FIELD GUIDE TO THE GRASSES
OF OREGON AND WASHINGTON

The authors and publisher gratefully acknowledge the generous financial support of the following organizations, which made possible the publication of the second edition of the *Field Guide to the Grasses of Oregon and Washington*.

 USDA Forest Service

 USDI Bureau of Land Management

FIELD GUIDE TO THE
Grasses of Oregon and Washington

Second Edition

Cindy Talbott Roché
Richard E. Brainerd
Barbara L. Wilson
Nick Otting
Robert C. Korfhage

Oregon State University Press
Corvallis

The John and Shirley Byrne Fund for Books on Nature and the Environment provides generous support that helps make publication of this and other Oregon State University Press books possible.

Cataloging in publication data is available from the Library of Congress.

ISBN 978-1-962645-26-3

♾ This paper meets the requirements of ANSI/NISO Z39.48-1992 (Permanence of Paper).

© 2025 Richard E. Brainerd, Robert C. Korfhage, Nick Otting, Cindy Talbott Roché, and Barbara L. Wilson

All rights reserved.
First edition published in 2019 by Oregon State University Press.
Second edition first published in 2025.

Printed in Korea

Title page: *Leucopoa kingii*. Photograph by Robert C. Korfhage.

Oregon State University Press
121 The Valley Library
Corvallis OR 97331-4501
541-737-3166 • fax 541-737-3170
www.osupress.oregonstate.edu

CONTENTS

Acknowledgments vii
    Illustration Credits viii
Introduction 1
    What Is (and Isn't) in This Book 3
    Grass Names 4
    Grass Identification 4
    Grass Structure and Vocabulary 5
    Grass Biology 11
    How to Use This Book 14
Keys 17
Species Accounts 95
    Bamboo 453
Glossary 463
References 469
Index 471
About the Authors 485

Native grasses form meadows over the floodplains of subalpine streams, storing and purifying the water. Photo by Robert Korfhage on the East Fork of the Wallowa River in the Eagle Cap Wilderness, Oregon.

Coastal habitats have been heavily impacted by the introduction of invasive grasses, including dune stabilizers like European beachgrass (*Calamagrostis arenaria*). Photo by Robert Korfhage on the north coast of Oregon.

ACKNOWLEDGMENTS

The first copies of the *Field Guide to Grasses of Oregon and Washington* became available in July 2019. In 2020 it won the Annual Literature Award, the highest award of the Council on Botanical and Horticultural Libraries. In four short years all of the copies sold and the guide went out of print. During that time the authors taught grass identification workshops, explored both states for new sightings of grasses, tracked down old collection records, and captured better photos. We have made a concerted effort to improve the distribution maps for Washington state. This second edition includes 18 additional taxa, one of which is a newly described species of *Agrostis* from the Oregon Coast Range. Three of the new species are members of the California floristic province found in southwestern Oregon. Two species of *Sporobolus* were rediscovered in Washington, and *Blepharidachne kingii* was found in southeastern Oregon, a range extension from Nevada. We decided it was time to acknowledge that *Miscanthus sinensis* and *Panicum virgatum* had naturalized. Also, we finally dove into the bamboo issue to identify which species appear to persist outside cultivation; this added another nine species.

In addition, the flood of name changes among grasses has not abated. Some changes are accepted in this book, others are not, and still others revert to previously familiar names. Our decisions are primarily in sync with the two major regional herbaria: Oregon State University (OSC) with the *Flora of Oregon* and the University of Washington (WTU) with the *Flora of the Pacific Northwest*. If name changes cause you major grief, just remember that even these names are probably temporary. We have expanded both the index and the glossary to reduce confusion between terms and help the learning process. The second edition of the *Field Guide* is finally done, with more than a little help from our friends. This help is what we credit here.

We received major financial support from the USDA Forest Service; Shawna Bautista of the Forest Service Region 6 ACES Program helped us navigate the ins and outs of federal funding. We also received funding from the Bureau of Land Management (BLM) with help from Oregon state botanist Sarah Canham.

At the Oregon State University Herbarium, we received help from Curator James Mickley, particularly in processing new collections, storing specimens, and acquiring loans from other herbaria. Katie Mitchell prepared the distribution maps from databases of the OregonFlora (Linda Hardison, project director; Stephen Meyers, taxonomic director; Thea Jaster, database manager) and the Consortium of Pacific Northwest Herbaria and the herbarium at the University of Washington (David Giblin, collections manager).

Robert Soreng, Smithsonian Institution research associate, assisted with all things *Poa*, ranging from determining which species are found in Oregon and Washington to identifying specimens and reviewing specimens, the photo plates, and species treatments. Peter Zika, research associate at the University of Washington Herbarium, suggested multiple improvements to the keys and descriptions, found numerous new records, especially in Washington, documented locations of other field locations, and checked identification of specimens at the University of Washington. Christina Veverka, Deschutes National Forest botanist, discovered the first *Poa wallowensis* in the Cascade Range. Peter Dunwiddie, affiliate curator, University of Washington Herbarium, assisted by reporting *Anemanthele lessoniona*, documenting field locations (*Polypogon maritimus* in Washington and removal of misidentified *Agrostis microphylla* in the San Juans). Frank Callahan, botanist, discovered three range extensions of California Floristic Province grasses in southwestern Oregon. Dain Sansome, owner of Valley Bamboo in Albany, Oregon, Ian Connor, PNW chapter president of the American Bamboo Society, and employees at Bamboo Garden Nursery in Portland taught us about bamboo species in Oregon and Washington.

Individuals who contributed to the previous edition include Kathy Ahlenslager, Curtis Bjork, Paula Brooks, Tim Butler, Ken Chambers, Jim Duncan, Jean Findley, Stu Garret, David Giblin, Ron Halvorson, Greg Haubrich, Lawrence Janeway, Ben Legler, Sarah Malaby, Don Mansfield, Caryn Meinicke, Mark Mousseaux, Beth Myers-Shenai, Mary Nicholson, Christine Ott-Hopkins, Wayne Rolle, Kathleen Sayce, Paul Slichter, Robert Soreng, Sue Vrilakas, Faye Weekley, Bryan Wender, Gene Yates, Peter Zika.

Illustration Credits

We are grateful to the photographers who provided images for this book. It is our intent to credit all photographs and photographers, and we apologize for any omissions or inaccuracies. All photographs not credited here were taken by the authors. Robert Korfhage digitally edited all of the photos. Cindy Roché prepared the line drawings for the introduction and for the bamboo section, and Barbara Wilson drew the fescue leaf cross sections.

Zoya Akulova-Barlow. *Ehrharta erecta* ligule, inflorescence, node; *Spartina alterniflora* habit, collar, culm, inflorescence; *Crypsis vaginiflora* habit, inflorescence; *Nassella lepida* inflorescence; *Rytidosperma penicillatum* habit, inflorescence, collar; *Phalaris californica* habit, inflorescence, ligule.

David and Diane Bilderback. *Poa unilateralis* ssp. *unilateralis* habit.

Curtis Bjork. *Festuca washingtonica* habit, habitat; *Podagrostis aequivalvis* habitat; *Coleanthus subtilis* habit; *Deschampsia bolanderi* habit.

Margo Bors. *Ehrharta erecta* habit.

Frank Callahan. *Muhlenbergia rigens* habit, inflorescence.

Lori Cheung. *Spartina anglica* habit, inflorescence. Courtesy of Olafson Environmental, Inc. (www.spartina.org)

Trent M. Draper. *Poa fendleriana* inflorescence, ligule.

Al Keuter. *Polypogon viridis* inflorescence and close-up, ligule.

Neal Kramer. *Aegilops triuncialis* spike; *Hainardia cylindrica* habit.

Cayman Lanzone. *Nassella lepida* habit.
Matt Lavin. *Alopecurus arundinaceus* habit, inflorescence; *Alopecurus carolinianus* habit, inflorescence; *Elymus scribneri* habit, inflorescence; *Bromus diandrus* habit; *Poa fendleriana* habit; *Poa reflexa* inflorescence, plant base.
Ben Legler. *Echinochloa muricata* inflorescence.
Max Licher. *Muhlenbergia minutissima* habit.
Steve Matson. *Hordeum depressum* habit.
Susan McDougall. *Calamagrostis tacomensis* inflorescence; *Podagrostis aequivalvis* habit.
Peter Meininger. *Puccinellia maritima* habit, inflorescence. Saxifraga Foundation – Images of European Biodiversity (www.freenatureimages.eu)
Keir Morse. *Melica geyeri* habit; *Nassella pulchra* floret and habit.
Bruce Newhouse. *Festuca viridula* habitat; *Holcus lanatus* closed inflorescence.
Richard Old. *Alopecurus myosuroides* habit, inflorescence, ligule; *Zizania palustris* root.
Amelia Ryan. *Blepharidachne kingii* habit.
Neil Ratzlaff. *Sphenopholis intermedia* inflorescence.
Lindsey Salmonson. *Poa curtifolia* habit.
Kathleen Sayce. *Bromus pacificus* inflorescence; *Poa unilateralis* ssp. *pachypholis* habit; *Nardus stricta* habit.
Kurt Schaefer. *Redfieldia flexuosa* habit, inflorescence, ligule, rhizome.
Quentin Skinner. *Bromus aleutensis* habit, ligule.
Paul Slichter. *Poa* × *multnomae* habit, inflorescence.
Robert Soreng. *Poa glauca* habit; *Poa arctica* habit.
Robert Steers. *Agrostis blasdalei* habit.
James H. Thomas. *Poa curtifolia* leaf closeup.
Trevor Van Loon. *Poa paucispicula* habit.
Washington State Department of Agriculture. *Spartina densiflora* habit, habitat, inflorescence; *Spartina patens* habit, habitat.
Gary Williams. *Spartina anglica* habit.
Gene Yates. *Poa alpina* inflorescence.
Peter Zika. *Corynephorus canescens* habit, habitat, inflorescences; *Miscanthus sinensis* habit; *Parapholis incurva* habit; *Poa infirma* habit; *Polypogon fugax* inflorescence; *Puccinellia simplex* habit; *Sporobolus vaginiflorus* habit.

Henderson's ricegrass (*Eriocoma hendersonii*) grows on rocky scablands in central Oregon and Washington. Photo by Robert Korfhage.

INTRODUCTION

Once upon a time, there were no grasses on earth. The fossil record of grasses begins about 145 to 100.5 million years ago, during the Cretaceous geologic period. Since then, the grass family (Poaceae) has expanded to include over 11,550 accepted species, making learning all of the grasses probably more than a lifelong project! Grassland environments constitute over 40% of the planet's nonpolar land area, and grasses are estimated to make up about 20% of global vegetation. Grasses are adapted to an amazingly wide variety of habitats, from cracks in freeways to pristine mountain fens covered with snow half the year. They dominate steppes, meadows, grassy balds, prairies, and savannas, forming the matrix within which a diversity of organisms thrives. They form the understory of woodlands and forests and mingle with sedges and rushes along streams, rivers, lakes, and marshes. They grow in salt flats and alkaline valleys, brackish coastal estuaries, and on shifting sand dunes, talus slopes, and cliff ledges. Some species thrive in deserts. Others survive in vernal pools where their rooting zone is either inundated with water or parched bone dry, depending on the season.

The dense fibrous roots of grasses prevent erosion and build up soil by trapping airborne silt and increasing underground organic matter; in fact, grass creates its own soil (known as mollisol), rich in organic matter from decayed fibrous roots. Along with other graminoids, grasses act as a filtration system to purify water and maintain the soil around marshes, lakes, streams, and canals. Native grasses provide critical habitat for a variety of native mammals, birds, amphibians, reptiles, insects, and more. Thus, native grasses are tremendously important for use in the restoration of degraded habitats. Grasses are ecological indicators. The grass species present at a site can reveal the site's substrate, moisture regime, climate, and history of disturbance. In the modern world, many grasslands have been severely disturbed or completely replaced by fields of cultivated plants, livestock grazing, and development.

Grasses feed herbivores ranging in size from ants to elephants, as well as domestic livestock. With the exception of their seeds, grasses are considered "low nutrient" food, but in vast savannas, grasses can provide the great quantities of forage needed to support grazing herds. In turn, herds of herbivores once supported another tier of life: predators, ranging from wolves and lions to humans, who later followed domesticated herds across the steppes. The traditional evolutionary story of an arms race between grasses and grazers is now known to be incorrect. The fossil record shows that grasses evolved during the Cretaceous period, when dinosaurs still roamed the Earth. Diversification among grasses began many millions of years before diversification in grazing mammals and the silica inclusions in grass epidermal cells are not hard enough to abrade tooth enamel. However, the expansion of grasslands in the drier climate of the Miocene greatly increased food supplies for grazers. The amount of soil grit ingested in this environment appears to be the driving force behind the evolution of longer teeth in grazers.

Grasses are essential for smaller animals as well. For example, Roemer's fescue is a larval host plant for the rare Mardon skipper butterfly, a candidate species for listing under the Endangered Species Act. In arid lands, grasses may provide not only food but also water because their roots extract soil moisture that birds and reptiles obtain directly from the insects and seeds.

Humans have always used grasses for food, fiber, and even shelter. During the Neolithic period, we developed a closer relationship with grasses when we started cultivating them for their seeds (grain). Early cultivated grasses included wheat, corn, rice, barley, oats, rye, teff, and a diversity of millets. The increased consumption of grass-based porridges, beer, and breads increased human populations. Easily stored and transported grain permitted the development of larger communities and eventually all of what we call civilization. Although seeds are by far the greatest contribution of grasses to human diets, we also use leaves of lemon grass (*Cymbopogon citratus*), shoots of various bamboos, and sugar from the culms of sugarcane (*Saccharum officinarum*). Citronella oil is used for cleaning products, insect repellents, aromatherapy, and soaps; it comes from citronella grass (*Cymbopogon nardus*). Vetiver, a fragrant essential oil used in the perfume industry, comes from the roots of the grass *Chrysopogon zizanioides* from India.

Grasses are also economically important as ornamentals for landscaping, lending beauty to parks and gardens. They can produce dense lawns. They provide the substrate for playgrounds, sports, and other recreational activities. Grasses are used in forensics; fire investigators use "grass-stem indicators" to help determine the point of origin and cause of fires.

However, many of the world's important weeds are grasses. Most of the weeds compete with our crops, reducing yields. Some agricultural weeds create special problems. Goatgrass and darnel are both crop-following weeds; their seeds are the right size and shape to get harvested with wheat and then planted with the next generation of the crop. In addition, goatgrass hybridizes with wheat, lowering the protein content of the crop in fields that include the hybrids. Darnel, or at least those strains of it infected by a certain fungus, can make flour poisonous. Several grasses are weeds in lawns and flower gardens; crabgrass has the worst reputation.

Many weedy grasses compete with native plants. In fact, some have become community dominants in otherwise natural habitats, pushing out native plants and the insects dependent on them. Cheatgrass is probably the worst wildland invader in arid lands east of the Cascade Range. It outcompetes native plant seedlings and promotes vast wildfires. In some habitats cheatgrass is being displaced in turn by medusahead, and now ventenata, which are not improvements for the native species, nor for livestock grazing. West of the Cascades, prairies are being invaded by introduced bentgrasses, among others, while falsebrome has exploded into forest understories, replacing native grasses. Thus, for good and bad, the grass family is probably the most economically and ecologically important of the world's plant families.

What Is (and Isn't) in This Book

This book differs from other field guides in that it covers all of the grass species known to occur in natural environments in Oregon and Washington. It includes native and naturalized species (ones introduced from elsewhere but reproducing successfully in their new environment), as well as common waifs (species that grow from seed blown off trucks, fallen from bird feeders, or dumped in yard waste, and that persist for only a few years). Major agricultural grasses (wheat, oats, and barley) are repeatedly found as waifs along roads or where straw was used as mulch, so they are included. We included native species that are apparently extirpated; some have not been collected in over 100 years.

We did not include some species that were collected around the ports of Portland and Seattle over a century ago and not recorded since. New in this edition are bamboos, which appear at the end of the species accounts in their own section. Approximately half of the taxa in this guide are native and the other half are introduced, with a few species that have both native and non-native varieties or there is still doubt as to whether they're native or not.

We had long debates about whether to include certain ornamentals that are spreading in landscape beds or plants that are aggressively invasive in neighboring states. We included several additional ornamental species in this edition, either with full treatments or just in the keys. We expect the number of grasses in Oregon and Washington to increase as new weeds appear, ornamentals escape and establish wild populations, and perhaps, rarely, as species new to science are described.

By "grasses" we mean members of the family Poaceae. Poaceae are monocots (plants with a single cotyledon leaf) that have flowers much reduced and hidden between two bracts (called lemma and palea). They have a specialized fruit with the embryo usually fused to its embryonic food supply.

We do not include other graminoids. The term "graminoid" refers to both grasses and grass-like plants, which have small green or brown wind-pollinated flowers. There are three main families of graminoids: grasses, rushes, and sedges. All grasses have stems with solid nodes and two-ranked leaves, one at each node. There are two parts to each leaf: the sheath, which encircles the culm, and the blade, which is commonly long and narrow. If you have only the leaves and stem of a plant, you can still tell whether or not it is a grass. According to Agnes Chase, "The only plants that may reasonably be mistaken for grasses are the sedges." (See references on p. 472 for Chase's book.) In sedges (Cyperaceae), the culms are not jointed and are often triangular. The leaves are three-ranked. In addition, their flowers are protected by one bract, not two as in grasses. Rushes (Juncaceae) in flower are easily identified by their tiny brown flowers that are lily-like and have six tepals (petals or sepals).

Many plants called "grasses" in common vernacular are not included here because this field guide is limited to members of the family Poaceae. Virtually any monocot with long, narrow, green leaves might be called a grass, but that doesn't make it one. Monocots that are called grasses, but aren't, include beargrass (*Xerophyllum*, Melanthiaceae; or *Nolina*, Asparagaceae), eelgrass (*Zostera*, Zosteraceae), seagrass (*Halophila*, Hydrocharitaceae), cultivated black mondograss (*Ophiopogon planiscapus* 'Nigrescens'; Asparagaceae), and Australia's grass tree (*Xanthorrhoea*, Asphodelaceae). Cottongrass (*Eriophorum*) is closer; it's a sedge (Cyperaceae). Grass widows (*Olsynium*) and blue-eyed grass (*Sisyrinchium*) are both in the iris family

(Iridaceae). Examples of dicot "grasses" include sabal snake grass (*Clinacanthus*, Acanthaceae), sour grass (*Oxalis*, Oxalidaceae), and grass of Parnassus (*Parnassia*, Celastraceae). Silk grass (*Pityopsis*) and viper's grass (*Scorzonera*) are composites (Asteraceae). Whitlow grass (*Draba*), peppergrass (*Lepidium*), and scurvy grass (*Barbarea*) are mustards (Brassicaceae). The name "snake grass" is often applied to *Equisetum* (Equisetaceae), also called horsetail and mare's tail; it is a fern ally. Some clues that a "grass" is not a true grass include succulent leaves (probably in the lily, asparagus, or orchid family), flowers with colorful petals, or berries. If the leaves are arranged edgewise to the stem as in irises, they're probably irises or relatives, though redtop panicgrass (*Panicum rigidulum*) has almost iris-like leaves. You'll get fooled sometimes, but the difference between true grasses and everything else will become clear with experience.

Because it covers all naturalized and native grasses and includes characters useful in the lab, this book can be more than a field guide. We hope it opens your eyes to seeing the splendor in the fascinating world of grasses.

Grass Names

Scientific names of plants tell us something about relationships. All members of a genus are supposed to be more closely related to each other than to plants in other genera. Therefore, as our understanding of the genetic, genealogical relationships among grasses improves, some names must change. That's been happening a lot in recent decades, to the annoyance of nearly everyone who works with grasses. In this book, we use names we hope might remain stable into the future, but that's no guarantee. Our nomenclature usually follows the *Flora of Oregon* (https://oregonflora.org/) and the *Flora of the Pacific Northwest* (Consortium of Pacific Northwest Herbaria). The synonyms are included in the index, where they are referred to the name we think is best.

Grass Identification

> *I have just made out my first grass, Hurrah, Hurrah! I must confess that Fortune favours the bold, for as good luck would have it, it was the easy* Anthoxanthum odoratum; *nevertheless it is a great discovery. I never expected to make out a grass in all my life. So Hurrah! It has done my stomach surprising good.*
> —Charles Darwin, in a letter to Sir Joseph Dalton Hooker, 15 June 1855

Grass identification can be difficult, but, as Darwin learned, it doesn't have to give you ulcers. Accurate grass identification is the key that opens the door to volumes of information on the plant itself, its role in the ecosystem, the history of the site, and appropriate objectives for land management. This book can help you get from the odd green or straw-colored thing in your hand to its name. You can start by browsing the photographs in the species accounts to match your specimens or use the formal identification key (p. 17). We recommend moving back and forth between them if the plant is at all difficult.

Why are grasses more difficult than some other plants? For one thing, grasses are diverse. This book covers 384 taxa in 105 genera that are known outside cultivation in Oregon and Washington, and 14 more are included in the keys (representing 7 additional genera). Many species look superficially alike, particularly among the larger genera such as *Agrostis*, *Bromus*, *Elymus*, *Festuca*, and *Poa*.

Second, grass parts are small. One of us (Barbara) can attest to the effectiveness of extreme myopia (nearsightedness) for seeing small things, but the great majority of us were born without that advantage and will have to settle for using tools. We recommend getting a 10× hand lens. A dissecting microscope is great but a bit cumbersome for field use. A pair of binoculars used backward makes a surprisingly effective hand lens, once the grass part is moved very close to the farther lens.

A third problem is that grass parts are highly modified compared to other plants. Just recognizing what you're looking at can be a challenge, and besides that, botanists have taken the opportunity to generate lots of specialized vocabulary. Those words are explained in the glossary (p. 463) and the most important ones are labeled on the plates illustrating each species.

Despite all the problems, grass identification is possible. Why? Because all grasses are built on the same basic plan. Once you figure out the spikelets, glumes, and lemmas of one grass, you can find those parts on the next one and the next one, until you encounter one that is odd even for grasses, but by and large, you can find the essential parts. In this book, we try to help you around the most obvious pitfalls. Grasses definitely get easier with practice, although there is the downside that as you become more experienced and confident, you become more likely to take on challenges like bentgrasses, fine-leaved fescues, annual bromes, or even bluegrasses! Grasses can provide a lifetime of mysteries, and triumphs.

Grass Structure and Vocabulary (READ ME!)

Grass leaves and flowers differ from those of most familiar plants, so botanists have created special terms for their parts. This can make grass identification confusing at first. But because all grasses have the same basic structures, a little time invested in learning the vocabulary is well worth the effort. Terms are defined in the glossary, p. 463. What follows is an illustrated description of basic grass structures and their associated terminology.

Plant Duration. Grass plants can be **annual**, growing and dying within one year, or **perennial**, lasting for two or more years. Annuals are generally classed as winter annuals (germinating in the fall and reproducing the following spring/summer) or spring annuals (germinating in the spring and reproducing in the fall). The distinction between annuals and perennials is usually clear, but a few perennials flower in the first year, and some annuals in mild winter zones may persist for a few years. Perennials can be recognized by the remains of the previous year's leaves at the base. Generally speaking, plants with small root systems that pull easily are annuals, and plants with robust roots that are difficult to pull are perennials. This rule fails with particularly robust annuals or diminutive perennials, and in sandy soils. Sometimes there are other clues like old flowering stems and leaves, or carcasses of dead plants from previous years. Be observant and willing to run through the key a second time if your first try doesn't yield satisfactory results.

VEGETATIVE STRUCTURES

Underground Parts. The underground parts of the grass plant are **roots**, **rhizomes**, and **bulb-like corms**. Most grasses have fibrous roots. Rhizomes are underground stems, usually growing horizontally rather than vertically. Rhizomes have nodes with scales that are reduced leaves. Roots have hairs, but never nodes or scales.

Rhizomes allow grasses to spread laterally; these are the sod-forming grasses. Some grasses such as the oniongrasses (*Melica*) and tall oatgrass (*Arrhenatherum*) develop a corm (bulb-like swelling) at the base of the culm. Rhizomes and corms are often "fleshy," making them attractive food for rodents.

Growth form, Shoots, and Stems. Sod-forming grasses have rhizomes or **stolons**. Like rhizomes, stolons are modified, more-or-less horizontal stems that allow the grass to expand by reproducing vegetatively. Unlike rhizomes, stolons run aboveground, sometimes over the top of other plants. **Bunchgrasses** are cespitose, with the shoots clustered together, either densely or loosely. Generally, bunchgrasses lack creeping stems; if present, rhizomes or stolons are short and tightly packed.

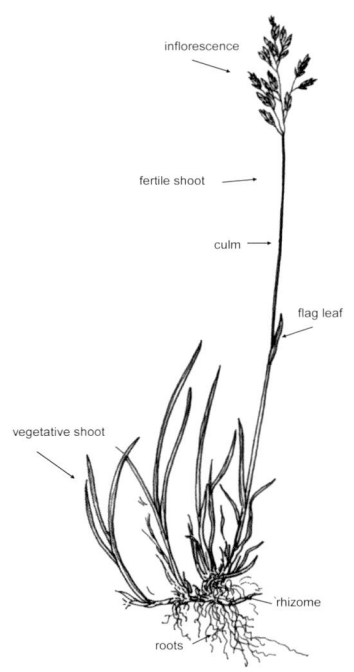

Figure 1. Grass diagram.

All these terms apply to perennials. There is some dispute as to whether an annual with multiple stems can be cespitose or should always be called **tufted**; we generally call them tufted. Many annuals and some perennials have **decumbent** stems that lie across the ground for a bit and then curve up. Decumbent stems often **root at the nodes** (effectively making them stolons), which are conveniently positioned close to the soil.

Grass shoots can be **vegetative**, without flowers, or **fertile**, with flowers. Vegetative shoots are also called **tillers** or **innovations**. The fertile shoot is topped by an **inflorescence**, a cluster of flowers or, in grasses, of spikelets (explained below). The stem of the fertile shoot is called a **culm**. Although there are some exceptions (bamboo), culms are generally annual stems that bear the inflorescence and then die. Stolons are a specialized type of culm that grows horizontally. Culms have nodes or joints, solid points at which the leaf sheath arises. The uppermost leaf on the culm, the one just below the inflorescence, is called the **flag leaf.**

Some grasses have tall vegetative shoots, but a special trick of many grasses is to keep the growing point, the stem tip where all growth is initiated, at or below ground level. This protects the all-important growing point from fire, grazing animals, and lawn mowers. In these protected shoots, the stem is still there but all telescoped down into a tiny nub, hidden inside the leaf bases. Each new leaf begins growth inside the sheath of the older leaf. It can be instructive to pull off one leaf (including the sheath) after another and see that when they're all gone, there is nothing left but a tiny nubbin. That nubbin may remain small for the entire life of the shoot, but in some species, it may expand in its second year, producing a fertile shoot.

In *Poa* and some other difficult-to-identify grasses, it is important to determine how new shoots originate on old shoots. Most bunchgrasses have **intravaginal shoots**, which grow up inside the lower leaf sheaths of the old shoot, parallel to it, and only gradually diverging from it. Rhizomatous grasses and some bunchgrasses have **extravaginal shoots**, which grow perpendicular to the old shoot at an early stage, bursting through lower leaf sheaths of the older shoot. And of course, some grasses have both intra- and extravaginal shoots! This can be a difficult character to assess, but sometimes it is critical for identification.

Leaves. The leaf consists of two parts, the **sheath**, which surrounds the stem, and the **blade**, which angles away from the stem and is often flat. The sheath is usually **open**, with margins overlapping, but in some species, it can be **closed**, with margins fused, forming a tube. The sheath may be closed only at the base and open above, or closed to the very top.

The region where the leaf sheath meets the leaf blade may offer characters useful for identification. The opening at the top is the **sheath mouth**. It may have a tuft of hairs or small flaps called **auricles**. Classic auricles are claw-like extensions that reach across the mouth, but we also apply the name to any little upright flap or bump, and to shelf-like flanges that reach around to connect with the blade base. The **collar** is the back (outside) part of the line connecting blade and sheath. The **ligule** is in front (inside) of this same junction. The ligule is parallel to the stem, at the base of the blade. It may be a small membranous flap, a line of hairs, or a small flap fringed with hairs. Ligules may serve to seal the space between the blade base and the culm or younger leaf, thus preventing water and fungal spores from getting inside the sheath and rotting it.

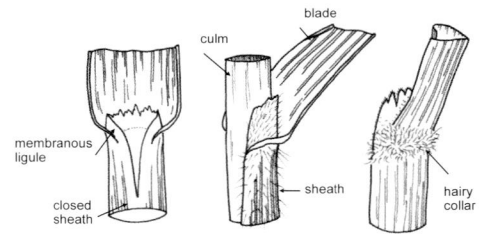

The leaf blade is usually flat but may be folded or rolled up, especially in the bud. Of course, botanists have developed special words like **involute**, **revolute**, **convolute**, and **conduplicate** to describe exactly how they are rolled or folded. The distinction is important: look for whether the back of the leaf is rounded (rolled) or sharply creased (folded). Check immature leaves if you can't tell on older ones.

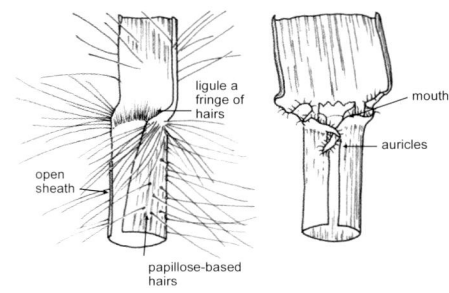

Figure 2. Collars.

Plant Surfaces and Hairs. A hairless surface is called **glabrous**. **Scabrous** surfaces have short rough points called **scabers**. You can feel the roughness when you rub your finger lightly along the surface. Surfaces without scabers or other rough bumps are called **smooth**. A **glaucous** surface is bluish or grayish due to a coating of wax. The wax can usually be rubbed off with your fingers and has a disturbing tendency to melt off in a plant press, so the specimen that looked so blue in the field may appear green when dried.

There are many words for hairs and hairiness, but we generally just use the word "hairy," sometimes modified as short-hairy, long-hairy, appressed-hairy, spreading-hairy, and so on. We may use a few other terms for hairs. A **puberulent** surface has very short hairs. A **pubescent** surface has more or less dense, short hairs. Stiff hairs can be called **bristles**. **Cobwebby hairs** are long, delicate, curly hairs that can be stretched out; they occur in some bluegrasses (*Poa*). Hairs growing out of a little swelling on the leaf surface are **papillose-based hairs**.

REPRODUCTIVE STRUCTURES

Inflorescences. The fertile culm is topped by an **inflorescence**, which bears the reproductive parts called spikelets. The grass inflorescence may be a **spike**, with all the spikelets attached directly to the main axis (the **rachis**) without a **pedicel** (stalk). It may be a **raceme**, with spikelets attached to pedicels that in turn attach directly to the main axis. Grass biologists admit that an inflorescence is a raceme only when the pedicels are relatively long. Otherwise, we use the term "spike" to cover both real spikes and racemes with short pedicels. This isn't as sloppy as it seems, because often spikelet arrangement can vary from completely sessile to short-pedicelled (subsessile) in the same "spike."

Figure 3 (top). Cobwebby hairs.
Figure 4. Papillose-based hairs.

More often, the grass inflorescence is a much-branched **panicle**. The primary panicle branches are the ones connected directly to the main axis. They often produce other branches (secondary, tertiary, etc.). Panicles may be open with the branches spreading away from the axis and the spikelets separated. In some open panicles, the lower branches are **reflexed**, oriented downward. Alternatively, the panicles may be contracted, with the branches mainly **ascending** (oriented upward) or **appressed** (lying parallel to the main axis or nearly so). If the primary panicle branches have sessile or nearly sessile spikelets, the inflorescence is called a **panicle with spike-like branches**, as in crabgrass (*Digitaria*) or bermudagrass (*Cynodon*). If all panicle branches are extremely short, the crowded inflorescence is a **congested panicle** or **spike-like panicle**, as in timothy (*Phleum*), foxtail (*Setaria*), or meadow foxtail (*Alopecurus*).

Some grasses, including bluestems (*Andropogon* and *Schizachyrium*), silvergrasses (*Miscanthus*), and ravenna grass (*Tripidium ravennae*), have very interesting (confusing) panicles divided into units called **rames**, which involve spikelet pairs consisting of a sessile and a pedicellate spikelet (or a cluster of spikelets) on panicle branches that break at the nodes, all usually hidden by a lot of hairs. The only good thing we can say about rames is that only a few grasses in Oregon and Washington have them. We're not going to try to explain them further; we use other traits to identify these grasses.

Spikelets. The basic unit of the grass inflorescence is the **spikelet**, a cluster of reduced flowers protected by **bracts** (modified leaves). At the base of the spikelet are the two bracts called **glumes**. Between or above the glumes are the **florets**, which are joined together by a rachilla. (Note that the central "stalk" of a spike is called the rachis, of which "rachilla" is the diminutive form. Think of a rachilla as the central "stalk" within a spikelet.) See the description below for florets.

Spikelets may be laterally compressed, cylindrical (terete), or dorsiventrally compressed. If you lay a **laterally compressed** spikelet on the table, it will lie on its side: that is, both glumes will be visible, one to the right and one to the left. Most of our grasses have laterally compressed spikelets. **Cylindrical** spikelets lie in a variety of positions, which is one way to determine that the spikelets are not compressed. If you lay a **dorsiventrally compressed** spikelet on the table, it lies with one of its glumes underneath, so that you can see just one glume and the back of a lemma.

Glumes have a **midrib**, the main vein running up the center of the glume back, and sometimes other **veins** (or **nerves**). If the glumes are strongly folded or the midrib is prominent, the midrib

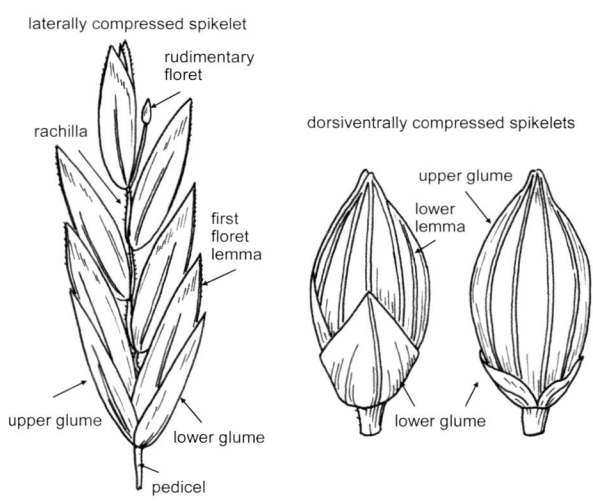

Figure 5. Laterally and dorsiventrally compressed spikelets.

may be called a keel. You can usually figure out the total number of veins on a glume by counting those on one side, multiplying by two (for those on the other side), and adding one for the midrib. This can be tricky when the glume is off-center (*Thinopyrum*, *Pascopyrum*) or if the veins are obscure. Glumes may or may not have **awns**, bristle-like tips formed by an extension of the midrib. Glumes are absent in a few aquatic grasses, such as rice (*Oryza*, *Zizania*) and cutgrasses (*Leersia*). Occasionally one glume is absent, as in ryegrasses (*Lolium*) and some panicgrasses (*Panicum* and *Dichanthelium*).

Florets. Each floret is composed of a reduced flower and the two bracts that enclose it, the lemma and the palea. The lemma is the larger one and usually the only one that is easy to see. It enfolds the palea, which usually has two keels (veins). The lemma has a **midrib** and often other veins or nerves. Many lemmas are more or less boat-shaped with a prominent midrib called a **keel**. The **marginal veins** are those closest to the lemma margin (edge). Between the keel and the marginal veins are the **lateral veins**. **Cilia** are hairs occurring on the margin of a structure, and lemmas may have them. The lemma tip may be **blunt**, **rounded**, **obtuse** (broadly angled), **acute** (forming a short point), or **acuminate** (drawn out into a long point). It may be **bifid** (divided into two **lobes** or **teeth** at the tip). The teeth, if present, may be long or very short. At the base of each lemma is a **callus**, a little hardened swelling or joint connecting the lemma to the rachilla.

Awns and bristles are thin and stiff. The term **awn** is used for an extension of a midrib or other vein of a glume, lemma, or palea. We use **bristle** for stiff hairs on a surface (*Echinochloa*) and for inflorescence structures made from modified panicle branches (*Setaria* or *Cenchrus*). Although glumes and lemmas usually have one awn each, formed by an extension of the midrib, in a few grasses such as goatgrasses (*Aegilops*) and threeawns (*Aristida*), awns originate from other nerves as well as the midrib. The rare paleas that are awned have two awns each, one from an extension of each vein (*Pleuropogon*). Awns of lemmas usually form at the lemma tip, but they may arise between two conspicuous or inconspicuous teeth or from lower on the lemma back, as a kind of branch from the midrib. They are usually straight, but they may be **bent** (**geniculate**), and if so, each straight part between the bends is called a **segment**. Awns may also be **twisted** along the main axis. **Outcurved** awns start straight and bend outward, away from the inflorescence axis, reaching their fully curved shape only at maturity. A point so short we don't want to call it an awn can be called a **mucro** or even a **short point**. A structure tipped with a mucro is **mucronate**. Measurements of glumes and lemmas do not include the awns, unless otherwise noted.

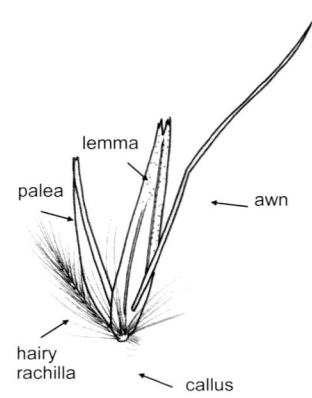

Figure 6. Florets.

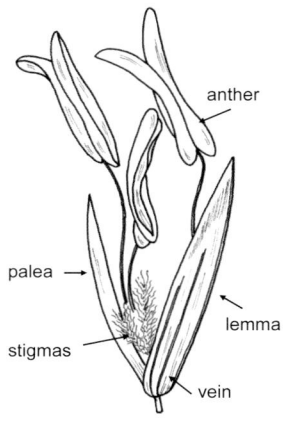

Flowers. The actual grass flower is much reduced, hidden between the lemma and palea. Florets can be pistillate, staminate, or bisexual. Pistillate florets contain three parts: **stigmas**, **styles**, and **ovary**. The feathery stigmas collect passing pollen from the air. The styles provide a passageway for the pollen to arrive at the ovary, where fertilization occurs. Staminate (male) flowers have only **stamens**, which have two parts: **anthers** and **filaments**. Anthers are filled with pollen and each one is connected to the floret by a thread-like whitish filament. Stamens commonly come in sets of three. Bisexual florets contain both the female and male structures. In most flowering plants, the most conspicuous parts we see are the petals, but in grasses the petals and sepals are reduced to two **lodicules**, nearly invisible structures that swell to push the lemma and palea apart at **anthesis** (flowering time), exposing the stigmas and anthers so pollen can be shared with, or accepted from, other individuals. Grass flowers at anthesis are a delight (unless you suffer from pollen allergies). Stigmas may be white, yellow, orange, magenta, or purple. Anthers, as well, can be colorful: yellow, white, or shades of purple or blue.

Figure 7. Purple stigma and yellow anthers of *Panicum*.

Seeds. The propagule in grasses is a **caryopsis**, or grain, a one-seeded fruit with its seed coat fused to the ovary wall. Some grasses, mostly in the tropics, have different kinds of fruits, but in our native flora only *Sporobolus* has fruits that aren't caryopses (they are achenes or utricles). The seed in *Sporobolus* is technically, a dry, indehiscent, one-seeded fruit that has developed from one flower having a single ovary, and the ovary wall becomes more or less bladdery or inflated. Thus, a utricle differs from an ordinary grass caryopsis by having an inflated ovary wall.

Disarticulation. Disarticulation is the term used for the separating of plant parts at maturity; the most well-known example is leaves falling from trees in the fall. The separation occurs along a natural abscission layer between the structures. Grass inflorescences disarticulate in a variety of places: along the rachis (*Elymus elymoides*), above the glumes (florets fall, leaving the glumes intact, *Bromus*), or below the glumes (spikelets fall as a unit, *Polypogon*), or the lemmas fall, leaving the paleas on the plant (*Eragrostis*). Some grasses disarticulate at culm nodes (*Triplasis*); some leaf blades disarticulate at maturity (*Crypsis, Blepharidachne*).

You can often determine where the spikelet disarticulates even before the parts fall naturally. Gently grasp the upper floret(s) and the pedicel, using fingers or forceps. Pulling gently should cause the spikelet to break where it normally would.

If you have managed to read this entire section, congratulations! You do not need to remember every detail, but once you've seen it, you can remember to come back and refer to this information as you need it later. And, while you are thinking about it, take a piece of tape and create a page tab for the glossary.

Grass Biology

ABOUT PHOTOSYNTHESIS AND WARM- VS. COOL-SEASON GRASSES

Early settlers to the Oregon Territory were appalled when the corn (*Zea mays*) that they had depended on in the Midwest barely set seed here. Fortunately for them, their wheat and oats thrived; until then, it seemed that farming might be a bust in the Pacific Northwest (PNW).

Why the difference between the outcomes for the crops? Corn is a warm-season (C4) grass, and wheat and oats are cool-season (C3) grasses. The C3 grasses do what we can call "normal" photosynthesis. They do fine as long as temperatures are mild, but their photosynthesis becomes inefficient at high temperatures. They shut down. Why? Because they depend on the ancient enzyme rubisco to trap ("fix") carbon dioxide and make it available for photosynthesis. Rubisco traps oxygen gas too, trapping more oxygen at higher temperatures. That was no problem when rubisco first evolved because free oxygen was rare. Now, however, it becomes a problem at high temperatures because, as temperatures rise, rubisco fixes more oxygen (less carbon dioxide), which slows photosynthesis. C4 grasses like corn thrive in hot weather, as long as there is adequate moisture. They use rubisco but hide it deeper inside the leaves, near the veins, and leave carbon dioxide fixation at the leaf surfaces to an efficient molecule (PEP-carboxylase) that can't fix oxygen.

Most PNW grasses are C3 grasses because PNW summers west of the Cascades are cool (relative to Kansas), not giving C4 grasses a chance. East of the Cascades, summers are hot but so dry grasses can't afford to do photosynthesis of either kind, but despite that, some native species find a way and flower late in the summer or in the fall (*Aristida* and *Sporobolus*, for example). In our climate, the majority of grass species have C3 photosynthesis, including all "normal" grasses like *Poa* and *Festuca*, and most of our crops. Warm-season photosynthesis has evolved repeatedly among other grasses, including bluestems, barnyard grasses, crabgrasses, *Setaria* and *Panicum*, as well as sorghum, millet, and corn, most of which are introduced grasses in the PNW.

MORE IRRESISTIBLE GRASS BIOLOGY (SEX AND OTHER REPRODUCTIVE OPTIONS)

Grasses are nearly all wind-pollinated. However, insect pollination occurs in some grasses. Usually it just supplements wind pollination, as in Dallisgrass (*Paspalum dilatatum*), but it is somewhat common among the few grasses of tropical and subtropical forests, where wind is not efficient at transferring pollen. Grasses have a few other tricks up their lemmas, too.

Botanists have, of course, come up with names for the various ways in which flower parts are distributed on grass plants. Florets that contain both male and female flowers are **bisexual** (also called perfect or hermaphroditic); this is common in many genera, such as *Bromus* and *Festuca*. If male and female flowers occur in separate florets, spikelets, or inflorescence parts on the same plant, the plant is **monoecious**. Sometimes male and female flowers are borne in different inflorescences on the same plant. Picture a corn plant with the staminate tassels (males) at the top and the pistillate ears (females) along the stalks. In wild rice (*Zizania*), male and female flowers occur in different parts of the same inflorescence: males at the bottom, females on top. If male and female flowers occur on separate plants, the plants are **dioecious** (saltgrass, *Distichlis*). Grasses aren't content with these simple alternatives. If some plants have only female flowers while others have only bisexual flowers, the population is **gynodioecious** (found in *Poa mansfieldii*). Flowers with female flowers and **vestigial** or aborted anthers are considered female (pistillate). However, aborted anthers may also be an indication that the plant is a hybrid.

To determine whether a flower is bisexual, male, or female, pry apart the lemma and palea and look inside. Early in flowering, the plume of the stigma and the anthers are obvious. Anthers may be obvious at anthesis (flowering), but later the anthers blow away. Look for fine threads; the filaments often persist much longer than the anthers they supported. As the grasses mature, the presence of a caryopsis is proof that at least a female flower was present. By this time, though, even the filaments are probably gone (assuming they were ever there). However, the anthers manage to persist in some grasses much longer than one might expect and thus can be a useful identifying trait. Be cautious measuring anthers that linger late into the fall; often these are aborted anthers and are much smaller than normal ones.

Empty lemmas and flowers without functional ovaries or anthers are **sterile**. You will encounter sterile florets or spikelets in a variety of settings. Often the terminal one or few florets in a spikelet are small and sterile. In oniongrasses (*Melica*), these sterile florets may be deformed in appearance and are called rudimentary florets or just rudiments. Sometimes sterile flowers occur at the base of the floret, below the fertile floret. That condition is seen in sweet vernalgrass (*Anthoxanthum*). Sometimes the sterile florets are vestigial, reduced to a nubbin, a bristle, or a tuft of hair; the first time you encounter this in reed canarygrass (*Phalaris arundinacea*) you are likely to think there is only one floret per spikelet. (The level of disbelief that the little tuft of hair is a floret will help you remember it when you see it again. Don't worry; you can identify it in our key whether you think there are three florets or only one.) In some grasses, such as dogtail (*Cynosurus*) and hooded canarygrass (*Phalaris paradoxa*), sterile spikelets are at least as common as the fertile ones and may aid in protection or dispersal of the seed.

Some populations of grasses are entirely female, such as two subspecies of *Poa cusickii*, which may leave you wondering how they produce seeds. **Apomixis** is the asexual production of seeds from the maternal tissue of the ovule, creating a clone. Apomixis has been reported in *Calamagrostis*, *Cenchrus*, *Chloris*, *Cortaderia*, *Hierochloë*, *Panicum*, *Paspalum*, *Poa*, and *Setaria*. Apomixis can be facultative (mixed with sexual reproduction) or obligate. Sometimes apomixis requires stimulation by a pollen grain but doesn't actually use it for fertilization.

Apomixis differs from selfing, in which pollen from the same flower (or plant) is used to produce a seed. Anytime you find really tiny anthers, you should suspect that some selfing is going on. Some genera have both obvious, outcrossing spikelets at the top of the culm and hidden, selfing (cleistogamous) spikelets inside leaf sheaths lower on the culm. Examples include cutgrass (*Leersia*), oatgrass (*Danthonia*), mannagrass (*Glyceria*), dropseed (*Sporobolus*), and sandgrass (*Triplasis*).

Sooner or later, you will encounter what look like leaves arising from the parts above the glumes where the florets should be. Do not panic. This is not uncommon in grasses and may represent either **vivipary**, which is germination of the seeds while still on the parent plant, or **pseudovivipary**, which is asexual reproduction. In cereal crops, vivipary is called preharvest sprouting. We've encountered leafy plantlets in late-season inflorescences of *Deschampsia cespitosa* and *Agrostis exarata*; these were probably seeds that germinated before dispersal. In pseudovivipary, the florets are replaced by leafy bulblets or plantlets that are clones of the mother plant. *Poa bulbosa* is the most common plant in our area that produces bulblets, but this method of reproduction is fairly common among alpine/arctic grasses, especially in the genera *Festuca* and *Poa*.

Grass sex options are fascinating and, in some cases, will actually help you identify a grass. Also, if you have some inkling of the possibilities of the arrangement of male and female parts in grass florets, spikelets, and inflorescences, you will appreciate that you cannot look at just one floret to answer the gender question in a key. You may need to look at multiple florets, multiple spikelets, and perhaps multiple plants.

How to Use This Book

This guide provides two pathways to grass names. You can use the formal dichotomous identification key, which works sequentially through a number of traits. Choose one of a pair of alternatives (lead 1a or lead 1b), go to the next lead below it (lead 2a or 2b), and so on. In theory you will automatically find the right name. In reality, choosing the right lead can be difficult. However, it will become easier with practice and as you learn how to recognize the characters. If you prefer, you can leaf through the pages of descriptions and photos. This "leaf method" is especially useful if you are trying to determine which group or genus your specimen belongs in, or if you already know the genus and are selecting a match from among two or three species. The photos were selected to illustrate important characteristics, which complement the description of traits, ecology, and similarity to other species. A distribution map shows where the species has been found in the past. We recommend using the key and photo/description pages together. This method is far superior to using either one alone, especially for species that may be only subtly different.

Experienced botanists routinely use keys to identify grasses, methodically working through the leads and examining the specimens carefully. With practice you will learn to recognize many genera and you'll be able to begin keying your specimen in the key for that genus. From teaching grass identification workshops for the past twenty years we've learned that students can become adept at recognizing characteristics of grasses and working through the keys. The technical terminology is unfamiliar for beginners, so we have included descriptions of grass structures and critical vocabulary (p. 5–11), and a glossary (p. 463). With some work and perseverance, you should be able to recognize many of the common and distinctive grasses in your area while gaining the skills to key some of the more difficult species that you encounter.

Read each lead carefully; skimming quickly is a reliable way to get yourself onto the wrong path. Over time, you will learn which characters are more reliable than others; for example, callus hairs are usually more consistent than plant height. And when two leads in the key offer overlapping measurements (e.g., 1–4 mm vs. 2–6 mm), look at where in the range your plant falls. It's generally better to go with the lead in which your plant's measurement is closer to the middle of the range. In this example, if your measurements (yes, take more than one!) hover around 4 mm, try the 2–6 mm option first, since 4 mm is in the middle of the 2–6 mm range but at the upper limit of the 1–4 mm range. Often there will be more than one character given in the lead and you can choose the lead with the best combination of characters for your plant. Important tip: whenever you have to make an uncertain choice, note that point in the key; if the first try takes you to a dead end, return to the lead where you were uncertain and try the other lead to see if it will take you to a correct identification. You may have to do this several times in difficult genera. Do not be discouraged. *Experienced botanists do this all the time!*

When collecting grass samples to identify, try to get at least parts of entire plants, including roots, basal leaves, culms, and inflorescences. If possible, get enough to tear apart during identification and still have ample material left for an herbarium specimen. It is preferable to have parts of more than one plant in the population, especially the inflorescences. If you are in a hurry and "top-snatch" a grass, you may not be able to tell whether it's an annual or perennial, whether it has rhizomes or stolons, or whether the basal branching is intra- or extravaginal (unless you thought ahead to note these features). Spending a little longer to bring back a good specimen will pay off in your ability to identify it, and its suitability for making a permanent record if it is an oddity.

The exception is for iNaturalist users, who use photos to document habitat, growth habit and larger features, but don't have time or a suitable camera to show small details of flowers or ligules. In this case, top-snatching would be useful to assist later identifcation.

There will be times that you are simply unable to key a specimen. In our experience, putting it away and trying again the next day sometimes works wonders. This guide is of limited usefulness for trying to identify a grass early in the season before any flowering parts develop, or late in the season when the bracts that enclose the flowers (glumes) are empty. But if you have an idea of what it might be, checking the photos might help you confirm an identification using the fragments.

With consistent use of this field guide, you will learn about habitat, habit, phenology, persistence of parts, and other features of species that provide clues to their identification. Over time, you will learn which grasses to expect in certain habitats and be able to compare basal leaf sheaths, ligules, or persistent glumes to images in the book. We hope you come to enjoy grasses as much as we do. Besides their ecological and economic importance, grasses themselves are fascinating. It's amazing to see how much diversity they exhibit from variations on a very simple body plan. Discovering new grasses is part of the adventure of exploring the amazing world around us.

Much of the high desert of eastern Oregon is actually shrub steppe, a grassland with a shrub overstory, intermingled with a diversity of forbs. Photo by Robert Korfhage in Deschutes County, Oregon.

KEYS

Keys to Genera of Grasses

Measurements of glumes, lemmas, and spikelets exclude the awns (if any) unless otherwise noted. Range descriptions focus on OR and WA; most species have ranges that extend beyond the Pacific Northwest (PNW).

KEY TO GROUPS

1a Culms woody; leaf blades distinctly narrowed to petiole-like bases (pseudopetioles) ....**Bamboos (p. 453)**
1b Culms delicate to tough; if woody, leaf blades not distinctly narrowed to petiole-like bases
    2a Leaf blades wide, up to 10–12 cm wide with prominent pale midvein; ligule membranous with a short fringe of hairs
        3a Leaf blade margins smooth; plants with both axillary pistillate spikes (corncobs) and terminal staminate panicles (tassels) .... ***Zea mays*** **ssp. *mays***
        3b Leaf blade margins serrate; plants with terminal panicles with bisexual spikelets.. ***Sorghum bicolor***
    2b Leaf blades various, but never as wide as 10 cm; ligules various
        4a Spikelets enclosed in a bur-like fascicle
            5a Burs spiny, painful to touch, held above leaves; plants tufted annuals; inflorescences bisexual.. ***Cenchrus longispinus***
            5b Burs not spiny, not painful to touch, hidden in leaf bases; plants rhizomatous perennials, dioecious; pistillate inflorescences short and bur-like, hidden in leaf bases, often overlooked; planted for xeriscaping (not treated) .... ***Bouteloua dactyloides***
        4b Spikelets not enclosed in a bur-like fascicle
            6a All or most florets producing bulblets instead of flowers and seeds; culms bulbous-based ....***Poa***
            6b Florets normally producing flowers and seeds; culms bulbous-based or not
                7a Inflorescence a spike, raceme, condensed spike-like panicle, dense head, or single spikelet; if a panicle, the branches not easily seen at arm's length even when the inflorescence is bent to the side ....**Group 1, p. 18**
                7b Inflorescence a panicle, not spike-like or head-like; branches easily seen at arm's length, at least when the inflorescence is bent to the side; branches sometimes spike-like

8a Inflorescence branches spikes or spike-like, the inflorescence sometimes digitate; spikelets often paired .... **Group 2, p. 26**
8b Inflorescence branches not spike-like, often branched again 2 or more times, sometimes raceme-like; spikelets usually not paired
   9a Florets 1 per spikelet .... **Group 3, p. 29**
   9b Florets 2 or more per spikelet (1 or more florets sometimes minute)
      10a Sheaths of upper leaves closed at least 25% their length
         .... **Group 4, p. 33**
      10b Sheaths of upper leaves open to the base, or nearly so
         .... **Group 5, p. 34**

Group 1. Inflorescence a spike, raceme, condensed spike-like panicle, head, or single spikelet; inflorescence branches none or obscure

1a Spikelets 1 per inflorescence
   2a Plants perennial .... *Danthonia unispicata*
   2b Plants annual
      3a Leaf sheaths closed nearly to the top; lemmas rounded .... ***Bromus***
      3b Leaf sheaths open to the base; lemmas keeled .... ***Festuca***
1b Spikelets 2 or more per inflorescence
   4a Plants short, <20 cm tall, growing in desert areas of SE OR; lemmas with long hairs that are conspicuous without breaking the spikelets apart
      5a Leaf tips stiff, sharp; plants short-stoloniferous, mat-forming; inflorescences mostly hidden in short clusters of leaves; lemmas only slightly lobed, with a tuft of hairs on each side, originating on the margin at mid-length; collected once in E OR but apparently not persisting (not treated) .... ***Munroa squarrosa***
      5b Leaf tips firm but not sharp; plants cespitose; inflorescences not hidden; lemmas deeply lobed, the lateral lobes wider than the central one and fringed with long hairs on one or both sides .... ***Blepharidachne kingii***
   4b Plants short or taller (to 350 cm), habitat and range various; lemmas hairy or not, hairs if present short or long, sometimes evident without breaking the spikelet apart
      6a Inflorescence a dense, spherical or ovoid, pale head with a mass of long, soft hairs that originate on narrow glumes or on bristles that arise below the spikelets
         7a Inflorescences 4–11.5 cm long, 5–7.5 cm wide; the few lowest spikelet clusters usually a bit separate from the others; perennial; rarely escaping, perhaps not persisting (not treated) .... ***Cenchrus longisetus***
         7b Inflorescences 1.5–3 cm long, 1–2 cm wide, compact; annual; rarely escaping, perhaps not persisting (not treated) .... ***Lagurus ovatus***
      6b Plants not as above; inflorescence not a dense, spherical or ovoid head with a mass of long, soft hairs that originate on narrow glumes or on bristles that arise below the spikelets
         8a Inflorescence 1-sided or with 1-sided clusters of spikelets
            9a Individual spikelets 20–40(50) mm, spreading to reflexed
               .... ***Pleuropogon***
            9b Individual spikelets 3–25 mm, if 20–25 mm, then spikelets appressed

10a Glumes absent or vestigial; leaves 0.5–1(2) mm wide, stiff and sharp-pointed; florets 1 per spikelet; spikelets 1 per node, sessile, evenly spaced along the axis .... **Nardus stricta**
10b Glumes present; leaves 0.5–14 mm wide, generally not stiff, not sharp-pointed; florets 1–5+ per spikelet; spikelets not arranged as above
    11a Glumes rounded at tip; base of mature inflorescence usually hidden in leaves; plants annual .... **Sclerochloa dura**
    11b Glumes pointed to awned at tip; base of mature inflorescence well-exserted from leaves; plants annual or perennial
        12a About half of spikelets fertile and normal-looking, the others reduced and strongly modified, often looking like a set of whitish, chaffy bracts; plants annual or perennial .... **Cynosurus**
        12b All spikelets similar, little modified, and usually fertile; plants perennial
            13a Upper part of callus with a band of dense soft hairs; range NE WA .... **Oryzopsis asperifolia**
            13b Upper part of callus without a band of dense soft hairs
                14a Sheaths of upper culm leaves closed 10–20% their length; spikelets not arranged like teeth of a comb; plants of coastal bluffs .... **Poa unilateralis**
                14b Sheaths of upper culm leaves open to the base; spikelets often arranged like the teeth of a comb in dense spike-like branches
                    15a Plants short, 1–30 cm, dioecious, strongly stoloniferous; staminate inflorescence branches 1–1.5 cm, usually held above leaves; pistillate inflorescences short and bur-like, hidden in leaf bases, often overlooked; planted for xeriscaping (not treated) .... **Bouteloua dactyloides**
                    15b Plants usually taller, bisexual, cespitose or rhizomatous, not stoloniferous; inflorescence branches 1–25 cm long, usually held above upper leaves
                        16a Habitat wetlands at the coast or E of the Cascades; glumes strongly scabrous or with very short, curved hairs on keels and sometimes also veins, the hairs slightly broader at the base; plants 15–250 cm tall .... **Spartina**
                        16b Habitat xeriscaping in E OR; upper glumes with moderately long hairs that originate in pimple-like, usually dark, rounded bases, each hair arising from a tiny socket and thus leaving a tiny donut-like ring if it falls off; plants 24–70 cm tall (not treated) .... **Bouteloua gracilis**
8b Inflorescence 2-sided or round in cross section; clusters of spikelets, if present, not 1-sided
    17a Inflorescence axis thickened; lower spikelets strongly appressed and completely or partially embedded in the axis
        18a Glumes with well-developed awns .... **Aegilops**
        18b Glumes awnless
            19a Corn. Leaves usually > 30 mm wide; plants with both axillary pistillate spikes (corncobs) and terminal staminate panicles (tassels) .... **Zea mays ssp. mays**

19b Not corn. Leaves < 30 mm wide; inflorescences bisexual, terminal
  20a Glumes 1 per spikelet, except the terminal spikelet with 2 glumes; spikelets oriented with the glumeless edge against the inflorescence axis
    21a Spikelets with 1 bisexual floret (sometimes with a second reduced, sterile floret) .... *Hainardia cylindrica*
    21b Spikelets with 2–20+ bisexual florets... *Lolium*
  20b Glumes 2 per spikelet on all spikelets; spikelets oriented with the flat side against the inflorescence axis; both glumes visible
    22a Spikelets with 3–12 florets; culms 50–200 cm .... *Thinopyrum*
    22b Spikelets with 1 floret; culms ≤ 35 cm
      23a Lemmas awnless...*Parapholis incurva*
      23b Lemmas awned, the awns 2–4 mm.............*Deschampsia bolanderi*
17b Inflorescence axis not clearly thickened; spikelets appressed to spreading, not embedded in the axis
  24a Spikelets subtended by scabrous hairs or bristles that are longer than the spikelets and arise on the pedicel or at the base of the spikelet; spikelets dorsiventrally flattened or round in cross section
    25a Bristles falling with the spikelets at maturity; disarticulation at the base of the reduced panicle branches; inflorescences 10–52 mm wide .... *Cenchrus*
    25b Bristles persistent on the inflorescence at maturity; disarticulation below the glumes; inflorescences (3)5–15(30) mm wide .... *Setaria*
  24b Spikelets not subtended by scabrous hairs or bristles that are longer than the spikelets (or if they seem to be, the bristles are modified glumes, not outgrowths of the pedicels, not arising at the base of the glumes); spikelets laterally flattened or round in cross section
    26a Spikelets attached directly to the main inflorescence axis, sessile or nearly so, 2–4(8) per node near the middle of the inflorescence; inflorescences spikes or condensed racemes .... **Group 1a, p. 24** (Tribe Triticeae)
    26b Spikelets either attached directly to the main inflorescence axis and only 1 per node, or attached to short branches and 1 or more per node; inflorescences condensed panicles (most species), racemes, or spikes
      27a Sheaths of upper culm leaves closed > 50% their length
        28a Lemmas awned
          29a Palea keels with awns or small triangular appendages at about midlength; plants perennial ....*Pleuropogon*
          29b Palea keels lacking awns or appendages at midlength; plants annual .... *Bromus*
        28b Lemmas awnless
          30a Plants annual; spikelets with 7–15 bisexual florets, disarticulating above the glumes; sheaths with dense, long, soft hairs .... *Bromus briziformis*
          30b Plants perennial; spikelets with 1–4 bisexual florets, disarticulating below the glumes; sheaths glabrous or sparsely short-hairy .... *Melica*

27b Sheaths of upper culm leaves open to the base, or closed < 50% their length
  31a Florets 1 per spikelet (rarely a few spikelets with 2–3 florets in *Muhlenbergia*)
    32a Lemmas with 3 awns .... **Aristida**
    32b Lemmas with 1 awn or awnless
      33a Spikelets 10–15 mm
        34a Leaves usually involute and 0.5–2.5 mm in diameter, sometimes flat and 4–8 mm wide; inflorescence a condensed panicle; perennial plants of coastal sands .... **Calamagrostis**
        34b Leaves flat, 6–15+ mm wide; inflorescence a spike; annual plants of disturbed, usually inland habitats...................... ***Triticum aestivum***
      33b Spikelets 0.8–10 mm
        35a Glume keels with coarse hairs
          36a Glumes awned
            37a Spikelets with a short, persistent stipe, disarticulating at the base of the stipe; panicle sometimes lobed............... **Polypogon**
            37b Spikelets lacking a stipe, initially disarticulating above the glumes, later also below the glumes; panicle a straight-sided cylinder .... **Phleum**
          36b Glumes awnless
            38a Ligules membranous; lemmas awned .... **Alopecurus**
            38b Ligules of hairs; lemmas awnless................................. **Crypsis**
        35b Glume keels lacking coarse hairs
          39a Upper glume (excluding awn) 3-5x as long as lemma body, long-tapered, somewhat swollen at base where distended by the floret, in and W of the Cascades .... ***Gastridium phleoides***
          39b Upper glume (excluding awn) <= 2.5 as long as lemma body, obtuse to acuminate, not swollen at base; range various
            40a Glumes clearly awned
              41a Spikelets disarticulating above glumes, floret falling alone; glumes 0.3–2× the lemma; lemma awn usually shorter than lemma body, sometimes lacking; plants perennials .... ***Muhlenbergia***
              41b Spikelets disarticulating below glumes and short, persistent stipe, falling as a unit; glumes 2–3x the lemma; lemma awn usually longer than lemma body, sometimes lacking; plants annuals .... **Polypogon**
            40b Glumes truncate to long-tapered or mucronate, not clearly awned
              42a Lemmas awned from the back
                43a Glumes fused to each other in the lower half; calluses glabrous .... ***Alopecurus myosuroides***
                43b Glumes free to the base; calluses with hairs, which may be inconspicuous
                  44a Paleas < half as long as lemmas; callus hairs usually inconspicuous, usually < 33% as long as lemmas (rarely to 50%); glumes 1.5–5(7) mm. .... **Agrostis**

      44b Paleas subequal to lemmas; callus hairs conspicuous, ≥ 33% as long as lemmas; glumes (4)5–7(9) mm .... ***Calamagrostis***
   42b Lemmas awned from the tip or awnless
      45a Ligules of hairs; lemmas awnless .... ***Crypsis***
      45b Ligules membranous; lemmas awned or awnless
         46a Lemmas stiffly membranous to hard, firmer than the glumes, tightly enfolding the palea and fruit; florets ± round in cross section; lemmas awned, awns early deciduous in some species
             .... **Group 3a, p. 32** (**Tribe Stipeae**)
         46b Lemmas thin and flexible, not or only slightly firmer than the glumes, not tightly enfolding the palea and fruit; florets usually laterally flattened; lemmas awnless or awned, awns if present not deciduous
            47a Paleas < half as long as lemmas; lemmas awnless; glumes not winged .... ***Agrostis***
            47b Paleas subequal to lemmas; lemmas awned or awnless; glumes winged or not
               48a Glumes 2–8 mm, often winged, longer than and hiding the florets; spikelets with 1 fertile floret subtended by 2 sterile, bristle-like or knob-like florets; lemmas 2–6.8 mm; culms (10)20–230 cm; leaves (0.5)2–20 mm wide; inflorescences 0.8–4 cm wide .... ***Phalaris***
               48b Glumes 1.5–3.2 mm, not winged, longer or shorter than but not hiding the florets; spikelets with 1 fertile floret not subtended by sterile florets; lemmas 1.5–4 mm; culms (3)5–150 cm; leaves 0.5–6 mm wide; inflorescences 0.1–3 cm wide .... ***Muhlenbergia***
31b Florets (1)2–20+ per spikelet
   49a Ligules of hairs; leaf sheaths with a tuft of hairs at the top; terminal inflorescences exserted; additional inflorescences completely hidden in upper leaf sheaths .... ***Danthonia***
   49b Ligules membranous (sometimes minutely ciliate); leaf sheaths usually lacking a tuft of hairs at the top; inflorescences not hidden in upper leaf sheaths (NOTE: *Anthoxanthum odoratum* has a tuft of hairs at the top of the sheath, but the ligule is membranous)
      50a Florets 3 per spikelet, with 1 firm fertile floret subtended by 2 softer, usually hairy, sterile florets (which may be reduced to tiny, soft bristles in *Phalaris*); inflorescences condensed panicles
         51a Sterile florets longer than fertile floret, awned .... ***Anthoxanthum***
         51b Sterile florets reduced, usually shorter than fertile floret, awnless .... ***Phalaris***
      50b Florets (1)2+ per spikelet, usually alike in shape; sterile florets, if any, above fertile florets; inflorescences spikes, racemes, or condensed panicles
         52a Glumes 1 per spikelet, except the terminal spikelet with 2 glumes; spikelets oriented with the glumeless edge against the inflorescence axis; spikelets sessile .... ***Lolium***

52b Glumes 2 on all spikelets, if apparently 1, then spikelets clearly stalked; spikelets oriented variously to the inflorescence axis but not with 1 edge tight against the axis
    53a Plants cespitose, 115–350 cm at flowering time, often glaucous; inflorescences often branched at lower nodes; blades 10–30 mm wide ....*Leymus condensatus*
    53b Plants not as above
        54a Glumes very narrow, awl-like, tapering from below midlength; spikelets paired, 1 spikelet of each pair sessile, the other on a pedicel to 15 mm; lemmas 10.5–15 mm with dense hairs 2–3 mm; plants rhizomatous; dunes and sandy soils in the Columbia Basin................. *Leymus flavescens*
        54b Plants not as above
            55a Inflorescences spikes; spikelets sessile or subsessile on inflorescence axis
                56a Spikelets long and narrow, (4.8)5–6× as long as wide; inflorescences nodding; mostly in and W of the Cascades ....*Brachypodium sylvaticum*
                56b Spikelets shorter and wider, 1–4× as long as wide; inflorescences nodding or erect; widespread .... **Group 1a, p. 24 (Tribe Triticeae)**
            55b Inflorescences condensed panicles or racemes; spikelets on short stalks, not attached directly to the main inflorescence axis
                57a Spikelets (15)20–40(50) mm; mostly in and W of the Cascades ....*Brachypodium sylvaticum*
                57b Spikelets 2.5–16 mm; range various
                    58a Upper glume 1.6–2.5+ times as long as the lower glume; florets apparently 2 per spikelet (actually 3 but the shorter, uppermost floret is hidden by the others), brown, hairy; awns bent, arising from lemma back ....*Anthoxanthum*
                    58b Upper and lower glumes more nearly equal in length; florets 2+ per spikelet, usually pale, glabrous or hairy; lemma awns absent or arising from the tip or the back of the lemma
                        59a Upper and lower glumes of different shapes (widest at different points along their lengths)
                            60a Upper glumes 1.5–2.9 mm, acute, rounded or truncate, sometimes hooded; lemmas 1.9–3 mm; plants of wet habitats; E of the Cascades; uncommon .... *Sphenopholis*
                            60b Upper glumes 2.5–5 mm, acute, not hooded; lemmas 2.5–6.5 mm; plants of drier habitats; widespread ....*Koeleria macrantha*
                        59b Upper and lower glumes ± alike in shape, though often differing in length
                            61a Lemmas awnless or with awns to 2 mm
                                62a Plants of coastal bluffs; panicle branches densely papillose and sparsely scabrous; inflorescences dull .... *Poa unilateralis*
                                62b Plants usually of inland habitats, if at the immediate coast then panicle branches covered with soft hairs; inflorescences usually shiny

  63a Panicle branches antrorsely scabrous, not covered with soft hairs; E of the Cascades... ***Graphephorum wolfii***
  63b Panicle branches covered with soft hairs; widespread......***Koeleria macrantha***
 61b Lemmas with awns 3–25 mm
  64a Lemmas awned from the tip... ***Festuca***
  64b Lemmas awned from the back
   65a Plants annual; spikelets 2.8–3.8 mm; lowlands in and W of the Cascades ....***Aira praecox***
   65b Plants perennial; spikelets 4–7 mm; widespread......... ***Trisetum spicatum***

Group 1a. Tribe Triticeae: Inflorescences spikes or spike-like racemes (rarely spike-like panicles), with 1–5+ sessile or subsessile spikelets per node

The wheatgrass tribe is characterized by spikes or spike-like racemes (rarely spike-like panicles in *Leymus*). The tribe has been rearranged to reflect genetic relationships. Most former *Agropyron* and all former *Sitanion* species have been placed in *Elymus*, making this genus particularly diverse. Several small groups of species are now treated as genera: *Leymus*, *Pascopyrum*, *Taeniatherum*, and *Thinopyrum*. Treatment of some taxa remains controversial.

 Assess the number of spikelets per node near the middle of the spike. If spikelet number is not obvious, then break the spike to isolate a node. Glumes and lemmas may be reduced and awn-like. Curvature of lemma awns increases with maturity; key plants with any curved lemma awns as strongly curved.

 Hybrids are common in the wheatgrass tribe, even between genera. Most are sterile, and only a few are mentioned in this field guide.

1a Inflorescence axis thickened; lower spikelets strongly appressed and completely or partially embedded in the axis
 2a Glumes with well-developed awns..***Aegilops***
 2b Glumes awnless... ***Thinopyrum***
1b Inflorescence axis not clearly thickened; spikelets appressed to spreading, not embedded in the axis
 3a Spikelets 2–8(15) per inflorescence node (sometimes to 35 per node including side branches in *Leymus*)
  4a Spikelets consistently 3 per node; inflorescence axis disarticulating at maturity (except *Hordeum vulgare*)
   5a Plants annual or perennial; central spikelet with 1 floret; lateral spikelets usually on pedicels; widespread....................................***Hordeum***
   5b Plants perennial; central spikelet usually with 2 florets; lateral spikelets sessile; E of the Cascades ....***Elymus elymoides* ssp. *hordeoides***
  4b Spikelets 2–8(15) per node at most nodes (sometimes to 35 per node including side branches in *Leymus*), but not consistently 3 per node; inflorescence axis disarticulating or not at maturity
   6a Plants annual; some awns longer than the inflorescence axis .... ***Taeniatherum caput-medusae***
   6b Plants perennial; awns, if present, shorter than the inflorescence axis
    7a Lemma awns > 3 mm; inflorescence axis disarticulating or not at maturity

8a Inflorescence axis disarticulating at maturity; awns outcurving at maturity...*Elymus*
8b Inflorescence not disarticulating; awns straight or outcurving at maturity
  9a Lemmas densely hairy with hairs 0.5–1 mm; glumes 1.5–4 mm wide; unstable coastal sands.. **Leymus mollis ssp. mollis**
  9b Lemmas usually glabrous or scabrous; glumes 0.6–2 mm wide; not on coastal sands..*Elymus*
7b Lemma awns 0–3 mm; inflorescence axis not disarticulating (tardily so in *Psathyrostachys*)
  10a Glumes awn-like or awl-like, tapering very gradually from base or from well below midlength
    11a Plants cespitose or rhizomatous, if cespitose then ligules 0.7–8 mm or spikelets 10–25 mm; inflorescence axis not disarticulating........*Leymus*
    11b Plants cespitose; ligules 0.2–0.3 mm; spikelets 7–10(12) mm; inflorescence axis tardily disarticulating; planted for forage, occasionally escaping E of the Cascades in central and NE WA (not treated).. **Psathyrostachys juncea**
  10b Glumes parallel-sided from base to above midlength, then tapering more abruptly
    12a Long-rhizomatous plants of unstable, sandy soils along the Snake and Columbia Rivers; inflorescences 1–3 cm wide ....*Leymus*
    12b Short-rhizomatous to loosely cespitose plants of more stable substrates; widespread; inflorescences 0.5–1 cm wide ....*Elymus*
3b Spikelets 1 per node at most inflorescence nodes
  13a Lemma marginal veins distally keeled and ciliate like the midrib, with stiff, short, separated hairs or small spines; lemmas otherwise glabrous; inflorescences nodding at maturity .... **Secale cereale**
  13b Lemma marginal veins not ciliate, generally not keeled, sometimes scabrous or hairy; lemmas glabrous or hairy; inflorescences erect or curved, rarely nodding at maturity
    14a Plants annual
      15a Lemmas 5–7.5 mm; spikes 1.3–2.4 cm (excluding awns); E of the Cascades ....**Eremopyrum triticeum**
      15b Lemmas 10–15 mm; spikes (3.5)6–18 cm (excluding awns); widespread.. **Triticum aestivum**
    14b Plants perennial
      16a Inflorescences very crowded, internodes usually < 3 mm; spikelets > 3× as long as the internodes; mainly E of the Cascades.........***Agropyron***
      16b Inflorescences less crowded, internodes 4–15+ mm; spikelets ≤ 3× as long as the internodes
        17a Glumes truncate to acute, very thick, stiff, and hard, awnless; lemmas usually awnless... ***Thinopyrum***
        17b Glumes obtuse to acuminate, thin and flexible, awned or awnless; lemmas awned or awnless
          18a Mature lemmas with outcurving awns
            19a Inflorescences disarticulating when mature; anthers 1–2 mm...*Elymus*

19b Inflorescences not disarticulating; anthers 2–8 mm
    20a Glumes narrowly lanceolate, the veins 1–3(4), not including the somewhat thickened margins; native in NE OR and SE WA, introduced elsewhere E of the Cascades .... *Elymus wawawaiensis*
    20b Glumes lanceolate to oblanceolate, the veins (3)4–5(7) (check several glumes); E of the Cascades and SW OR .... *Pseudoroegneria spicata*
18b Mature lemmas with straight awns or awnless
    21a Plants cespitose, sometimes loosely so; lemmas glabrous to sparsely hairy
        22a Anthers 4–8 mm; spikes loose and open, spikelets usually 1.1–1.5× as long as internodes of inflorescence axis; inflorescence axis usually plainly visible; E of the Cascades and SW OR .... *Pseudoroegneria spicata*
        22b Anthers 0.7–2.5(3.5) mm; spikes compact, spikelets usually 1.5–3× as long as internodes of inflorescence axis; inflorescence axis partly concealed by spikelets to plainly visible; range various...*Elymus*
    21b Plants strongly rhizomatous; lemmas glabrous to densely long-hairy
        23a Spikelets 2(3) per node at most or all nodes in middle of spike, 1 per node near base and tip of spike; glumes very narrow, awl-shaped or awn-shaped .... *Leymus triticoides*
        23b Spikelets 1 per node at most or all nodes in all parts of spike; glumes broader, not awl-shaped or awn-shaped
            24a Blades 5–10 mm wide, usually flat; lower sheaths usually with spreading hairs..*Elymus*
            24b Blades normally strongly involute or considerably < 6 mm wide; lower sheaths usually glabrous
                25a Glumes oblong-lanceolate, widest at or above midlength, 3–9-veined, tips not twisted; glume midvein straight; lemmas glabrous to copiously hairy; range various ....*Elymus*
                25b Glumes lanceolate, widest near base, 3–5-veined, tips acuminate, often twisted; glume midvein curving slightly to 1 side; lemmas glabrous to moderately pubescent; usually E of the Cascades .... *Pascopyrum smithii*

Group 2. Inflorescences with spike-like or raceme-like branches

1a Spikelets dorsiventrally compressed
    2a Lower glume much smaller than the upper, exposing the florets; upper lemmas usually thick, leathery or hardened, resembling a turtle shell; inflorescences not conspicuously hairy, hairs if present not longer than the spikelets; lemmas awnless or awned (if lemmas awned, leaves lacking ligules); spikelets paired or not .... **Group 2a, p. 28** (Tribe Paniceae)
    2b Both glumes well-developed, hiding the florets; upper lemmas hyaline to membranous, not thick, not leathery or hardened, not resembling a turtle shell; inflorescences conspicuously hairy, hairs often longer than the spikelets; lemmas awned; ligules present; spikelets paired
        3a Inflorescences terminal, dense, with > 15 branches; both spikelets of each spikelet pair pedicellate, similar in shape and size, neither spikelet reduced; plants cespitose, 60–200 cm tall; exotic ornamental rarely escaping in W OR and WA... *Miscanthus sinensis*

3b Inflorescences axillary and terminal, open, each with 1–2 branches; spikelet pairs with one sessile spikelet and one pedicellate spikelet, the pedicellate spikelet reduced; plants cespitose, sometimes short-rhizomatous; 30–150 cm tall; rare along the Columbia River in central WA
....*Schizachyrium scoparium* **var.** *scoparium*
1b Spikelets laterally compressed or ± round in cross section
   4a Glumes semicircular, winged, the wings inflated and the veins prominently raised; plants annual; wetlands, often growing in standing water
....*Beckmannia syzigachne*
   4b Glumes not semicircular, generally not winged, veins not prominently raised; plants annual or perennial; habitat various
     5a Inflorescences digitate, with spike-like branches arranged in 1 or more whorls
       6a Lemmas awned ....*Chloris verticillata*
       6b Lemmas awnless
         7a Spikelets with 1 floret (sometimes with a 2nd rudimentary floret); inflorescence branches 1.4–6 cm × 1–2 mm, arranged in a single terminal whorl..*Cynodon dactylon*
         7b Spikelets with 5–9(11) florets; inflorescence branches 1–16 cm × 3–14 mm, arranged in a terminal whorl, sometimes with a single branch attached ≤ 3 cm lower on the inflorescence axis; 2 species (*Eleusine indica, E. tristachya*) collected on ballast in Portland in the early 1900s; *E. indica* more recently documented near Vancouver, B.C., and to be expected in our area (not treated) ....*Eleusine*
     5b Inflorescences racemose, branches spike-like or raceme-like, not arranged in whorls
       8a Spikelets very flat and very tightly crowded on 1-sided branches, arranged like the teeth of a comb
         9a Plants short, 1–30 cm, dioecious, strongly stoloniferous; staminate inflorescence branches 1–1.5 cm, usually held above leaves; pistillate inflorescences short and bur-like, hidden in leaf bases, often overlooked; planted for xeriscaping (not treated)
....*Bouteloua dactyloides*
         9b Plants usually taller, bisexual, cespitose or rhizomatous, not stoloniferous; inflorescence branches 1–25 cm long, usually held above upper leaves
           10a Habitat wetlands at the coast or E of the Cascades; glumes strongly scabrous or with very short, curved hairs on keels and sometimes also veins, the hairs slightly broader at the base; plants 15–250 cm tall..*Spartina*
           10b Habitat arid grasslands; upper glumes with moderately long hairs that originate in pimple-like, usually dark, rounded bases, each hair arising from a tiny socket and thus leaving a tiny donut-like ring if it falls off; plants 24–70 cm tall; planted for xeriscaping (not treated)................................*Bouteloua gracilis*
       8b Spikelets not crowded, not arranged like the teeth of a comb
         11a Each spikelet with a tuft of long hairs at the base (at top of pedicel just below glumes, not on the lemma callus) ......*Miscanthus sinensis*

11b Spikelets glabrous or hairy but not with a tuft of long hairs at the base
  12a Ligules of hairs; leaf sheaths often with a tuft of hairs at the top; inflorescences sometimes enclosed in the uppermost leaf sheath even when mature .... ***Sporobolus***
  12b Ligules membranous or ciliate, not entirely of hairs; leaf sheaths lacking a tuft of hairs at the top; inflorescences normally emerging from the uppermost leaf sheath before maturity
    13a Lemmas (1)3-veined, awnless or awned; ligules 2–8 mm .... ***Diplachne fusca*** (with 2 subspecies, see p. 56)
    13b Lemmas (3)5–9-veined, awned; ligules ≤ 1 mm .... ***Festuca***

Group 2a. Tribe Paniceae: spikelets dorsiventrally flattened; lower glume reduced; lower lemma resembling upper glume; lower floret staminate or sterile; upper floret fertile. Upper lemma often hard, often shiny. Leaf sheaths open. Inflorescences panicles (sometimes condensed and spike-like)

1a Spikelets enclosed in a spiny, bur-like fascicle .... ***Cenchrus longispinus***
1b Spikelets not enclosed in a spiny, bur-like fascicle
  2a Inflorescences condensed, cylindrical, spike-like panicles; each spikelet subtended by 1–many scabrous bristles
    3a Bristles falling with the spikelets at maturity; disarticulation at the base of the panicle branches; inflorescences 10–52 mm wide .... ***Cenchrus***
    3b Bristles persistent on the inflorescence; disarticulation below the spikelets; inflorescences 5–15 mm wide, to 30 mm wide in *S. italica* .... ***Setaria***
  2b Inflorescences not as above; spikelets not subtended by bristles, though papillose-based hairs may occur on the pedicel
    4a Ligules absent; glumes and lower lemmas usually awned, sometimes awnless .... ***Echinochloa***
    4b Ligules present, membranous, sometimes ciliate, or of hairs and hard to see; glumes and lemmas awnless
      5a Inflorescence branches spike-like, usually not branched again; spikelets in 2 rows along 1 side of each branch, ± appressed to the branch axis
        6a Plants annual, rarely short-lived perennials; lemma of fertile (upper) floret membranous, flexible at maturity; spikelets paired or in 3s, on pedicels of unequal length .... ***Digitaria***
        6b Plants perennial; lemma of fertile floret leathery to hardened, rigid at maturity; spikelets solitary or paired, subsessile or on short pedicels of equal length .... ***Paspalum***
      5b Inflorescence branches not spike-like, often branched again 2 or more times; spikelets not in 2 rows along 1 side of each branch
        7a Inflorescence branches 1-sided, the secondary branches ± appressed to the primary branches and to each other; pedicels < 2 mm; SW OR .... ***Panicum rigidulum* ssp. *rigidulum***
        7b Inflorescence branches not 1-sided, the secondary branches diverging from the primary branches; at least some pedicels 2–20 mm; widespread

8a Basal leaves not forming rosettes, similar to cauline leaves; plants annual or perennial..***Panicum***
8b Basal leaves forming rosettes, shorter and wider than the cauline leaves; plants perennial...***Dichanthelium***

Group 3. Inflorescence a panicle, branches not spike-like; florets 1 per spikelet

1a Glumes lacking; habitat wetlands
  2a Culms 1–5(10) cm; panicle branches in umbel-like clusters; lower Columbia River and Columbia Gorge...***Coleanthus subtilis***
  2b Culms 20–500 cm; panicle branches not in umbel-like clusters; range various
    3a Leaf margins and surfaces sharply scabrous, apt to cut skin and stick to clothing; plants to 150 cm; spikelets with all florets bisexual ....***Leersia oryzoides***
    3b Leaf margins and surfaces smooth or scabrous but not sharp and not sticking to clothing; plants 20–500 cm; spikelets of the lower part of panicle staminate, narrow, and falling early; spikelets of the upper part of panicle pistillate, wider, persisting longer..***Zizania***
1b Glumes 2, usually well-developed, the lower glume sometimes much smaller than the upper; habitat wetlands or uplands
  4a Spikelets dorsiventrally flattened
    5a Spikelets with 1 floret; spikelets not paired or in 3s; Eurasian weed of winter wheat in ID; to be expected in E WA and OR (not treated) ....***Milium vernale***
    5b Spikelets appearing to have one floret, but with a sterile or vestigial lower floret below the upper fertile floret; spikelets solitary, paired or in 3s
      6a Lower glume much smaller than the upper, exposing the lower floret; spikelets solitary or in pairs ....**Group 2a, p. 28 (Tribe Paniceae)**
      6b Both glumes well-developed, hiding the florets; spikelets paired or in 3s (if 3, the 2 staminate spikelets may fall early)
        7a Inflorescences plumose; upper surface of leaf blade densely long-hairy near the base, the hairs obscuring the blade surface for 2+ cm; spikelets subtended by numerous, conspicuous callus hairs as long as the spikelets..***Tripidium ravennae***
        7b Inflorescences not plumose; upper surface of leaf blade sparsely hairy or glabrous near the base, the hairs not obscuring the blade surface; spikelets subtended by inconspicuous callus hairs much < the spikelets...***Sorghum***
  4b Spikelets ± laterally flattened, sometimes round in cross section
    8a Glumes with a flat or inflated wing-like keel
      9a Inflorescences open, branches easily discernible; glumes semicircular; glume wings inflated and the veins prominently raised ....***Beckmannia syzigachne***
      9b Inflorescences dense, branches obscure; glumes not semicircular; glume wings not inflated; the veins not prominently raised .....***Phalaris***
    8b Glume not with a flat or inflated wing-like keel

10a Glumes and lemmas awnless (opposite lead on p. 31)
- 11a Ligules of hairs; leaf sheaths often with a tuft of hairs at the top; inflorescence sometimes enclosed in the uppermost leaf sheath at maturity
  - 12a Lemma base clearly hairy-tufted, hairs 1+ mm long, about 25–85% as long as lemma, lemma otherwise glabrous; inflorescence not enclosed in uppermost leaf sheath at maturity .... ***Calamovilfa longifolia* var. *longifolia***
  - 12b Lemma base not hairy-tufted, lemma hairs if present < 1 mm; inflorescence sometimes enclosed in uppermost leaf sheath at maturity .... ***Sporobolus***
- 11b Ligules membranous, sometimes ciliate; leaf sheaths lacking a tuft of hairs at the top; inflorescence normally emerging from the uppermost leaf sheath before maturity
  - 13a Lemmas stiffly membranous to hard, not keeled, tightly enfolding the palea and fruit; florets ± round in cross section
    - 14a Lemmas 2–2.3 mm, glabrous, awnless; calluses glabrous; plants annual; Eurasian weed of winter wheat in ID; to be expected in E WA and OR (not treated) .... ***Milium vernale***
    - 14b Lemmas 2–7 mm, glabrous or hairy, awned, awns early deciduous in some species; calluses glabrous or hairy; plants perennial .... **Group 3a, p. 32** (**Tribe Stipeae**)
  - 13b Lemmas thin and flexible, often keeled, not tightly enfolding palea and fruit; floret usually laterally flattened
    - 15a Lemma base clearly hairy-tufted; if lemmas evenly hairy, then the hairs at least 1 mm
      - 16a Spikelets 8–13 mm, subsessile; inflorescences dense; plants of coastal sand dunes .... ***Calamagrostis***
      - 16b Spikelets 3–10 mm, usually clearly stalked; inflorescences dense to open; plants not of coastal sand dunes
        - 17a Lemmas awned from the tip, sometimes awnless; paleas well-developed, mostly subequal to lemmas .... ***Muhlenbergia***
        - 17b Lemmas awnless or awned from the back; paleas often reduced or lacking, sometimes subequal to lemmas
          - 18a Glumes 1.5–4 mm; plants (5)10–120 cm; leaves 1–8(10) mm wide .... ***Agrostis***
          - 18b Glumes 4–8 mm; plants (40)150–230 cm; leaves 5–20 mm wide .... ***Phalaris***
    - 15b Lemma base not hairy-tufted; if lemmas evenly hairy, the hairs < 1 mm
      - 19a One or both glumes < lemma
        - 20a Leaf sheath closed ~ 50% its length; leaves 2–13 mm wide, panicles open; SE OR .... ***Catabrosa aquatica***
        - 20b Leaf sheath open to the base; leaves 0.6–6 mm wide; panicles condensed or open; range various .... ***Muhlenbergia***
      - 19b Both glumes equaling or > lemma
        - 21a Spikelets shed as a unit, the glumes not persisting on the plant; if leaves > 6 mm wide, then panicles open
          - 22a Leaf blades (1)7–20 mm wide; panicles open, nodding; widespread .... ***Cinna latifolia***

22b Leaf blades 1–5 mm wide; panicles condensed, sometimes lobed or somewhat interrupted; Columbia Gorge and W of the Cascades .... ***Polypogon viridis***

21b Spikelets not shed as a unit, disarticulating above the glumes, the glumes persistent on the plant; if leaves > 6 mm wide, then panicles usually condensed

    23a Glumes (4)4.5–8.1 mm; wider leaf blades > 10 mm wide .... ***Phalaris***

    23b Glumes 1.5–4.3(5) mm; leaf blades 2–8(10) mm wide

        24a Paleas ≥ 67% as long as lemmas; rachilla usually prolonged 0.1–1.9 mm beyond the base of the floret, sometimes absent or very difficult to see; montane to alpine (if in doubt go to the *Agrostis* key where *Podagrostis* species are keyed using different traits) .... ***Podagrostis***

        24b Paleas 0–67% as long as lemmas; rachilla not prolonged beyond the base of the floret; lowlands to alpine .... ***Agrostis***

10b Glumes or lemmas awned

  25a Lemmas with 3 awns .... ***Aristida***

  25b Lemmas with 1 awn or awnless

    26a Lemmas stiffly membranous to hard, firmer than the glumes, tightly enfolding the palea and fruit; florets ± round in cross section; lemma awns 3–225 mm, early deciduous in some species .... **Group 3a, p. 32 (Tribe Stipeae)**

    26b Lemmas thin and flexible, not or only slightly firmer than the glumes, not tightly enfolding the palea and fruit; florets usually laterally flattened; lemma awns absent or to 16 mm, not deciduous

        27a One or both glumes awned or abruptly short-pointed

            28a Spikelets disarticulating below glumes and the short, persistent stipe, falling as 1 unit; glumes 2–3× as long as the lemma; lemma awn usually longer than lemma body, sometimes lacking .... ***Polypogon***

            28b Spikelets disarticulating above glumes, floret falling alone; glumes 0.3–2× as long as the lemma; lemma awn usually shorter than lemma body, sometimes lacking

                29a Lemmas 5-veined or veins too faint to see; palea < 67% as long as lemma .... ***Agrostis***

                29b Lemmas 3-veined; palea generally as long as lemma .... ***Muhlenbergia***

        27b Glumes obtuse, acute, or tapered to tip, not awned

            30a Lemma awns > 2× as long as the spikelets

                31a Lemma awns subterminal .... ***Apera***

                31b Lemma awns terminal .... **Group 3a, p. 32 (Tribe Stipeae)**

            30b Lemma awns lacking or < 2× as long as the spikelets

                32a Lemmas awned from near midlength or below

                    33a Lemmas with conspicuous callus hairs, easily visible with the naked eye or a 10× hand lens .... ***Calamagrostis***

                    33b Lemmas glabrous or with inconspicuous callus hairs

                        34a Paleas ≤ 67% as long as the lemma; rachilla not prolonged beyond the base of the floret as a bristle .... ***Agrostis***

34b Paleas 67–100% as long as the lemma; rachilla prolonged beyond the base of the floret as a bristle, though sometimes for only 0.1 mm (sometimes absent or very difficult to see)
    35a Calluses hairy, the hairs 0.5–4.5 mm; branches spikelet-bearing throughout; culms (10)35–100+ cm .... *Calamagrostis*
    35b Calluses glabrous or with hairs to 0.5 mm; branches spikelet-bearing only in the outer half; culms 5–50 cm .... *Podagrostis*
32b Lemmas awned at or near the tip, abruptly short-pointed, or awnless
  36a Plants annual, delicate; panicle branches widely spreading .... *Ventenata dubia*
  36b Plants perennial, not especially delicate; panicle branches widely spreading or appressed
    37a Inflorescences erect, the branches appressed; leaves 0.5–1.6(4.2) mm wide; spikelets disarticulating above the glumes .... *Muhlenbergia*
    37b Inflorescences nodding, the branches spreading; leaves (1)7–20 mm wide; spikelets disarticulating below the glumes .... *Cinna latifolia*

Group 3a. Tribe Stipeae: floret 1, ± round in cross section, generally awned though awn may be deciduous

Nearly all needlegrasses were formerly placed in the worldwide genus *Stipa*. Recent studies have shown that true *Stipa* species are not native to North America, and our needlegrasses and ricegrasses are now placed in *Eriocoma* and several other genera.

1a Panicle branches 10–20+ per node at lower and central nodes; florets ~2 mm; awns ≤ 8 mm, deciduous; exotic ornamental from New Zealand, rarely escaping W of the Cascades and perhaps not persisting (not treated) .... *Anemanthele lessoniana*
1b Panicle branches 1–4 per node; florets ≥ (1.5)2.5 mm; awns 3–225 mm, deciduous or persistent
  2a Lemmas narrowing near the tip into a cylindrical or cup-like crown topped by a row of hairs and surrounding the base of the awn; lemma margins strongly overlapping their entire length at maturity; lemmas often minutely papillose, especially distally, ± textured at 20× .... *Nassella*
  2b Lemmas narrowing near the tip but not extended into a cylindrical or cup-like crown; lemma margins overlapping or not; lemmas not papillose, ± smooth at 20×
    3a Lemma awns 3–20 mm, deciduous or persistent, straight or bent once
      4a Uppermost culm leaf blades reduced, 2–12 mm long; basal leaves overwintering, remaining green; NE WA .... *Oryzopsis asperifolia*
      4b Uppermost culm leaf blades not reduced, > 12 mm long; basal leaves not overwintering; range various, including NE WA
        5a Lemma margins not overlapping, separate their whole length; awns persistent .... *Piptatheropsis exigua*
        5b Lemma margins overlapping; awns readily deciduous .... *Eriocoma*
    3b Lemma awns 12–225 mm, persistent; bent once or twice, if < 20 mm bent twice
      6a Hairs on basal segment of awn 3–8 mm; SE OR .... *Pappostipa speciosa*
      6b Hairs on basal segment of awn 0–2 mm; range various

7a Awns 75–225 mm; calluses 2–4 mm, sharply pointed; E of the Cascades .... ***Hesperostipa comata* ssp. *comata***
7b Awns 12–60 mm; calluses 0.2–1.5 mm, blunt or sharply pointed; range various ....***Eriocoma***

Group 4. Inflorescence a panicle; florets 2+ per spikelet; leaf sheaths closed ≥ 25%

1a Spikelets in dense 1-sided clusters; panicles often with 1–2 remote lower branches; culm bases flattened; plants cespitose .... ***Dactylis glomerata***
1b Spikelets not borne in dense 1-sided clusters; other characters various
    2a Culms bulbous-based
        3a Sheaths of upper culm leaves closed to the top or nearly so .... ***Melica***
        3b Sheaths of upper culm leaves closed about 25% their length .... ***Poa***
    2b Culms not bulbous-based or culm bases not available
        4a Leaf sheaths generally closed 100% their length; upper 1–4 florets sterile, forming a rudiment composed of empty lemmas enclosing one another and lacking paleas; auricles lacking; plants perennial .... ***Melica***
        4b Leaf sheaths closed 25–90% their length or more but with at least a V-shaped opening near the top; upper florets normal, sometimes reduced and staminate or pistillate, but not forming a rudiment composed of empty lemmas enclosing one another, paleas always present in each lemma; auricles sometimes present; plants perennial or annual
            5a Lemmas awned
                6a Plants annual .... ***Bromus***
                6b Plants perennial
                      7a Spikelets 15–80 mm; lemma awns usually arising from between lemma teeth; leaf sheaths of fertile culms closed > 90% their length (with a V-shaped opening at the top) .... ***Bromus***
                      7b Spikelets 7–17 mm; lemma awns terminal or nearly so; leaf sheaths of fertile culms usually split by growing culms .... ***Festuca***
            5b Lemmas awnless
                8a Lemma veins faint to conspicuous, not equally spaced their entire length, usually curved and converging but not actually meeting near the tip; leaf tips prow-shaped or not
                9a Spikelets 16–40(45) mm; tips of ovary and seed minutely hairy; callus never with cobwebby hairs; sheaths closed > 90% their length .... ***Bromus***
                9b Spikelets 2–17 mm; tips of ovary and seed glabrous (except *Festuca viridula*); callus sometimes with cobwebby hairs; sheaths closed ≤ 90% their length
                      10a Leaf tips gradually tapered to a point; glumes ≤ the lowest lemmas; callus lacking hairs; seed attachment scars linear .... ***Festuca***
                      10b Leaf tips prow-shaped, usually conspicuously so; glumes often ≥ the lowest lemmas; callus often hairy, the hairs sometimes cobwebby; seed attachment scars round to oval .... ***Poa***

8b Lemma veins conspicuous, equally spaced their entire length, not converging; leaf tips prow-shaped
    11a Spikelets sessile or subsessile on 1 side of the inflorescence axis; annual on disturbed upland sites; E of the Cascades...................................**Sclerochloa dura**
    11b Spikelets pedicellate in open or contracted panicles, not restricted to 1 side of the inflorescence axis; plants usually perennial, growing in wetlands; range various
        12a Spikelets with (1)2(3) florets; lower glumes veinless; spikelets open; lower floret sessile, upper floret borne on a relatively long, exposed rachilla internode; SE OR...**Catabrosa aquatica**
        12b Spikelets with 2–16 florets; lower glumes 1-veined; spikelets compact; rachilla internodes relatively short and often concealed by the strongly overlapping florets; range various ....**Glyceria**

Group 5. Inflorescence a panicle; florets 2+ per spikelet; leaf sheaths open, or closed < 25%

1a Spikelets dorsiventrally flattened
    2a Lower glume much smaller than the upper, exposing the florets; spikelets solitary or in pairs..**Group 2a, p. 28** (**Tribe Paniceae**)
    2b Both glumes well-developed, hiding the small lemmas; spikelets paired or in 3s (if 3, the 2 staminate spikelets may fall early)
        3a Inflorescences plumose; upper surface of leaf blade densely long-hairy near the base, the hairs obscuring the blade surface for 2+ cm; spikelets subtended by numerous, conspicuous callus hairs as long as the spikelets ....***Tripidium ravennae***
        3b Inflorescences not plumose; upper surface of leaf blade sparsely hairy or glabrous near the base, the hairs not obscuring the blade surface; spikelets subtended by inconspicuous callus hairs, much < the spikelets ....***Sorghum***
1b Spikelets laterally flattened or round in cross section
    4a Mature glumes leathery, shiny, thicker than fertile lemmas................***Sorghum***
    4b Mature glumes and lemmas alike in texture or lemmas thicker than glumes
        5a Plants usually > 2 m; inflorescences large, fluffy panicles 15–130 cm; culms (5)10–15+ mm thick; lemmas or rachillas with long, silky hairs
            6a Plants strongly rhizomatous; leaf margins not sharply scabrous, not apt to cut the skin; lemmas glabrous or long-hairy
                7a Lemmas glabrous, awnless, with long-acuminate tips; rachilla internodes long-silky-hairy; leaf blades 15–40 cm long × 2–4 cm wide, lacking wedge-shaped brown area at base
...........................***Phragmites australis*** (with 2 subspecies, see p. 77)
                7b Lemmas long-hairy, awned from between 2 long, awn-like teeth; rachilla internodes glabrous; leaf blades 30–100 cm long × 2–7(9) cm wide, with wedge-shaped brown area at base ....***Arundo donax***
            6b Plants densely cespitose; leaf margins sharply scabrous, likely to draw blood; lemmas long-hairy, especially below midlength ....***Cortaderia***

5b Plants differing from the above in some way; inflorescences various but not both fluffy and 15–130 cm; plants usually < 2 m; culms usually < 10 mm thick; lemmas and rachillas often lacking long silky hairs
  8a Inflorescences 1-sided, dense
    9a Spikelets paired, the upper spikelet of each pair fertile, the lower spikelet sterile, reduced and strongly modified, often looking like a set of whitish, chaffy bracts; plants annual or perennial, (5)9-75(100)cm tall; range various .... ***Cynosurus***
    9b Spikelets not paired, similar in appearance, all normally fertile; plants annual, 2-15(30) cm tall; E of the Cascades .... ***Sclerochloa dura***
  8b Inflorescences not both 1-sided and dense
    10a Top of leaf sheath with a distinct tuft of long hairs (look closely; they may be inconspicuous)
      11a Plants strongly rhizomatous, dioecious; leaves stiffly 2-ranked; alkaline areas E of the Cascades or saline areas near the coast .... ***Distichlis spicata***
      11b Plants cespitose, bisexual; leaves neither stiff nor strongly 2-ranked; habitat and range various
        12a Palea keels conspicuously densely hairy, the tufts of hairs visible to the naked eye; culms disarticulating at maturity; leaf sheaths somewhat inflated; sandy soils along the lower Columbia River .... ***Triplasis purpurea* var. *purpurea***
        12b Palea keels not conspicuously densely hairy; culms not disarticulating at maturity; leaf sheaths not inflated; habitat and range various
          13a Lemmas awned
            14a Lemmas with distinct tufts of hairs near the callus base and elsewhere on the lemma, hairs often in 1 or more transverse rows above the callus and/or at midlength ....***Rytidosperma***
            14b Lemmas with tufts of hairs only at the base of the callus, otherwise glabrous, or with hairs evenly distributed or mostly marginal .... ***Danthonia***
          13b Lemmas awnless
            15a Glumes longer than the adjacent floret; mainly coastal .... ***Danthonia decumbens***
            15b Glumes shorter than the adjacent floret; range various
              16a Spikelets with 3–40 florets; rachilla not prolonged beyond the distal floret; collars not marked with a line or ridge; leaf blades not disarticulating from sheaths; plants annual or perennial, often with saucer-shaped glands or glandular bands; widespread .... ***Eragrostis***
              16b Spikelets with (1)2–5 florets; rachilla often prolonged beyond the distal floret and terminating in a much-reduced, rudimentary floret; collars marked with a line or narrow ridge; leaf blades eventually disarticulating from sheaths; plants perennial, lacking glands; introduced ornamental rarely escaping W of the Cascades .... ***Molinia caerulea***

10b Top of leaf sheath lacking a distinct tuft of long hairs, or if hairy then similar hairs present elsewhere on the sheath
  17a Plants annual; lemma awns easily visible without dissecting the spikelets, bent, arising from lemma tip or below midlength; panicles open and much-branched at maturity
    18a Plants robust; lemma awns 15–45 mm; glumes 14–33 mm; spikelets pendent at maturity .... ***Avena***
    18b Plants delicate; lemma awns 2–9(15) mm; glumes 1–12 mm; spikelets spreading to erect at maturity
      19a Spikelets 1.5–3.5 mm .... ***Aira***
      19b Spikelets 4–15 mm
        20a Glume veins (3)5–9, conspicuous; lemmas 5–10 mm; spikelets with 2–3 florets; lowest floret staminate with awn arising from the lemma tip, upper floret(s) bisexual, with awns arising from below midlength; ligules 1–8 mm, acute to obtuse, usually lacerate .... ***Ventenata dubia***
        20b Glume veins 1–3, inconspicuous; lemmas 1.5–3 mm; spikelets with 2 bisexual florets; both lemmas in each spikelet with awn arising from below midlength; ligules (0.5)2–3(4.7) mm, acute to acuminate, entire .... ***Deschampsia danthonioides***
  17b Plants annual or perennial, not as above in one or more ways
    21a Spikelets with 3 florets, the upper floret fertile, the lower florets sterile or staminate, sometimes modified and greatly reduced (NOTE: If in doubt try both leads)
      22a Glumes clearly < the florets; auricles ciliate; lemma of second floret rugose in distal 50%, exceeding the upper floret; established in coastal N CA, recently found near Reedsport in Douglas County, OR, and likely to occur elsewhere .... ***Ehrharta erecta***
      22b Glumes slightly < to > the florets; auricles lacking; lemma of second floret not rugose in distal 50%, exceeding the upper floret or not, sometimes modified and much reduced; widespread
        23a Lower florets vestigial, 0.2–4.5 mm, from much < to 75% as long as the upper, fertile floret; plants without a sweet odor when crushed .... ***Phalaris***
        23b Lower florets staminate, 3–5 mm, > the upper, fertile floret; plants with a sweet odor when crushed .... ***Hierochloë***
    21b Spikelets with (1)2-many fertile florets, the lower floret(s) fertile, the upper florets sometimes reduced and sterile but similar to the lower
      24a Spikelets with 2(3) florets; the awn of the lower lemma distinctly different from the awn of the upper lemma, or 1 of the lemmas awnless
        25a Plants annual, delicate; leaves and nodes glabrous .... ***Aira***
        25b Plants perennial, not delicate; leaves and nodes glabrous or pubescent
          26a Plants velvety-hairy throughout or only on culm nodes; lower lemma awnless; spikelets disarticulating below the glumes .... ***Holcus***

26b Plants not velvety-hairy; culm nodes glabrous or pubescent; lower lemma awned, the awn 10–20 mm; spikelets disarticulating above the glumes ....***Arrhenatherum elatius***

24b Spikelets with (1)2–many florets; lemmas awnless or with awns that are similar (but may vary in length)
  27a Lemmas (6)7–20 mm; spikelets nearly sessile
    28a Lemmas awned from between 2 acute to aristate, apical teeth; awns twisted and geniculate, flattened; range various........................***Danthonia***
    28b Lemmas awned from the tip; awns straight, terete; E of the Cascades ....***Leymus***
  27b Lemmas 2–6(7) mm; spikelets stalked or sessile
    29a Some lemmas with awns ≥ 1 mm
      30a Awns arising at or below middle of lemma
        31a Plants delicate annuals
          32a Glumes 1.5–3.5 mm; lemma tips with 2 slender teeth ....***Aira***
          32b Glumes 3.5–9 mm; lemma tips acute to acuminate, not toothed ....***Deschampsia danthonioides***
        31b Plants perennial
          33a Awns bent, with a ring of short projections at the bend, the distal segment club-shaped; introduced ornamental, escaped and forming small persistent populations in the Puget Sound area and in BC..***Corynephorus canescens***
          33b Awns straight or bent, but not as above; range various
            34a Callus hairs about half as long as the lemmas; ligules truncate to obtuse, 1.5–3.5 mm...........***Vahlodea atropurpurea***
            34b Callus hairs < half as long as the lemmas; ligules acute to acuminate or rounded, 1.5–13 mm
              35a Lemmas scabrous or puberulent, not shiny; introduced to S BC from E North America; to be watched for in N WA (not treated)...***Avenella flexuosa***
              35b Lemmas smooth, glabrous, shiny..................***Deschampsia***
      30b Awns arising above middle or at tip of lemma
        36a Lower glumes > lowest lemmas...................... ***Graphephorum wolfii***
        36b Lower glumes ≤ lowest lemmas
          37a Panicle branches soft-hairy; leaf tips ± prow-shaped ....***Koeleria macrantha***
          37b Panicle branches glabrous to scabrous; leaf tips tapering gradually to a point, not prow-shaped
            38a Awns usually arising > 1 mm below lemma tip; inflorescences shiny; branches often branched, not spike-like; plants perennial..***Graphephorum***
            38b Awns arising ≤ 1mm below lemma tip or from tip; inflorescences not shiny, branches spike-like or not; plants annual or perennial
              39a Ligules 2–8 mm; panicle branches +/- spike-like .... ***Diplachne***

39b Ligules 0.1–1.5 mm; panicle branches usually not spike-like
- 40a Leaves with claw-like or clasping auricles or with expanded, flaring, rounded area at base of blade; plants perennial .... **Schedonorus**
- 40b Leaves without auricles or these reduced to tiny erect flaps, without expanded, flaring, rounded area at base of blade; plants annual or perennial .... **Festuca**

29b Lemmas awnless or with awns < 1 mm.
- 41a Lemmas wider than long, awnless .... **Briza**
- 41b Lemmas longer than wide, awnless or with a short awn
  - 42a Culms solid; panicles very open; lemmas 1–6 per spikelet, often dark gray, with 3 veins or veins obscure; glumes with 1(2–3) veins
    - 43a Florets 2–6 per spikelet; calluses with hairs to 1.5 mm; culms 50–130 cm; sand dunes and wind-blown soil in E WA......... ***Redfieldia flexuosa***
    - 43b Florets 1–2(3) per spikelet; calluses glabrous; culms 10–60 cm; moist alkaline meadows, wet ditches; widespread E of the Cascades .... **Muhlenbergia**
  - 42b Culms hollow; panicles open to condensed; lemmas 2–many per spikelet, usually whitish, green, or straw-colored; glumes often with 3+ veins
    - 44a Lemma tips blunt (sometimes acute); lower glumes usually ≤ 50% as long as the lowest lemmas; lemma veins usually parallel, not converging toward lemma tips; wetlands
      - 45a Plants rhizomatous; lemma veins prominent; freshwater wetlands, streams, and lakes.. **Torreyochloa**
      - 45b Plants cespitose, not rhizomatous; lemma veins usually faint; freshwater, saline, or alkaline wetlands
        - 46a Panicle branches erect to ascending, spike-like; spikelets sessile or nearly so in 2 rows; freshwater to alkaline wetlands .... **Diplachne**
        - 46b Panicle branches ascending, spreading, or descending, not spike-like, secondary and tertiary branches often present; spikelets pedicellate and not in 2 rows; alkaline to saline wetlands .... **Puccinellia**
    - 44b Lemma tips acute to acuminate (sometimes blunt); lower glumes usually > 50% as long as lowest lemmas; lemma veins usually not parallel, converging toward lemma tips; wetlands or uplands
      - 47a Plants emergent in shallow water, rhizomatous
        - 48a Spikelets with 3–4 florets; lemmas 4–9 mm; calluses densely hairy, the hairs 1–1.5 mm; culms 6–8 mm thick at the base; E of the Cascades in S OR..................................... ***Scolochloa festucacea***
        - 48b Spikelets with 4–8 florets; lemmas 2–3.6 mm; calluses glabrous; culms 1–4.8 mm thick at the base; range various .... **Torreyochloa**
      - 47b Plants of wetlands or uplands, not emergent; rhizomatous or cespitose
        - 49a Florets 3, the lower 2 sterile or staminate, the upper 1 fertile; leaves 2–15 mm wide, with a sweet odor when crushed .... **Hierochloë**

49b Florets 2–12, all fertile, or the upper 1–few sometimes sterile; leaves 0.4–12 mm wide, without a sweet odor when crushed
  50a Upper and lower glumes of different shapes (widest at different points along their lengths)
    51a Upper glumes 1.5–2.9 mm, acute or rounded to truncate, sometimes hooded; lemmas 1.9–3 mm; spikelets disarticulating below the glumes; panicle axis and branches scabrous, not soft-hairy; E of the Cascades .... **Sphenopholis**
    51b Upper glumes 2.5–5 mm, acute, usually not hooded; lemmas 2.5–6.5 mm; spikelets disarticulating above the glumes; panicle axis and branches soft-hairy; widespread.. ***Koeleria macrantha***
  50b Upper and lower glumes nearly alike in shape, though often differing in length
    52a Panicle axis and branches soft-hairy .... ***Koeleria macrantha***
    52b Panicle axis and branches glabrous, scabrous, or strigose, not soft-hairy
      53a Leaves with claw-like or clasping auricles, or with expanded, flaring, rounded area at base of blade .... ***Schedonorus***
      53b Leaves without auricles or expanded, flaring, rounded area at base of blade
        54a Leaf tips usually prow-shaped; upper surface of leaf blade with a thinner, translucent band along each side of the midrib; calluses sometimes with cobwebby hairs; sheaths of upper culm leaves closed at least 10% their length .... ***Poa***
        54b Leaf tips tapered gradually to the tip, not prow-shaped or only minutely so; upper surface of leaf blade lacking thinner translucent band along each side of the midrib; calluses lacking cobwebby hairs; sheaths of upper culm leaves open to the base or nearly so (sometimes torn open by growing culms)
          55a Leaves 0.5–1.5 mm wide, generally folded or rolled .... ***Festuca***
          55b Leaves 1.5–10+ mm wide, flat, sometimes rolling lengthwise when dry
            56a Plants dioecious (pistillate florets with vestigial anthers, staminate florets with vestigial pistils); panicles condensed and spike-like; alpine; Steens Mt., SE OR.................. ***Leucopoa kingii***
            56b Plants with bisexual florets; panicles dense to open; habitat and range various
              57a Lower glumes < the lowest lemmas; florets 3–10 (or as few as 1 or 2 in dwarf alpine plants) .... ***Festuca***
              57b Lower glumes generally > the lowest lemmas; florets 2(3 .... ***Graphephorum wolfii***

Genus Descriptions and Keys to Species

NOTE: If the genus has only one species, it is keyed only in the key to genera (above) and not in a key to species within the genus. Exceptions: *Arrhenatherum elatius*, *Diplachne fusca*, and *Phragmites australis* each have two subspecies in the PNW and keys for those genera are provided below.

Aegilops, Goatgrass

Annuals with open leaf sheaths and membranous ligules. Spikes with 1 spikelet per node, the spikelets embedded in the inflorescence axis; disarticulation at the base of the spike or in the inflorescence axis. At the base of the spike are (0)1–3 inconspicuous, rudimentary spikelets. The thin inflorescence axis associated with these spikelets is a point of weakness where the spike will break off. Above these are (1)2–7(12) thick, fertile spikelets called lower spikelets. Most species also have smaller, sometimes sterile, terminal spikelets. Glumes and lemmas are often awned or toothed.

1a Glumes of lower spikelets 1-awned or with a long tooth; glumes of terminal spikelets 1-awned; spikes cylindric; widespread .... ***A. cylindrica***
1b Glumes of lower spikelets 2–3-awned; glumes of terminal spikelets 3-awned; spikes ± broadened at the base; SW OR .... ***A. triuncialis* var. *triuncialis***

Agropyron, Wheatgrass

Cespitose perennials with open leaf sheaths, auricles usually present, and membranous ligules. Spikes with crowded spikelets; disarticulation above the glumes. Spikelets usually > 3× longer than the internodes; with 3–16 florets; glumes and lemmas sometimes awned.

This once-sprawling genus was recently split into multiple genera (*Elymus*, *Leymus*, *Pascopyrum*, *Pseudoroegneria*, *Thinopyrum*, and others; see Group 1a key to the Triticeae, p. 24). *Agropyron* species have been introduced in much of western North America for forage and erosion control in disturbed shrub steppe. Crested wheatgrass (*A. cristatum*) quickly establishes and grows into vigorous bunches, competes well with cheatgrass (*Bromus tectorum*), and tolerates heavy grazing and drought. With these traits it outcompetes native plant species. A number of cultivars have been developed by crossing three Old World species (*A. cristatum*, *A. fragile*, and *A. desertorum*). Trying to distinguish them can seem futile because they are undergoing a process of merging and "de-speciation" on this continent, and many intermediate forms occur. We recognize two species in the PNW; crested wheatgrass (*A. cristatum*) is widespread and commonly planted; Siberian wheatgrass (*A. fragile*) is uncommon outside of experimental plantings and has been documented twice in the PNW, in SE OR and SE WA.

1a Lemmas usually awned, awns 1–6 mm; spikelets diverging from the rachis at 30–95°; spikes narrowly to broadly lanceolate, rectangular, or ovate in outline; common .... ***A. cristatum***
1b Lemmas awnless, sometimes mucronate; spikelets diverging from the rachis < 30(35)°; spikes linear to narrowly lanceolate in outline; rare .... ***A. fragile***

Agrostis, Bentgrass

Perennials or annuals, cespitose, rhizomatous, or stoloniferous. Leaf sheaths open; ligules membranous. Inflorescences condensed to open panicles; disarticulation above the glumes and below the florets, sometimes also at the panicle base. Spikelets laterally compressed, with 1 floret; rachilla not prolonged beyond the floret. Glumes 1(3)-veined, longer than the florets. Calluses glabrous or hairy. Lemmas 3–5-veined, awnless or awned, the awns arising from near the lemma base to near the tip, often bent. Paleas 0–67% as long as the lemmas.

One of the most distinctive characteristics about the genus *Agrostis* is how undistinctive (and undistinguishable!) most of the species are. They are separated based on tiny characters that arguably are not field characters and will leave you standing out in a field wishing you had a portable microscope. Palea length relative to lemma length is an important character. The trick is to actually see the palea, which can range from transparent to nonexistent. Anther length, awn length, ligule shape, growth habit, and habitat are all important to assess. Many of our native species grow in specialized habitats such as dunes, vernal pools, and montane wet meadows and bogs. A few are narrow endemics. Five introduced species (*A. canina, A. capillaris, A. castellana, A. gigantea,* and *A. stolonifera*) thrive in low-elevation grasslands or wetlands and are aggressive invaders of native habitats.

The genera *Apera* and *Podagrostis*, closely related to *Agrostis*, were at one time included in the genus; thus they are included in this key. In these genera the rachilla is prolonged beyond the base of the floret, but is nearly invisible.

1a Plants annual; awns often 2–5× as long as lemma bodies (sometimes lacking or < 2× as long as lemmas in *A. exarata*)
    2a Panicles open, branches without spikelets at the base for 5+ mm
    .... ***Apera spica-venti***
    2b Panicles ± contracted, branches usually spikelet-bearing to the base and hidden by the spikelets
        3a Awns 0–3.5 mm, < 2× as long as lemma bodies .... ***A. exarata***
        3b Awns (0)3–12 mm, 2–5× as long as lemma bodies
            4a Glumes unequal, 1–2.5 mm; lemmas firm; paleas 75–100% as long as lemmas; rachilla prolonged 0.2–0.6 mm beyond base of floret
            .... ***Apera interrupta***
            4b Glumes subequal, 2.5–7 mm; lemmas delicate; paleas 0–15% as long as the lemmas; rachilla not prolonged
                5a Lemmas 2.5–4 mm; lemma tips with 2 teeth to 1.5 mm; awns (5)8–10 mm; leaf blades 1–4.5 cm; SW OR, apparently extirpated
                .... ***A. hendersonii***
                5b Lemmas 1.5–2.3 mm; lemma tips with 2 teeth to 0.5 mm; awns 3.5–8 mm; leaf blades 3–15 cm; widespread W of the Cascades and in the Columbia Gorge... ***A. microphylla***
1b Plants perennial; awns lacking or to 2× as long as lemma bodies (sometimes > 2× as long in *A. castellana*)
    6a Panicles dense and often ± spike-like at anthesis; lower panicle branches appressed to ascending, usually hidden by the spikelets, 0.3–5 cm
        7a Leaf blades 0.5–2 mm wide, usually inrolled or folded; panicles 0.2–1.2(2) cm wide

8a Anthers 0.7–2 mm; coastal bluffs and dunes; recently documented at two sites in Curry County, OR .... ***A. blasdalei***
8b Anthers 0.4–1 mm; montane to alpine; widespread, not coastal
9a Paleas 0–0.2 mm, much < lemmas; lemmas delicate .... ***A. variabilis***
9b Paleas 0.9–1.6 mm, about 75% as long as lemmas; lemmas firm .... ***Podagrostis humilis***
7b Leaf blades usually 2–10 mm wide, usually flat in life (may roll or fold if wilted before pressing); panicles 0.4–4(6) cm wide
10a Lower panicle branches 2–5 cm; plants stoloniferous or rhizomatous; anthers 0.7–2 mm
11a Plants stoloniferous; paleas 0.7–1.4 mm; moist to wet meadows, shorelines, ditches .... ***A. stolonifera***
11b Plants rhizomatous; paleas 0–0.2 mm; habitat often drier: dunes, meadows, open woods .... ***A. pallens***
10b Lower panicle branches 0.3–2(5) cm; plants cespitose, sometimes short-rhizomatous if buried in sand; anthers 0.3–0.7 mm
12a Back of glumes fine-scabrous throughout; paleas usually 0.5–0.7 mm, to about 33% as long as the lemmas; ligules mostly 1–3 mm; panicles very dense, often partially included in upper sheaths; sandy soils and cliffs; range strictly coastal .... ***A. densiflora***
12b Back of glumes usually ± glabrous except finely scabrous on the keel; paleas 0–0.5 mm, about 20% as long as the lemmas; ligules usually 2.5–4 mm; panicles less dense, usually exserted from upper sheaths; various wet habitats; widespread, including coastal .... ***A. exarata***
6b Panicles open, diffuse, or ± contracted at anthesis, not spike-like; lower panicle branches erect to spreading, readily visible, 0.5–12 cm
13a Callus hairs prominent, (0.8)1–2 mm; paleas lacking; anthers 1.5–2.3 mm; W OR.. ***A. hallii***
13b Callus hairs inconspicuous, to 0.2–0.8 mm; paleas lacking or present; anthers 0.3–2.3 mm; range various
14a Longer lemma awns ≥ 3.6 mm; anthers 0.8–2 mm
15a Paleas 0.6–1.1 mm, ≥ 50% as long as the lemmas; lemmas of terminal spikelet of a branch generally with awns to 5 mm; lemmas of proximal spikelets awnless or with shorter awns .... ***A. castellana***
15b Paleas 0–0.2 mm, much < lemmas; all lemmas awned, the awns ± similar in length
16a Inflorescences 10–25 cm, 1–11 cm wide; lemmas 2.3–3.1 mm; leaf blades 0.8–5 mm wide; Coast Range of NW OR or Columbia River Gorge
17a Plants short-rhizomatous, sod-forming; leaf blades 0.8–3 mm wide; anthers (1.2)1.5–2 mm; Coast Range of NW OR .... ***A. swalalahos***
17b Plants cespitose, not sod-forming; leaf blades 3–5 mm wide; anthers 1–1.3 mm; Columbia River Gorge .... ***A. howellii***
16b Inflorescences (2)3–10 cm, 1–7 cm wide; lemmas 1–2.6 mm; leaf blades 0.5–3 mm wide; range various
18a Anthers 0.5–0.8 mm; plants cespitose; subalpine to alpine; N WA .... ***A. mertensii***

18b Anthers 1–1.5 mm; plants stoloniferous, appearing loosely cespitose; lowlands to montane; mostly W of the Cascades..**A. canina**
14b Longer lemma awns ≤ 3.5 mm or awns lacking; anthers 0.3–2.3 mm
  19a Anthers (0.8)1–2.3 mm, ± filling the lemmas
    20a Rachilla prolonged 0.5–1.9 mm beyond the base of the floret; glumes 2.3–4.4 mm; lemmas firm; paleas 2–3 mm; mts. of N WA ....***Podagrostis aequivalvis***
    20b Rachilla not prolonged beyond the base of the floret; glumes 1.7–3.6 mm; lemmas delicate; paleas 0–1.4 mm; widespread
      21a Paleas 0.6–1.4 mm, 40–67% as long as the lemmas; introduced
        22a Ligules of the upper leaves longer than wide, 2–7.5 mm; lower panicle branches often spikelet-bearing to the base
          23a Inflorescences 8–30 cm, usually open after anthesis, usually reddish (sometimes green to tan in shaded plants); longest lower panicle branches 4–9 cm; shoots erect to decumbent-based; rhizomes present, stolons absent; culms 20–120 cm ......**A. gigantea**
          23b Inflorescences 3–20 cm, usually contracted after anthesis, usually light green to tan; longest lower panicle branches 2–6 cm; shoots generally sprawling; stolons present, rhizomes absent; culms 8–60 cm ....**A. stolonifera**
        22b Ligules of the upper leaves usually shorter than wide, 0.3–3 mm; lower panicle branches with spikelets confined to the distal 30–50%
          24a Calluses glabrous or with a few hairs to 0.1 mm; panicles stiffly erect, 3–20 cm; lemmas glabrous, usually awnless or with awns to 2 mm ....**A. capillaris**
          24b Calluses hairy, hairs to 0.6 mm; panicles somewhat lax, 10–30 cm; lemmas occasionally pubescent on the lower half; awns to 5 mm often present on lemma of the terminal spikelet of a branch ....**A. castellana**
      21b Paleas 0–0.2 mm, 0–20% as long as the lemmas; native
        25a Plants rhizomatous; primary panicle branches branched from below midlength, spikelet-bearing to the base; habitat moist to ± xeric, usually drier than that of *A. oregonensis*....................**A. pallens**
        25b Plants cespitose; primary panicle branches branched mainly above midlength, spikelet-bearing on the distal half only; wet meadows, bogs, often growing in water or saturated soil ....**A. oregonensis**
  19b Anthers 0.3–0.8 mm, much smaller than the lemmas
    26a Lemmas firm, with prominent veins; paleas 75–100% as long as lemmas; rachilla prolonged 0.1–0.6 mm beyond the floret base; montane to alpine .... ***Podagrostis thurberiana***
    26b Lemmas delicate, without prominent veins; paleas lacking or much < lemmas; rachilla not prolonged; habitat various
      27a Lemma awns (2)3–4.4 mm, usually bent; subalpine to alpine; N WA ....**A. mertensii**
      27b Lemma awns absent or to 3 mm, straight; lowland to alpine; widespread
        28a Wider leaf blades usually 2–7 mm wide, 5–30 cm long, flat; leaves mostly cauline to mostly basal; plants rhizomatous or cespitose

29a Plants rhizomatous; primary panicle branches branched from below midlength, spikelet-bearing to the base; habitat moist to ± xeric, usually drier than that of *A. oregonensis* .... ***A. pallens***
29b Plants cespitose; primary panicle branches branched mainly above midlength, spikelet-bearing on the distal half only; wet meadows, bogs, often growing in water or saturated soil .... ***A. oregonensis***
28b Wider leaf blades ≤ 2 mm wide, 1–14 cm long, flat or involute; leaves mostly basal; plants cespitose
  30a Lower panicle branches 4–12 cm; panicles at maturity often detaching at the base and forming a tumbleweed; plants 15–90 cm .... ***A. scabra***
  30b Lower panicle branches 1–4 cm; panicles at maturity not detaching and forming a tumbleweed; plants 8–40 cm .... ***A. idahoensis***

Aira, Hairgrass

Small introduced annuals. Leaf sheaths open; ligules membranous. Panicles open or contracted, sometimes spike-like. Spikelets with 2 florets; glumes > the florets; calluses with short hairs; lemmas awned from the lower back or awnless; lemma apex with a slender bifid tip.

1a Inflorescences condensed, narrow, branches appressed; pedicels shorter than spikelets .... ***A. praecox***
1b Inflorescences open, branches spreading; pedicels longer than spikelets
  2a Spikelets solitary; lower lemmas usually awnless; pedicels usually 2–8× as long as the spikelets; spikelets 1.7–2.5 mm; uncommon .... ***A. elegans***
  2b Spikelets ± clustered; both lemmas usually awned; pedicels usually 1–2× as long as the spikelets; spikelets 2–3.5 mm; common .... ***A. caryophyllea***

Alopecurus, Meadow Foxtail

Annuals and perennials, native and introduced; cespitose or rhizomatous. Leaf sheaths open; ligules membranous. Inflorescences dense, spike-like, cylindrical panicles; disarticulation below the glumes. Spikelets strongly laterally compressed, oval in outline, with 1 floret. Glumes 3-veined, often fused in lower half, keeled, the keels ciliate at least basally. Lemmas slightly < to slightly > glumes, 3-veined, awned, the awns arising from near the base to midlength, bent or straight. Paleas lacking.

1a Spikelets (excluding awns) ≥ 4 mm
  2a Plants perennial
    3a Lemma tips acute; lemmas green to purplish or straw-colored at maturity; awns exserted 2.2–5.5 mm from the glumes; glume tips not outcurved at maturity; widespread .... ***A. pratensis***
    3b Lemma tips obtuse to truncate; lemmas often lead-gray at maturity; awns hidden by glumes or exserted up to 3 mm; glume tips distinctly outcurved at maturity; few scattered populations E of the Cascades, probably underreported .... ***A. arundinaceus***
  2b Plants annual
    4a Glumes glabrous except keels short-ciliate, 4.5–7.5 mm; upper sheaths inconspicuously inflated; anthers 2.4–4.1 mm .... ***A. myosuroides***

  4b Glumes pubescent, 3–5 mm; upper sheaths conspicuously inflated; anthers 0.7–1.8 mm ....***A. saccatus***
 1b Spikelets (excluding awns) < 4 mm
  5a Awns 0.7–3 mm, straight, hidden in the glumes or exceeding them by < 1 mm, the inflorescences thus with a neat, trim appearance ....***A. aequalis* var. *aequalis***
  5b Awns 3–10 mm, bent, exceeding glumes by (1.2)1.5–6 mm; inflorescences ± shaggy in appearance
   6a Upper leaf sheaths conspicuously inflated; spikelets 3–5 mm; awns 6–10 mm, exceeding glumes by 3–6 mm .... ***A. saccatus***
   6b Upper leaf sheaths not or slightly inflated; spikelets 1.9–3.5 mm; awns 3–6.5 mm, exceeding glumes by 1.2–4 mm
    7a Plants perennial, rooting at lower nodes; anthers (0.9)1.4–2.2 mm ....***A. geniculatus***
    7b Plants annual, not rooting at the nodes; anthers 0.3–1 mm ....***A. carolinianus***

Anthoxanthum, Vernalgrass

Eurasian annuals or cespitose perennials. Leaf sheaths open; ligules membranous, sometimes short-ciliate. Panicles dense, spike-like; disarticulation above the glumes. Spikelets with 3 florets: 2 sterile lower florets, and a bisexual upper floret; lemmas of lower florets with golden brown hairs, rounded bifid tips, and bent awns arising from the back; lemmas of upper florets firmer and awnless. Glumes unequal, the longer one > the florets.

1a Plants annual; upper glumes usually < 7 mm; longer lemma awns exserted 2–3 mm beyond the tip of the upper glume; leaf blades 1–2.7(5) mm wide; inflorescences 1–4 cm......................................***A. aristatum* ssp. *aristatum***
1b Plants perennial; upper glumes usually > 7 mm; longer lemma awns exserted 0–2 mm beyond the tip of the upper glume; leaf blades 3–10 mm wide; inflorescences 3–14 cm...***A. odoratum***

Apera, Windgrass

Introduced annuals similar to *Agrostis* and *Podagrostis*. The rachilla is prolonged beyond the base of the floret as an inconspicuous bristle.

1a Anthers 0.3–0.5 mm; panicle branches appressed, short, spikelet-bearing to within 2 mm of the base; widespread... ***A. interrupta***
1b Anthers 1–2 mm; panicle branches spreading, naked for the basal 5 mm or more; few historical records... ***A. spica-venti***

Aristida, Three-awn

Tufted annuals or cespitose perennials. Leaf sheaths open; ligules a fringe of hairs or membranous with ciliate margins. Inflorescences panicles or racemes; disarticulation above the glumes. Spikelets with 1 floret; glumes often > the florets, sometimes awned; lemmas hardened, terete, enclosing the palea; lemmas with 3 awns arising from the tip.

1a Lower glumes (9)12–22(28) mm, 3–7-veined, with awns 1–13 mm; anthers usually 1(3); plants annual; W of the Cascades in OR, but present in ID and expected along the Snake River .... ***A. oligantha***
1b Lower glumes 4–13 mm, 1(2)-veined, acuminate or with awns to 2 mm; anthers 3; plants annual or perennial; E of the Cascades
    2a Plants annual; central lemma awns 7–15 mm; inflorescences dense panicles; rare waif in central OR, perhaps not persisting (not treated) .... ***A. adscensionis***
    2b Plants perennial; central lemma awns 40–100(140) mm; inflorescences open racemes or sparsely branched panicles; E of the Cascades .... ***A. purpurea* var. *longiseta***

Arrhenatherum elatius, Tall Oatgrass

Tall introduced perennial with large, narrow panicles bearing 2-flowered spikelets. The lower floret staminate and with a stout, twisted, strongly geniculate awn from midback; the upper bisexual and with a short, straight, subterminal awn.

1a Basal internodes of culm not bulb-like, not or only slightly swollen, 2–4 mm thick at the widest point .... ***A. e.* ssp. *elatius***
1b Basal internodes of culm bulb-like and swollen, 5–10 mm thick at the widest point .... ***A. e.* ssp. *bulbosum***

Avena, Oats

Introduced annuals. Leaf sheaths open; ligules membranous. Inflorescences open panicles; disarticulation above the glumes. Spikelets with 2–3 florets; glumes > florets; lemmas awned from the back; awns stout, strongly bent and twisted.

1a Lemma tips with 2 elongated awn-like teeth 2–4 mm; mainly W of the Cascades .... ***A. barbata***
1b Lemma tips acute, narrowed, teeth if present not elongated and awn-like, ≤ 1.5 mm; W and E of the Cascades
    2a Callus glabrous; upper lemmas awnless or with straight to slightly twisted awn much < the awn of lowest lemma; spikelets not disarticulating, remaining attached to the plant even when mature .... ***A. sativa***
    2b Callus hairy; upper lemmas with bent, twisted awn usually subequal to the awn of the lowest lemma; mature spikelets disarticulating above the glumes .... ***A. fatua***

Briza, Quaking Grass

Cespitose annual or perennial. Leaf sheaths open; blades flat; ligules membranous, sometimes decurrent on the sides. Inflorescences open panicles with spikelets that dangle on slender pedicels; disarticulation above the glumes. Spikelets laterally flattened, but glumes and lemmas broadly rounded over the back, and spreading nearly 180°. Glumes often purple, contrasting with green to straw-colored lemmas. Lemmas inflated, awnless.

1a Spikelets 10–20(27) mm; W of the Cascades .... ***B. maxima***
1b Spikelets (2)3–5.5 mm

2a Plants annual; lemmas 1.6–2 mm; ligules 4–13 mm; W of the Cascades .... ***B. minor***
2b Plants perennial; lemmas 3–4 mm; ligules to 0.5 mm; NE OR .... ***B. media***

Bromus, Brome

Annuals, biennials, or perennials, usually cespitose, sometimes rhizomatous. Leaf sheaths closed nearly to the top, usually hairy; ligules membranous, blades usually flat. Inflorescences panicles, sometimes racemes or with a single spikelet in depauperate plants; disarticulation above the glumes and below the florets. Spikelets terete or laterally compressed. Glumes unequal, shorter than the spikelets. Lemmas keeled or rounded on the back, tips entire to toothed, usually awned, the awns terminal or subterminal.

Bromus is a diverse genus with both important native species and some of our most significant exotic invasives. All of our annual bromes are introduced, but most of our perennials are native. Some species interbreed, and intermediate forms are fairly common. Culm diameter is measured near the middle of one of the lowest elongated internodes. Distance from the base of the lemma awn to the lemma apex is measured on the distal florets in a spikelet. Lemma lengths are measured on the lowest florets in a spikelet. Lemma, spikelet, and inflorescence branch lengths do not include awns.

1a Plants perennial or biennial
  2a Lemmas keeled on the back, V-shaped in cross section at least near the tip; spikelets ± flattened
    3a Lemmas lacking awns, or with awns to 4 mm; spikelets very strongly flattened; lemma veins prominent; disturbed areas; W of the Cascades .... ***B. catharticus* var. *catharticus***
    3b Lemmas awned, the awns 4–17 mm; spikelets flattened but not so strongly as in *B. catharticus*; lemma veins prominent or obscure; habitat and range various
      4a Lower glumes 1(3)-veined, conspicuously hairy; lemmas usually only slightly keeled, with long, soft hairs, especially toward the base; moist habitats near the coast... ***B. pacificus***
      4b Lower glumes 3–7(9)-veined, glabrous to hairy but less conspicuously so than in *B. pacificus*; lemmas usually strongly keeled, glabrous or hairy but without long, soft hairs .... ***B. sitchensis*** (with 5 varieties)
        5a Panicles dense, the branches erect and mostly shorter than the spikelets; culms 20–70 cm; coastal sands and headlands; Lincoln County, OR , south to S CA .... ***B. s.* var. *maritimus***
        5b Panicles loose to compact, at least some panicle branches longer than the spikelets, erect, ascending, spreading, or reflexed; culms 30–180 cm; habitat various, including coastal sands; widespread
          6a Some lower panicle branches 10-20 cm (measured from culm to tip of terminal spikelet), spreading to reflexed, with 1–2(3) spikelets borne at the tip; ligules 3–8 mm; sheaths glabrous to sparsely pilose; lemmas usually glabrous (sometimes short-hairy); culms generally ≥ 3 mm diameter at midlength; mostly W of the Cascades... ***B. s.* var. *sitchensis***

6b Lower panicle branches usually < 10 cm (measured from culm to tip of terminal spikelet), erect, ascending, spreading, or reflexed, with 1–5 spikelets variously arranged, borne at the tip or sometimes to near the base; ligules 1–5(6) mm; sheaths and lemmas variously hairy or glabrous; culms < or > 3 mm diameter at midlength; range various
    7a Panicle branches erect to ascending, generally with 1–2(3) spikelets borne at the tip; ligules (1)3.5–5 mm; leaf blade veins relatively narrow, mostly < one-third as broad as the area between them; spikelets with 3–6 florets; culms often > 3 mm thick at midlength; coastal NW WA .... ***B. s.* var. *aleutensis***
    7b Panicle branches erect, ascending, spreading, or reflexed, with 1–5 spikelets borne variously, often not limited to the tip; ligules 1–3(6) mm; leaf veins relatively wide, mostly at least half as broad as the area between them; spikelets with 4–11 florets; culms usually < 3(4) mm thick at midlength; widespread
        8a Most awns 8–17 mm; panicle branches mostly spreading to reflexed .... ***B. s.* var. *carinatus***
        8b Most awns 4–7 mm; panicle branches mostly erect to ascending .... ***B. s.* var. *marginatus***
2b Lemmas rounded on the back, C-shaped in cross section; spikelets ± cylindrical
  9a Plants strongly rhizomatous; lemmas awnless or with awns to 3 mm; widespread, mostly E of the Cascades .... ***B. inermis***
  9b Plants not rhizomatous; lemmas usually with awns (1)2–12 mm
    10a Inflorescence narrow and dense, branches ascending to erect, usually shorter than the spikelets
      11a Inflorescence nodding; longer lemma awns 6–12 mm; ligules 2–6 mm; widespread in shady habitats .... ***B. vulgaris***
      11b Inflorescence erect; lemma awns 2–8 mm; ligules up to 3 mm; montane to subalpine; in and E of the Cascades, and SW OR
        12a Awns 2–5 mm; anthers 2–3.5 mm; leaf sheaths glabrous .... ***B. suksdorfii***
        12b Awns (4)5–8 mm; anthers 3–5 mm; some leaf sheaths hairy, at least at the collar .... ***B. orcuttianus***
    10b Inflorescence wider, more open, lower branches reflexed to ascending, usually longer than the spikelets
      13a Most lower glumes with 1 vein (occasionally some with 3 veins)
        14a Glumes glabrous (occasionally pubescent)
          15a Lemma awns 2–4(5) mm; anthers 1.5–2 mm; in and E of the Cascades .... ***B. ciliatus***
          15b Lemma awns 4–12 mm; anthers 2–5 mm; range various
            16a Inflorescences nodding, the branches appressed to ascending; lemma awns (4)6–12 mm; ligules 2–6 mm; widespread .... ***B. vulgaris***
            16b Inflorescences erect, the branches widely spreading when mature; lemma awns (4)5.5–8 mm; ligules 1–3 mm; in and E of the Cascades, and SW OR .... ***B. orcuttianus***
        14b Glumes hairy

17a Inflorescences nodding; panicle branches appressed to ascending; lemma awns (4)6–12 mm; ligules 2–6 mm; widespread .... ***B. vulgaris***
17b Inflorescences upright; panicle branches widely spreading when mature, the lower ones sometimes reflexed; lemma awns 3.5–8 mm; ligules 1–4 mm
    18a Culms with 2–4 nodes; panicles 7–13.5 cm; in and E of the Cascades, and SW OR.. ***B. orcuttianus***
    18b Culms with (5)6–8 nodes; panicles 10–25 cm; W of the Cascades, mainly coastal .... ***B. pacificus***
13b Most lower glumes with 3 veins (occasionally some with 1 vein)
    19a Upper glumes 5-veined; panicle branches ± flexuous; SW OR to S Willamette Valley (1 historic record each in Hood River County, OR, and Klickitat County, WA) ....***B. laevipes***
    19b Most upper glumes 3-veined (some may be 5-veined); panicle branches ± stiff; in and E of the Cascades.. ***B. orcuttianus***
1b Plants annual
  20a Lemmas keeled, V-shaped in cross section; spikelets very strongly flattened; W of the Cascades..***B. catharticus* var. *catharticus***
  20b Lemmas ± rounded over the back, C-shaped in cross section; spikelets not strongly flattened; range various
    21a Lemmas tapered very gradually to attenuate tips, the lemma teeth acuminate to acuminate-aristate; lemma awns 8–65 mm; first glume 1(3)-veined; second glume usually 3(5)-veined
      22a Lemmas 20–35 mm with awns 30–65 mm; spikelets 25–70 mm ....***B. diandrus***
      22b Lemmas 9–20 mm with awns 8–30 mm; spikelets 10–50 mm
        23a Panicles ± open and loose, erect to drooping; panicle branches often > the spikelets, slender, mostly spreading to reflexed
          24a Spikelets 10–20 mm; lemma awns 10–18 mm; usually at least 1 panicle branch with 4–8 spikelets.................................***B. tectorum***
          24b Spikelets 20–35 mm; lemma awns 15–30 mm; panicle branches usually with 1–2(3) spikelets .... ***B. sterilis***
        23b Panicles strongly contracted, erect; panicle branches usually much < the spikelets, stout, ascending to erect
          25a Panicles very dense; panicle branches 0.1–1 cm, usually hidden by the spikelets; spikelets 18–25 mm; lemmas 10–13(15) mm ....***B. rubens***
          25b Panicles less dense; some panicle branches 1–3+ cm, not hidden by the spikelets; spikelets 30–50 mm; lemmas 12–20 mm; collected on ballast in Portland in the early 1900s; other PNW records dubious (not treated)........................***B. madritensis***
    21b Lemma tips ± rounded to acute, usually shallowly bidentate; lemma awns 1–13 mm or lacking; first glume 3(5)-veined; second glume 5(9)-veined
      26a Lemmas usually awnless or with awns scarcely 1 mm, strongly inflated; spikelets 15–30 mm, with 9–15 florets; ligules pubescent; E of the Cascades (a few historic records W of the Cascades).......... ***B. briziformis***

26b Lemmas with awns ≥ 2 mm, not strongly inflated, occasionally awnless, but then plants otherwise not as above
    27a Inflorescence consisting of 2 or more spikelets
        28a Lower glumes 7–10 mm; upper glumes 8–12 mm; panicle branches conspicuously sinuous; awns erect to weakly spreading, 10–16 mm; seldom collected in PNW ....*B. arenarius*
        28b Lower glumes 4–7 mm; upper glumes 5–9 mm; panicle branches sometimes sinuous; awns erect to strongly divergent, (0)3–10 mm
            29a Awns originating ≥ 1.5 mm below the lemma tips (check several lemmas from middle of spikelet), 8–13(16) mm; spikelets 15–70 mm
                30a Lemmas with hyaline margins 0.3–0.6 mm wide, slightly angled above the middle; spikelets 20–40 mm with 6–12 florets; branches often with more than 1 spikelet..*B. japonicus*
                30b Lemmas with hyaline margins 0.6–0.9 mm wide, strongly angled above the middle; spikelets 15–70 mm with 8–30 florets; branches usually with 1 spikelet... *B. squarrosus*
            29b Awns originating < 1.5 mm below lemma tips, (0)3–10 mm; spikelets 10–20(30) mm
                31a Panicles usually dense, erect to strongly ascending, the branches usually shorter than the spikelets; lemmas thin and papery at maturity, usually hairy over the back (occasionally glabrous); lemma veins thickened and raised at maturity; anthers 0.6–1.5(2) mm ....*B. hordeaceus*
                31b Panicles more open, ascending to spreading, the branches usually longer than the spikelets; lemmas leathery or firm at maturity, usually glabrous over the back; lemma veins not or only slightly thickened and raised at maturity; anthers 0.7–2(3) mm
                    32a Lemmas 8–11.5 mm, margins bluntly angled; anthers 0.7–2(3) mm; leaf blades 2–4 mm wide................................*B. commutatus*
                    32b Lemmas 6.5–8.5 mm, margins rounded and inrolled at maturity; anthers 1–2 mm; leaf blades (2)4–12 mm wide ....*B. secalinus*
    27b Inflorescence consisting of a single spikelet
        33a Lemma veins thickened and raised when dry; lemmas papery in texture, usually hairy over the back (occasionally glabrous); anthers 0.6–1.5(2) mm..*B. hordeaceus*
        33b Lemma veins not or only slightly thickened and raised; lemmas leathery or firm, usually glabrous over the back; anthers 0.7–2(3) mm ....*B. commutatus*

Calamagrostis, Reedgrass, Beachgrass

Cespitose or rhizomatous perennials. Leaf sheaths open; ligules membranous. Inflorescences open to contracted panicles; disarticulation above the glumes. Spikelets with 1 floret; rachillas prolonged beyond the floret, usually hairy. Glumes ± equal, ≥ florets. Calluses with hairs 0.2–6.5 mm, 20–150% of the lemma length. Lemmas with 4 lateral veins that often extend into minute teeth; awnless or awned from the back with a straight to geniculate awn; awns sometimes hidden by callus hairs. Paleas nearly = to slightly > lemmas. *Agrostis* and *Trisetum* can be confused with some *Calamagrostis*. *Calamagrostis* plants generally are larger and coarser than *Agrostis*, with firmer glumes, proportionately longer paleas, and prolonged rachillas. *Trisetum* species have 2–3 florets per spikelet. Recent work has placed the genus *Ammophila* (beachgrass) in *Calamagrostis*.

1a Lemmas awnless or with a short terminal point, plants of coastal sands
    2a Ligules 10–35 mm, acute, bifid or lacerate; common .... ***C. arenaria* ssp. *arenaria***
    2b Ligules 1–4.6 mm, truncate to obtuse; uncommon ....***C. breviligulata* ssp. *breviligulata***
1b Lemmas awned from the back
    3a Lemma awns geniculate, at least the longer ones exserted (1.5)2–11 mm beyond the glume tips
        4a Lemma awns 10–16 mm, exserted 7–11 mm beyond the glume tips; panicles open, branches spreading; Columbia Gorge..................***C. howellii***
        4b Lemma awns 2.9–8.5(10) mm; panicles open or congested, branches spreading to erect; range various
            5a Wider leaf blades 6–13 mm wide; culms (47)60–120(150) cm; Cascades of central WA .... ***C. tweedyi***
            5b Leaf blades 0.4–5(7) mm wide; culms (20)30–80(95) cm; range various
                6a Panicles open; leaf blades 0.4–1.7 mm wide; Mt. Hood and Mt. Jefferson, OR.. ***C. breweri***
                6b Panicles congested, narrow; leaf blades (1.5)2–5(7) mm wide; range various
                    7a Upper leaf surface with dense short hairs; lower leaf surface strongly scabrous; inflorescence branches with short spreading hairs; glume keels scabrous; awns (4.5)6–7(9) mm; Olympic Mts. and North Cascades of WA, and Wallowa Mts., OR ....***C. purpurascens***
                    7b Upper leaf surface glabrous, scabrous, or sparsely hairy; lower leaf surface glabrous or scabrous; inflorescence branches scabrous; glume keels smooth or slightly scabrous; awns 2.9–8.5(10) mm; range various
                        8a Glumes 3.2–5.8 mm; lemmas 2.9–4.2 mm; awns 2.9–6.6 mm, exserted 0.5–2.3 mm; Steens Mt., OR .... ***C. utsutsuensis***
                        8b Glumes (4)6–6.5(7) mm; lemmas 4–5(5.5) mm; awns 7–8.5(10) mm, exserted 3–5 mm; mts. of W WA and N OR .... ***C. tacomensis***
    3b Lemma awns straight or sometimes geniculate, but not exserted > 1(1.5) mm beyond the glume tips
        9a Plants coastal; collars thickened; collar veins obscure or lacking; leaf blades thick and tough, usually (2)4–10(20) mm wide ....***C. nutkaensis***
        9b Plants inland or occasionally near the coast, especially in the Olympic Mts.; collars not thickened; collar veins evident; leaf blades not particularly thick or tough, 2–5(8) mm wide
            10a Callus hairs usually 20–50(60)% as long as the lemma

11a At least some collars hairy, the hairs longer than those of adjacent blade or sheath .... ***C. rubescens***
11b Collars glabrous or only puberulent, the hairs, if any, no longer than those of the adjacent blade or sheath
    12a Upper leaf surface densely short-hairy; lower leaf surface strongly scabrous; subalpine to alpine; Olympic Mts. and North Cascades of WA and Wallowa Mts., OR ....***C. purpurascens***
    12b Upper leaf surface scabrous to smooth, sometimes sparsely hairy; lower leaf surface scabrous to smooth; habitat various, including alpine; range various
        13a Foliage glaucous; anthers (1)1.6–2.5 mm; Steens Mt., OR .... ***C. utsutsuensis***
        13b Foliage green, not glaucous; anthers (1)1.3–3.5 mm; range various, but not Steens Mt., OR
            14a Anthers (1)1.3–2(2.6) mm; spikelets (3)4–4.5(5.5) mm; lemmas 2.5–3.5(4) mm; E of the Cascades................................... ***C. rubescens***
            14b Anthers 2–3.5 mm; spikelets (4)4.5–6(7) mm; lemmas (3.5)4–5(6) mm; SW OR.. ***C. koelerioides***
10b Callus hairs usually (50)70–120% or more as long as the lemma
    15a Panicles relatively congested, rarely > 2 cm wide when pressed, branches erect or ascending, not readily distinguished; lemma awns stout ....***C. stricta*** (with 2 subspecies, p. 176)
    15b Panicles almost always open, mostly > 2 cm wide when pressed, branches tending to spread, rarely obscured; awns delicate ....***C. canadensis*** (with 2 varieties, see p. 169)

Cenchrus, Sandbur, Fountaingrass, Feathertop

Annuals or perennials, habit various. Culms solid or hollow. Leaf sheaths open; ligules membranous, ciliate, or of hairs. Inflorescences spike-like panicles, the branches much reduced, forming clusters called fascicles. Fascicle axes 0.2–7.5(28) mm, with 3–many scabrous bristles. In some taxa the bristles may be partially fused into painfully spiny burs. Spikelets 1–12 per fascicle, each with 2 florets, but the lower sterile or staminate and often hard to distinguish. Lower glume sometimes absent. Lemmas scarious, membranous, or coriaceous (leathery).

    Cenchrus has recently been expanded to include *Pennisetum*. Three of our former *Pennisetum* species are commonly grown as ornamentals and rarely escape into the wild. They are keyed here but not given full treatment.

1a Spikelets enclosed in a spiny, bur-like fascicle..................... ***Cenchrus longispinus***
1b Spikelets not enclosed in a spiny, bur-like fascicle
    2a Inflorescence broadly cylindrical to subglobose, 50–75 mm wide; spikelets 9–14 mm; rarely escaping ornamental (not treated)........ ***Cenchrus longisetus***
    2b Inflorescence narrowly cylindrical, 35–52 mm wide; spikelets 4.5–7 mm
        3a Midculm leaves 2–3.5 mm wide, convolute or folded, the midveins noticeably thickened; plants cespitose; rarely escaping ornamental (not treated) .... ***Cenchrus setaceus***
        3b Midculm leaves 3–9 mm wide, flat, the midveins not noticeably thickened; plants rhizomatous; rarely escaping ornamental (not treated) ....***Cenchrus orientalis***

Cortaderia, Pampas Grass

Very large, densely cespitose perennials; plants dioecious; culms to 7 m, woody. Leaf sheaths open; blades flat to folded, edges dangerously serrate; ligules 1–2 mm, of hairs. Inflorescences large, dense, plumose panicles; disarticulation above the glumes. Spikelets ± laterally compressed, usually unisexual, with 3–7 florets. Glumes unequal, nearly as long as spikelets, 1-veined, calluses hairy; lemmas 3–5-veined, with straight awns; pistillate lemmas with long silky hairs; staminate lemmas less hairy or glabrous.

Pampas grasses are huge bunchgrasses with large, pale, fluffy panicles. Two species occur in North America and the PNW. Both have escaped from ornamental plantings and are invasive in coastal and inland habitats W of the Cascades. In keying *Cortaderia* it is helpful to examine unbroken, fresh or rehydrated lemmas under a dissecting scope to determine the shape of the lemma tip. The folded lemma can be flattened by moving a horizontal needle toward the lemma tip.

1a Plants staminate, spikelets with well-developed anthers; base of inflorescence usually not held above the leaves at maturity .... ***C. selloana***
1b Plants pistillate, spikelets lacking well-developed anthers; base of inflorescence held above the leaves at maturity or not
    2a Lemmas tapering very gradually to the awn (lemma tips sometimes broken off); lemma awns > 1 mm, exceeding lemma hairs by 5–12 mm; glume midribs white; panicles ± bright white; base of inflorescence usually not held above leaves at maturity; culms 2–4× as long as inflorescences ..... ***C. selloana***
    2b Lemmas acute or tapering abruptly or gradually to the awn; lemma awns usually < 0.5(1) mm, exceeding lemma hairs by ≤ 5 mm; glume midribs white or purple; panicles white or pinkish to dull purplish; base of inflorescence usually held above leaves at maturity; culms 4–7× as long as inflorescences .... ***C. jubata***

Crypsis, Pricklegrass

Annuals. Leaf sheaths open, often inflated; blades sometimes disarticulating from the sheaths; ligules of hairs. Inflorescences dense spike-like or head-like panicles, often partially enclosed in upper sheaths. Spikelets strongly laterally compressed with 1 floret enclosed in the glumes. Glumes and lemmas 1-veined, strongly keeled; lemmas awnless, or sometimes with a short point.

Pricklegrasses resemble small *Alopecurus* and grow on receding shores of lakes, reservoirs, and streams. Unlike *Alopecurus*, pricklegrasses have harsh-feeling, scabrous inflorescences, ligules of hairs, and awnless lemmas. Pricklegrasses are dispersed by migrating waterfowl. They are native to Europe, Asia, and N Africa.

1a Inflorescences ± cylindric, generally 7–8× longer than wide, exserted from uppermost sheath; spikelets 1.5–2.7 mm; widespread .... ***C. alopecuroides***
1b Inflorescences ± shortly ovoid, ≤ 5× longer than wide, bases partially to mostly enclosed by the uppermost sheath; spikelets 2.5–3.2 mm; range various
    2a Collars glabrous; glumes unequal, glume margins glabrous; leaf blades not deciduous, present on mature plants; few records in S OR .... ***C. schoenoides***
    2b Collars hairy; glumes subequal, lower glume margins hairy; leaf blades deciduous, often missing on mature plants; few records E of the Cascades .... ***C. vaginiflora***

Cynosurus, Dogtail

Weedy introduced grasses with open sheaths, membranous ligules, and dense, strongly 1-sided, spike-like panicles. Spikelets in pairs, 1 fertile, the other sterile with narrow, empty glumes and lemmas. Range mainly W of the Cascades.

1a Inflorescences narrow, spike-like, < 10 mm wide; awn of fertile lemma < 1 mm; plants perennial .... ***C. cristatus***
1b Inflorescences broadly rounded-triangular, head-like, > 10 mm wide; awn of fertile lemma 3–10 mm; plants annual .... ***C. echinatus***

Danthonia, Oatgrass

Cespitose perennials, sometimes short-rhizomatous. Leaf sheaths open, with a tuft of spreading hairs at top; ligules a fringe of hairs. Inflorescences panicles, racemes, or solitary spikelets. Spikelets with 3+ florets; glumes ≥ florets; calluses hairy; lemmas (5)7–11-veined, tips with 2 acute to awn-like teeth, awned from between the teeth (except *D. decumbens*), awns bent, twisted. Most species have self-pollinated spikelets hidden in leaf sheaths and tend to self-pollinate even the obvious florets. In some species, the culm falls apart at the nodes when mature. Oatgrasses are recognized by tufts of hairs at the top of the sheath and large spikelets with bent awns. See also *Rytidosperma*.

Lemma body length is measured from the base of the lemma to the base of the awn. Length of the lemma teeth is measured from the base of the awn to the tips of the teeth.

1a Lowest branches with pedicels generally > the spikelets; mature inflorescence branches usually spreading or reflexed; upper leaf blades abruptly spreading to reflexed; widespread... ***D. californica***
1b Lowest branches with pedicels generally ≤ the spikelets; mature inflorescence branches ascending to erect; upper leaf blades often ascending to erect, sometimes spreading; range various
   2a Lemmas awnless, or with an inconspicuous, short, straight point between the short apical teeth; near the coast .... ***D. decumbens***
   2b Lemmas with a twisted, geniculate awn; range various
      3a Inflorescences with 1(4) spikelets, if > 1 spikelet, the inflorescence a raceme with the pedicels attached to the inflorescence axis; lemmas 5.5–11 mm; mature culms disarticulating at the nodes; in and E of the Cascades, and SW OR... ***D. unispicata***
      3b Inflorescences with (4)5–10(18) spikelets, usually panicles with pedicels attached to inflorescence branches; lemmas 2.5–6 mm; mature culms disarticulating at the nodes or not; range various
         4a Glumes (11)13–17 mm; calluses of middle florets longer than wide, concave dorsally; seeds 1.5–2(2.3) mm; lemma bodies nearly glabrous over the back, densely hairy on the margins; mature culms not disarticulating at the nodes; in and E of the Cascades, Olympic Mts., and SW OR .... ***D. intermedia***
         4b Glumes (8)9–12 mm; calluses of middle florets about as long as wide, convex dorsally; seeds (2)2.3–3 mm; lemma bodies usually pilose (sometimes glabrous) over the back, margins hairy on lower 50%; mature culms disarticulating at the nodes; OR and WA Cascades, Olympic Mts., N WA; uncommon in OR................................... ***D. spicata***

Deschampsia, Hairgrass

Cespitose annuals or perennials. Leaf sheaths open, glabrous; ligules membranous, long and decurrent, usually acute to acuminate. Inflorescences open to condensed panicles or spikes, disarticulation above the glumes. Spikelets with 1–2(3) florets, rachillas hairy and prolonged > 0.5 mm beyond the base of the upper floret. Glumes > the lowest floret, sometimes > upper floret. Calluses with hairs to 50% as long as lemmas. Lemmas smooth, shiny, erose to toothed, awned from near the base to about the middle of the back, or from between 2 apical lobes; awns straight to bent, sometimes twisted.

Deschampsia species grow in seasonally wet habitats. Three of our species are cespitose, with 2 small, well-separated florets hidden by the glumes, and hairy rachillas that are prolonged beyond the base of the upper floret. Examine the lowest lemma of a spikelet to determine lemma characters. Recent work has placed *Scribneria bolanderi* in *Deschampsia*. Its inflorescences are spikes with spikelets that have only 1 floret.

1a Inflorescences spikes; spikelets 1-flowered, sunken into the inflorescence axis .... ***D. bolanderi***
1b Inflorescences open to condensed panicles, not spikes; spikelets 2(3)-flowered, not sunken into the inflorescence axis
    2a Plants annual; leaves 0.3–1(1.5) mm wide, mainly cauline, not forming a basal tuft; lemma awns strongly bent; glumes 3.5–9 mm ..... ***D. danthonioides***
    2b Plants perennial; leaves 0.2–6 mm wide, mainly basal, forming a tuft; lemma awns straight or slightly bent; glumes 2.7–7.5 mm
        3a Inflorescences ≤ 2 cm wide; basal leaf blades ≤ 1(2) mm wide; glumes light green; anthers 0.3–0.5(0.7) mm .... ***D. elongata***
        3b Inflorescences 4–30 cm wide; basal leaf blades 1–4 mm wide; glumes often purplish; anthers 1.5–3 mm ....***D. cespitosa***

Dichanthelium, Panicgrass, Rosette Grass

Cespitose perennials with much-branched culms and winter rosettes of ovate to lanceolate leaves that are well-differentiated from longer, narrower culm leaves. Leaf sheaths open; ligules a fringe of hairs, 1–5 mm. Panicles open, sparse, of 2 types: primary panicles form in spring and early summer with outcrossing flowers; secondary panicles form in summer and fall with closed, self-pollinating flowers. Disarticulation is below the glumes. Spikelets dorsiventrally compressed, elliptic to lanceolate; florets 2, the lower sterile or staminate, the upper bisexual. Lower glumes < spikelets; upper glumes about = spikelets; lower lemmas similar to upper glumes; lower paleas absent; upper lemmas ± hard and leathery, shiny, glabrous, awnless; upper paleas striate.

Dichanthelium is a genus of cool-season grasses with C3 photosynthesis. It is closely related to *Panicum*, a warm-season grass with C4 photosynthesis that also differs in not having winter rosettes.

1a Spikelets 1.5–2.0 mm; upper glumes 7-veined; panicles well-exserted from leaf sheaths ....***D. acuminatum* ssp. *fasciculatum***
1b Spikelets 2.7–3.5 mm; upper glumes 9-veined; panicles sometimes partially hidden in leaf sheaths .... ***D. oligosanthes* ssp. *scribnerianum***

Digitaria, Crabgrass

Sprawling, often mat-forming annuals or perennials. Culms often decumbent and rooting at the lower nodes. Leaf sheaths open; blades flat; ligules membranous. Panicles with narrow, 1-sided, spike-like branches, attached close together on the inflorescence axis. Spikelets 2 per node, 1 on a stalk, the other sessile or nearly so, borne in 2 rows on 1 side of panicle branches, dorsiventrally compressed, tips pointed, with 2 florets, lower floret sterile; disarticulation below the glumes.

Crabgrasses have panicles with thin, spike-like branches that are arranged like the spokes of an umbrella. They are weedy, late-season grasses with C4 photosynthesis. They differ from *Cynodon dactylon* in having dorsiventrally compressed spikelets.

1a Leaf sheaths, collars, and often blades with many long, papillose-based hairs; upper glume 33–60% as long as the spikelet .... **D. sanguinalis**
1b Leaf sheaths, collars, and blades nearly glabrous but with sparse papillose-based hairs on upper margins of sheaths and lower margins of blades; upper glume 75–100% as long as the spikelet... **D. ischaemum**

Diplachne fusca, Sprangletop

Cespitose annuals or perennials; leaf sheaths open; ligules membranous. Inflorescences panicles with spike-like branches; spikelets arranged in 2 rows along the branch with 6–20 florets per spikelet; lemmas awnless or awned.

1a Uppermost leaf blades exceeding the panicles; panicles usually partially enclosed in the uppermost sheaths; mature lemmas 3.5–5 mm, often smoky white with a dark smudge in the basal half............................ **D. f. ssp. *fascicularis***
1b Uppermost leaf blades exceeded by the panicles; panicles usually completely exserted; mature lemmas 2–3.6 mm, usually dark green to dark gray, lacking a dark spot; rare waif... **D. f. ssp. *uninervia***

Echinochloa, Barnyard Grass

Annuals. Leaf sheaths open, laterally compressed, blades flat with a prominent midrib; ligules usually absent. Inflorescences panicles with spike-like branches that may be further branched. Spikelets subsessile, crowded, dorsiventrally compressed, ± flat on 1 side, rounded on the other, with 2(3) florets; disarticulation below glumes. Glumes unequal, awnless or short-awned; lower glumes ~ half as long as spikelet; upper glumes as long as spikelet. Lower lemmas similar to upper glumes in length and texture, awnless or awned; upper lemmas tough, rounded over the back, smooth and shiny, awnless.

1a Upper lemmas gradually tapering to a stiff tip; lemmas awnless or with awns to 6(10) mm.. **E. muricata var. *microstachya***
1b Upper lemmas abruptly narrowed to a withering green tip; lemmas often with awns to 50 mm, sometimes awnless .... **E. crus-galli**

Elymus, Wildrye, Wheatgrass

Cespitose or rhizomatous perennials. Leaf sheaths open; ligules membranous, often ciliolate. Inflorescences spikes with 1–3(5) spikelets per node; internodes 1.5–26 mm. Spikelets laterally flattened, with 1–11 florets, usually disarticulating above the glumes, but sometimes below the glumes or in the inflorescence axis; glumes and lemmas often awned. See also *Leymus*.

Elymus and many of the other genera in the Triticeae are prone to producing sterile hybrids, often between genera. If you find an abnormal form that doesn't key, it may be one of these hybrids.

1a Spikelets 1 at all or most nodes
  2a Inflorescence axis disarticulating when mature; plants cespitose; anthers 1–1.6 mm; subalpine and alpine; few records in central and N Cascades, WA .... *E. scribneri*
  2b Inflorescence axis not disarticulating; plants cespitose or rhizomatous; anthers 0.5–8 mm
    3a Lemma awns curving outward at maturity
      4a Glumes narrowly lanceolate with 1–3(4) veins; native in NE OR and SE WA; planted widely E of the Cascades .... *E. wawawaiensis*
      4b Glumes lanceolate to oblanceolate with (3)4–5(7) veins (check several); E of the Cascades and SW OR .... *Pseudoroegneria spicata*
    3b Lemmas awnless or with straight awns
      5a Plants strongly rhizomatous; anthers 2.5–7 mm; spikelets mostly > 1.25× as long as internodes; lemmas glabrous to densely hairy
        6a Leaf blades 6–10 mm wide, usually flat; leaf veins unequal in width, with broader primary veins separated by narrower secondary veins; glumes distally keeled with scabrous midveins, relatively flat and smooth near base; widespread........................ ....*E. repens*
        6b Leaf blades 1.5–6 mm wide, usually involute; leaf veins subequal in width; glumes rounded or keeled throughout their length, sometimes distally scabrous; E of the Cascades (rare records W of the Cascades are waifs).................. *E. lanceolatus* (with 3 subspecies)
          7a Lemmas densely hairy, hairs flexible, some hairs ≥ 1 mm; E of the Cascades in N OR and S WA .... *E. l.* ssp. *psammophilus*
          7b Lemmas glabrous or with stiff hairs < 1 mm; widespread E of the Cascades
            8a Lemmas with hairs, not scabrous .... *E. l.* ssp. *lanceolatus*
            8b Lemmas mostly glabrous, sometimes scabrous distally, margins sometimes hairy proximally ....*E. l.* ssp. *riparius*
      5b Plants cespitose, occasionally loosely so; anthers 0.5–8 mm, if > 2.5 mm, then spikelets mostly 1–1.25× as long as internodes; lemmas glabrous to sparsely hairy
        9a Anthers 3.5–8 mm; spikelets mostly 1–1.25× as long as internodes; E of the Cascades and SW OR..................... *Pseudoroegneria spicata*
        9b Anthers 0.5–2.5 mm; spikelets mostly 2–3× as long as internodes; range various
          10a Glumes 3(5)-veined; hyaline margins of glumes wider on one side of the glume than the other, the wider margins 0.3–1 mm wide, widest in the distal third; lemma awns 0.5–3 mm; high montane to alpine; Wallowa Mts., OR, and Olympic Mts. and N Cascades, WA ....*E. violaceus*

10b Glumes 3–7-veined; hyaline margins of glumes usually ± equal in width on each side of the glume, 0.1–0.5 mm wide, widest at or slightly beyond midlength; lemmas awnless or with awns to 40 mm; lowlands to subalpine; widespread E and W of the Cascades..........***E. trachycaulus* ssp. *trachycaulus***

1b Spikelets 2–3(5) at all or most nodes

11a Both glumes awn-like to awl-like, 10–135 mm including awns, sometimes split lengthwise, flexuous to outcurving from near the base; inflorescence axis disarticulating at maturity

12a Glume awns split into 3–9 divisions; lemma awns about 0.2 mm wide at the base; inflorescence axis internodes 3–5.5 mm; widespread except W WA ....***E. multisetus***

12b Glume awns entire or split into 2(3) divisions; lemma awns about 0.4 mm wide at base; inflorescence axis internodes 3–10(15) mm; in and E of the Cascades, SW OR, and Olympic Mts., WA
.... ***E. elymoides*** (with 3 subspecies)

13a Inflorescence with 3 spikelets per node, the central spikelet usually with 2 fertile florets, the florets of the lateral spikelets not fertile, rudimentary to awn-like; lemma awns 15–30 mm; E of the Cascades
....***E. e.* ssp. *hordeoides***

13b Inflorescence usually with 2 spikelets per node, each spikelet usually with (1)2–4(5) fertile florets; lemma awns 15–120 mm

14a One or more of the spikelets at most nodes appearing to have 3 glumes, 1 of these actually the reduced lemma of a sterile, lowermost floret; glume awns entire or split into 2–3 divisions; awns 15–85 mm; paleas usually with the veins extended as bristles; widespread ....***E. e.* ssp. *elymoides***

14b No spikelets appearing to have 3 glumes, the lowermost floret of each spikelet well-developed; glume awns entire, 50–125 mm; paleas rarely with veins extended as bristles; widespread generally at higher elevations, E of the Cascades, Olympic Mts., and SW OR
.... ***E. e.* ssp. *brevifolius***

11b One or both glumes broader, not awn-like to awl-like, (4.5)7–40 mm including awns, entire, straight, or outcurving from well above the base; inflorescence axis not disarticulating at maturity

15a Glume awns 0–10 mm; lemma awns straight or outcurving at maturity; inflorescences nodding or erect

16a Inflorescences usually strongly nodding; distal lemma margins with hairs > those elsewhere on the lemma; lemma awns ± outcurving; W WA and NW OR...***E. hirsutus***

16b Inflorescences usually erect or sometimes slightly curved to the side; distal lemma margins glabrous or with hairs similar to those elsewhere on the lemma; lemma awns usually straight; range various

17a Glume bases ± round in cross section, bowed out, hard, without evident veins in the lower 2–3 mm; glume awns 0–3(5) mm; NE WA ....***E. curvatus***

17b Glume bases flat, not bowed out, not hard, with evident veins; glume awns 0–9 mm; widespread....... ***E. glaucus*** (with 2 subspecies)

18a Lemma awns (5)10–25(35) mm; glume awns 0.5–5(9) mm; widespread ....*E. g.* ssp. *glaucus*
18b Lemma awns (0)1–5(7) mm; glume awns 0–2 mm; SW OR and Olympic Mts., WA, usually near the coast.. *E. g.* ssp. *virescens*
15b Glume awns 10–25 mm; lemma awns outcurving at maturity; inflorescences nodding; E WA and along the Snake River in E OR; introduced in W WA and seeded elsewhere ....*E. canadensis* var. *canadensis*

Eragrostis, Lovegrass

Cespitose or mat-forming annuals and perennials. Sheaths open; ligules a fringe of hairs, often with longer hairs on the collar and upper sheath. Inflorescences are panicles, with a distinctive bulge (pulvinus) in the axils of the branches. Spikelets with 2–60 florets; glumes < the lowest floret. Lemmas with 3 veins not converging at the lemma tip, awnless. Seeds tiny and variable in shape, surface texture, and whether there is a groove on the side opposite the embryo. (You will need a microscope to see these features.) Various distinctive glands are found in some species on the leaves, sheaths, panicle branches, pedicels, and lemmas. Lemmas of some species fall off before the paleas, sometimes enclosing the caryopsis, sometimes leaving it behind in the palea; the palea drops off later, sometimes leaving a persistent rachilla.

1a Plants perennial, densely cespitose; anthers 0.6–1.2 mm; escaped ornamental; uncommon but spreading E and W of the Cascades .... *E. curvula*
1b Plants annual, tufted or creeping; anthers 0.2–0.5 mm
    2a Stems stoloniferous, widely creeping, rooting at the nodes, forming loose mats; muddy, sandy, or gravelly shorelines; widespread............... *E. hypnoides*
    2b Stems erect to decumbent, not creeping, not forming mats; habitat various
        3a Saucer-like glands or glandular areas present somewhere on the plant (check leaf blade veins and margins, leaf sheaths, culms, panicle branches, and the keels of glumes and lemmas)
            4a Panicles 0.5–2 cm wide, branches appressed or ascending; spikelets light yellowish, sometimes marked with red; glands absent from lemmas and glumes, present on leaf blade bases and panicle branches; rare in E OR and WA .... *E. lutescens*
            4b Panicles 2–18 cm wide, branches spreading or appressed to ascending; spikelets green, dark gray, or reddish-purple; glands sometimes present on lemmas and/or glumes, often also on culms, sheaths, blades, or inflorescence branches; range various
                5a Leaf blade margins, glume keels, and lemmas without glands; culms, sheaths, and ventral blade surfaces sometimes glandular; uncommon; widespread... *E. pilosa* var. *pilosa*
                5b Leaf blade margins with conspicuous glands; glume keels, lemmas, and other parts usually glandular; widespread especially E of the Cascades
                      6a Spikelets 2–4 mm wide; lemmas 2–2.8 mm .... *E. cilianensis*
                    6b Spikelets usually ≤ 2 mm wide; lemmas 1.4–1.8 mm ..... *E. minor*
        3b Glands lacking
            7a Lower glumes 0.3–0.6(0.8) mm, < half as long as the lowest floret; lowest inflorescence branches whorled; paleas falling soon after lemmas; uncommon; widespread ....*E. pilosa* var. *pilosa*

7b Lower glumes 0.5–1.7 mm, sometimes ≥ half as long as the lowest floret; lowest inflorescence branches usually solitary or paired, sometimes whorled; paleas persisting on spikelet rachis after the lemmas fall
    8a Seeds with a shallow ventral groove on the side opposite the embryo; leaf blades 2–7(9) mm wide; culms 10–130 cm; widespread ....*E. mexicana* ssp. *virescens*
    8b Seeds rounded on the ventral side; lacking a shallow groove opposite the embryo; leaf blades 1–4.5 mm wide; culms 10–80 cm ....*E. pectinacea* (with 2 varieties)
        9a Pedicels appressed, diverging ≤ 20° from the panicle branches; widespread .... *E. p.* var. *pectinacea*
        9b Pedicels spreading to widely divergent, usually diverging 20–60° from the panicle branches; SW OR .... *E. p.* var. *miserrima*

Eriocoma, Needlegrasses and Ricegrasses

Cespitose perennials. Leaf sheaths open; ligules membranous, sometimes ciliate. Panicles often contracted, sometimes open; disarticulation above the glumes. Spikelets with 1 floret; glumes longer than the floret; florets round in cross section, sometimes laterally compressed, fusiform to elliptic; lemmas with terminal awns that are persistent or deciduous.

    NOTE: The callus extends very nearly to the base of the palea, but callus hairs obscure the junction and can make the callus seem shorter than it is. When in doubt, remove the hairs. If you need to measure the palea in a mature floret, cut the floret crosswise with a razor blade just above the callus. This should free the palea so you can compare its length to the lemma.

    Eriocoma lettermanii has been reported repeatedly from OR, but we have not seen any correctly identified specimens. The name *Eriocoma* dates back to 1818 when the name was bestowed on Indian ricegrass by British botanist Thomas Nuttall.

    Once upon a time most of the needlegrasses and ricegrasses of western North America were placed in *Stipa* and *Oryzopsis*. Research from the 1970s to the early 2000s led to these genera being split and recombined into multiple genera, including several that occur in our area: *Achnatherum*, *Hesperostipa*, *Nassella*, *Oryzopsis*, *Pappostipa*, and *Piptatheropsis*. More recent work has resulted in more splitting. *Achnatherum* is now considered to be native to the Mediterranean region and Eurasia, and our *Achnatherum* species have been given the genus name *Eriocoma*. Not everyone agrees with this arrangement and some long for the days of one genus to rule them all—*Stipa*! However, there is evidence that *Eriocoma* is still polyphyletic and we may expect more splits in the future!

1a Ligules of upper culm leaves 3–8 mm; leaves 0.5–2 mm wide, curling as they age; E of the Cascades .... *E. thurberiana*
1b Ligules of upper culm leaves 0.1–2 mm; leaves 0.4–5 mm wide, usually not curling; range various
    2a Lower awn segments glabrous, scabrous, or with hairs < 0.5 mm; awns persistent or deciduous
        3a Lemmas glabrous, shiny, dark brown, elliptic; awns deciduous; E of the Cascades

4a Panicles drooping, branches flexuous and often diverging; Crook and Wallowa Counties, OR .... ***E. wallowaensis***
4b Panicles erect, branches straight, ascending to appressed; Kittitas and Yakima Counties, WA, and Crook and Wasco Counties, OR .... ***E. hendersonii***

3b Lemmas ± hairy, dull to shiny, pale or dark, oval or cylindric; awns deciduous or persistent; range various

5a Panicle branches spreading to strongly divergent; E of the Cascades
6a Awns 15–25 mm, persistent, bent twice; panicle branches spreading; spikelets drooping; ligules of basal leaves 0.1–0.5 mm; NE WA and NE OR.. ***E. richardsonii***
6b Awns 3–6 mm, deciduous, straight or bent once; panicle branches strongly divergent; spikelets not drooping; ligules of basal leaves 1.5–4 mm; widespread E of the Cascades.................................. ***E. hymenoides***

5b Panicle branches appressed to strongly ascending; range various
7a Lemmas densely long-hairy, longest hairs > 2 mm; E of the Cascades in OR
8a Awns persistent, 13–15 mm, bent twice........................ ***E. pinetorum***
8b Awns rapidly deciduous, 4–11 mm, straight or bent once .... ***E. webberi***

7b Lemmas sparsely hairy, longest hairs ≤ 2 mm; range various
9a Lemmas with a thick, stiff, apical lobe, about 0.1 mm; awn appearing off-center on the lemma tip; paleas 75–100% as long as the lemmas; florets slightly laterally compressed; ligules of upper culm leaves to 2.5 mm; leaf blades 0.5–2.5 mm wide .... ***E. lemmonii* ssp. *lemmonii***
9b Lemma apical lobe thin, membranous, 0.1–0.4 mm; awn centered on the lemma tip or very nearly so; paleas 30–67% as long as the lemmas; floret round in cross section; ligules of upper culm leaves to 1.5(2) mm; leaf blades 0.8–5 mm wide
10a Calluses blunt, the glabrous area on the dorsal side about as long as wide, with a straight to rounded boundary between the glabrous tip and the hairy portion; leaves 1.2–5 mm wide; widespread except absent in NW OR and SW WA .... ***E. nelsonii* ssp. *dorei***
10b Calluses sharp, the glabrous area on the dorsal side longer than wide, with an acute boundary between the glabrous tip and the hairy portion; leaves 0.8–5 mm wide; in and E of the Cascades
11a Leaves 1.2–5 mm wide; awns 19–45 mm, lower segment scabrous; calluses somewhat less sharp, with a shorter extension of the glabrous area into the hairy part of the callus; habitat generally mesic to moist........... ***E. nelsonii* ssp. *nelsonii***
11b Leaves 0.8–2 mm wide; awns 18–55 mm, lower segment scabrous to short-hairy; calluses sharper, with a longer extension of the glabrous area into the hairy part of the callus; habitat generally dry..... ***E. occidentalis* ssp. *californica***

2b Lower awn segments with spreading hairs, at least some of the hairs ≥ 0.5 mm (check several awns); awns persistent

12a Lemma tip hairs ≤ the lower awn hairs; hairs of lower awn segment with a tidy appearance, gradually and evenly becoming shorter toward the first bend; leaves 0.3–2 mm wide.. ***E. occidentalis* ssp. *pubescens***
12b Lemma tip hairs > most of the lower awn hairs; hairs of lower awn segment with an untidy appearance, with longer hairs scattered among shorter hairs, the hairs not gradually and evenly becoming shorter toward the first bend; leaves 0.8–3 mm wide
    13a Calluses 0.5–0.7 mm, glabrous area on dorsal side rounded to acute distally; palea 50–75% as long as the lemma; leaves 1–3 mm wide; SE and NE OR; rare ....*E. nevadensis*
    13b Calluses 0.8–1.2 mm, glabrous area on dorsal side narrow and acute distally; palea 40–60% as long as the lemma; leaves 0.8–2 mm wide; widespread .... *E. occidentalis* ssp. *californica*

Festuca, Fescue

Tufted annuals, or cespitose or rhizomatous perennials. Plants with intra- or extravaginal branching. Leaf sheaths open or closed; ligules membranous, usually ciliate; blades flat, folded, or involute. Inflorescences open or contracted panicles, occasionally reduced to racemes. Spikelets laterally compressed with (1)2–11(17) florets; disarticulation above the glumes. Glumes usually shorter than florets; lower glumes 1(3)-veined; upper glumes 3(5)-veined. Lemmas 5(7)-veined, often minutely 2-toothed, usually awned; awns terminal or subterminal, usually straight, sometimes bent or curved.

    Festuca is a diverse and difficult genus of temperate, arctic, and alpine regions. The taxonomy of the fine-leaved fescues is controversial because different species look similar but differ consistently in cryptic traits such as ovary pubescence or leaf anatomy.

    The limits of the genus *Festuca* are controversial. The annual species are sometimes segregated in *Vulpia* but have recently been returned to *Festuca*. Tall fescue and meadow fescue (formerly *F. arundinacea* and *F. pratensis*) are now in the genus *Schedonorus* but are close to *Lolium* and could probably be included in that genus, or both *Lolium* and *Schedonorus* could be included in *Festuca*; in a moment of indecisiveness we decided to treat all three genera as distinct. Spike fescue (formerly *F. kingii*) is now placed in the genus *Leucopoa*. Some of our fine-leaved fescues have been included in *F. ovina* in past treatments, but this was incorrect, and wild populations of *F. ovina* are not present in the PNW.

    Among the fine-leaved fescues, leaf sclerenchyma fibers (here referred to as "fiber bundles") under the epidermis on the outer/lower side of the leaf are diagnostic. Fiber bundles may be broad or narrow (> 2× or < 2× as wide as thick, respectively) and occur in 1 of 4 patterns: (1) at each margin and at the midrib; (2) at the margins, midrib, and some or all of the lateral veins; (3) in a continuous or sometimes broken band that is either uniform or variable in thickness; or (4) extending through the leaf from the lower to upper surface like pillars. Fibers are viewed by making thin cross sections of the middle third of a leaf in a drop of water and viewing the cut surface under a microscope. With practice, some traits can often be observed in the field (with sharp scissors and a hand lens). Obviously, this is a difficult character to assess.

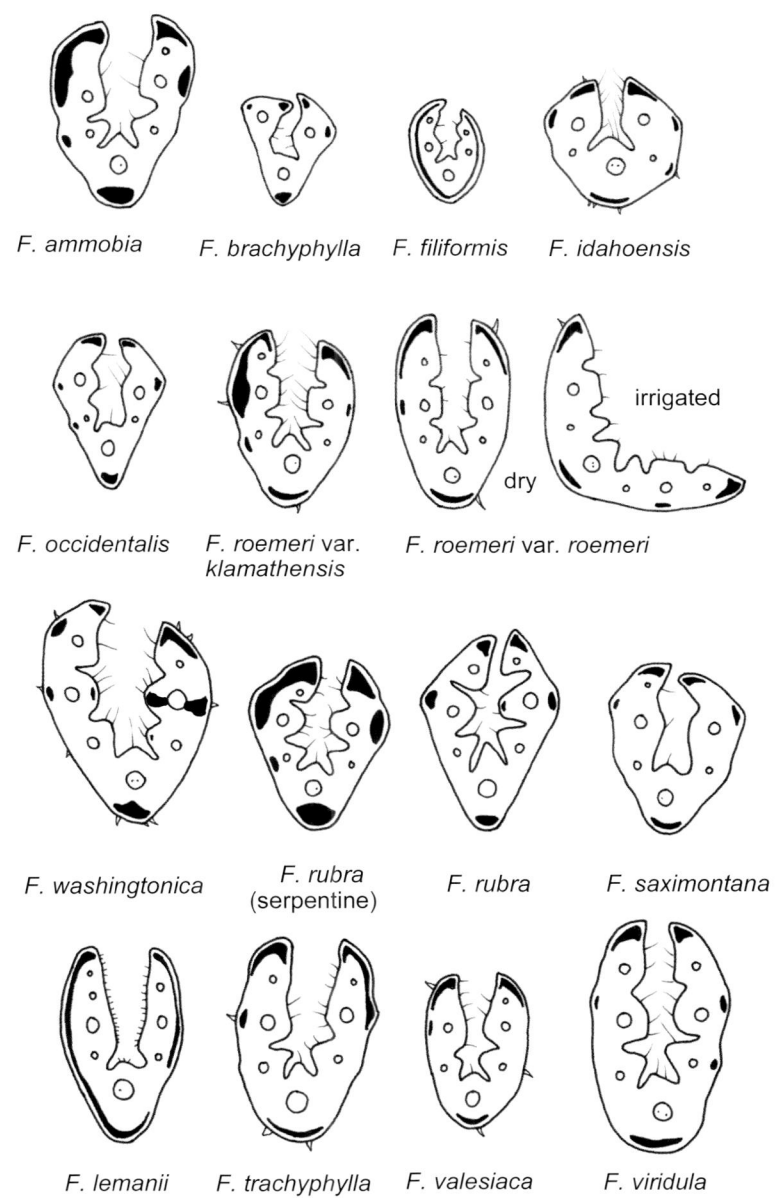

Figure 8. Leaf cross sections of fine-leaved fescues in Oregon and Washington cut from the middle third of innovation leaves (not culm leaves). Black shapes = fiber bundles, which appear white in fresh sections. Clear circles = vascular bundles (veins). Ventral grooves = indentations on the inner surfaces of the leaves. Note the hairs on the inner surface of the leaf and scabers on the outer surface.

Certain fine-leaved, cespitose fescues introduced for erosion control and xeriscaping may be confused with each other: *F. lemanii*, *F. nigrescens*, *F. trachyphylla*, and *F. valesiaca*. Sometimes the best way to differentiate them is with leaf cross sections (see illustrations on p. 63).

In keying *Festuca*, lemma length is measured on the lowest floret in a spikelet, at or after anthesis. Lemma and spikelet lengths do not include awns. Relative lengths of upper and lower glumes should not be assessed on the terminal spikelet of a branch. Sheaths are often split open by a growing shoot, so closure should be assessed on the second-from-top leaf of a vegetative shoot. Do not assess closure on culm leaves. Leaf cross sections should be cut from the middle third of innovation leaves (the leaves of nonflowering basal shoots), not from culm leaves.

1a Plants annual
    2a Inflorescence branches and/or spikelets spreading to reflexed; pulvini well-developed ....*F. microstachys*
    2b Inflorescence branches and spikelets ascending to erect; pulvini absent or poorly developed
        3a Lower glumes < 50% as long as the upper glumes, 0.1–2.5 mm......*F. myuros*
        3b Lower glumes ≥ 50% as long as the upper glumes, 1.7–6.5 mm
            4a Florets usually (5)7–12 per spikelet, spikelets closely overlapping; awns usually < 4(9) mm, ≤ lemmas; panicle branches 1–2 per node ....*F. octoflora*
            4b Florets usually 1–5(8) per spikelet; spikelets loosely overlapping; awns usually > 4 mm, ≥ the lemmas; panicle branches 1 per node
                5a Paleas usually slightly > lemma bodies; pulvini poorly developed but present at base of panicle branches and often the spikelets; plants immature (mature plants key at lead 2a) ....*F. microstachys*
                5b Paleas ≤ lemma bodies; pulvini lacking in the inflorescence; plants immature or mature.. *F. bromoides*
1b Plants perennial
    6a Widest leaves > 2 mm wide, flat or loosely rolled
        7a Leaves with claw-like or clasping auricles, or with expanded, flaring, rounded area at base of blade; lemma awns < 4 mm, much < the lemma bodies, or absent; ovary apices glabrous .... ***Schedonorus*** (go to key on p. 90)
        7b Leaves without auricles or expanded, flaring, rounded area at base of blade; lemma awns < or > lemma bodies, or absent; ovary apices usually hairy (except *F. rubra* and *F. roemeri*)
            8a Plants rhizomatous
                9a Glumes translucent; sheaths of basal leaves not shredding, pale with veins similar in color; leaf sheaths open, the margins overlapping; plants dioecious; dry, alpine; Steens Mt., SE OR...... ***Leucopoa kingii***
                9b Glumes not translucent, sometimes with hyaline margins; sheaths of basal leaves shredding lengthwise, brown with contrasting white veins; leaf sheaths closed about 75% their length, the margins fused; plants bisexual; moist to mesic, lowlands to high montane; widespread .... *F. rubra*
            8b Plants loosely to densely cespitose or growth habit unknown

10a Leaves mostly basal
- 11a Collars usually hairy at least on the margins; inflorescences 14–30 cm with branches 5–12 cm, spreading to drooping; W OR .... **F. californica**
- 11b Collars glabrous; inflorescences (2)10–25(30) cm with branches 3–7 cm, appressed to reflexed; range various
  - 12a Lower leaf sheaths pale, not contrasting with pale veins, intact to tearing but not shredding; leaves of vegetative shoots oval, rounded, or L-shaped in cross section with fiber bundles > twice as wide as thick; upland grasslands; W Cascades and W, and in the Columbia Gorge .... **F. roemeri** (with 2 varieties)
    - 13a Hairs on inner surface of leaf blades sparse, shorter than the thickness of the blade when dried; western Cascades and W, including near the coast, but not in the Klamath region of SW OR .... **F. r. var. roemeri**
    - 13b Hairs on inner surface of leaf blades dense, longer, approximately as long as the thickness of the blade when dried; Klamath region of SW OR to CA, not coastal .... **F. r. var. klamathensis**
  - 12b Lower leaf sheaths dark brown, contrasting with pale veins, shredding; leaves of vegetative shoots diamond-shaped, oval, or round in cross section with fiber bundles < twice as wide as thick; habitat and range various
    - 14a Lower surface of leaves glabrous to pubescent but not scabrous, or uniformly scabrous; ovary apex glabrous; leaf sclerenchyma usually not extending around major lateral veins; leaf blades 0.3–2.5 mm wide; plants green or glaucous; widespread................................. **F. rubra**
    - 14b Lower surface of leaves densely scabrous on the veins; ovary apex ± pubescent; leaf sclerenchyma often extending around major lateral veins, forming pillars from upper to lower surface of leaf; leaf blades 1.5–3 mm wide; plants green; in and E of the Cascades in WA and S BC.. **F. washingtonica**

10b Leaves mostly cauline
- 15a Inflorescence a spike-like condensed panicle; plants dioecious; alpine; Steens Mt., SE OR .... **Leucopoa kingii**
- 15b Inflorescence ± open; plants bisexual; lowland to montane forest and savanna; range various
  - 16a Calluses hairy, longer than wide, (0.3)0.4–0.6 mm long; upper leaf surfaces densely short-hairy; lemma awns 10–15 mm, much > lemma bodies; mesic to dry forest; W of the Cascades .. .. **F. subuliflora**
  - 16b Calluses glabrous, wider than long, 0.1–0.3 mm long; upper leaf surfaces glabrous, scabrous, or pubescent but not densely short-hairy; lemma awns 1.5–10(17) mm, < or > lemma bodies; range various.
    - 17a Lemma veins 0(1–3); lemma awns (2.5)5–10(17) mm, often curved; leaves cauline, 3–10 mm wide, glabrous (or nearly so) on upper surface; moist to mesic forest; widespread.............................. **F. subulata**
    - 17b Lemma veins (3)5 (sometimes obscure); lemma awns (1)2–5(8) mm, straight; leaves basal and cauline or cauline only, 1.8–6 mm wide, pubescent on upper surface; dry, open forest. SW OR.............. **F. elmeri**

6b Leaves < 2 mm wide, folded lengthwise
   18a Collars hairy at least on the margins; lemmas 7.5–11 mm; W OR .... *F. californica*
   18b Collars glabrous or if hairy then lemmas < 7 mm; range various
      19a Lemmas (6.2)7–8.5(10) mm, awnless or with awns to 1.5 mm; leaf blades usually scabrous over the entire surface, stiff when fresh; plants very densely cespitose; leaf sheaths from previous years persistent, forming a dense tuft at the base of the plant; dry uplands; E of the Cascades ....*F. campestris*
      19b Lemmas 2.3–10 mm, awnless to long-awned; leaf blades scabrous, smooth, or hairy, not stiff when fresh; plants cespitose to rhizomatous; leaf sheaths from previous years often persistent but usually not forming a dense tuft at the base of the plant; habitat and range various
         20a Lower cauline leaves with blades greatly reduced, usually 1–2 cm long; apex of ovary and seed densely hairy; montane to alpine.......... *F. viridula*
         20b Lower cauline leaves with blades normally developed; apex of ovary and seed glabrous or hairy; lowlands to alpine
            21a Lemmas awnless or with awns < 10% the length of the lemma bodies
               22a Lemmas 2.3–4.0(4.4) mm; anthers 1.5–2.2 mm; folded leaves 0.3–0.6 mm wide; leaf sheaths open to the base; leaf sclerenchyma forming a continuous band; rare introduction W of the Cascades.. *F. filiformis*
               22b Lemmas (3.5)4–9 mm; anthers (2)3–4 mm; folded leaves ≥ 0.5 mm wide; leaf sheaths closed ≥ half their length; leaf sclerenchyma in separate bundles
                  23a Inflorescences dense; plants short, often < 20 cm; leaf fiber bundles large; coastal; sands and bluffs ....*F. ammobia*
                  23b Inflorescences more open; plants usually taller; leaf fiber bundles small except sometimes in plants of extremely dry or serpentine habitats; coastal or not................................... *F. rubra*
            21b Lemmas awned; awns > 10% the length of the lemma bodies
               24a Plants rhizomatous or stoloniferous; vegetative shoots in part extravaginal... *F. rubra*
               24b Plants cespitose, lacking rhizomes or stolons, or habit unknown; vegetative shoots usually intravaginal
                  25a Mature anthers 0.4–1.8 mm; plants ≤ 60 cm, often dwarfed; lemmas 2–6 mm; lemma awns ≤ half as long as the lemma bodies; plants alpine
                     26a Anthers 0.4–0.8(0.9) mm when dry, 0.9–1.2 mm when fresh or rehydrated, usually retained long past anthesis ....*F. brachyphylla* (with 2 subspecies)
                       27a Culms about twice as long as the vegetative shoot leaves; lemmas 3–4(4.5) mm; North Cascades, WA, and Wallowa Mts. and Steens Mt., OR
                         .... *F. b.* ssp. *coloradensis*

27b Culms usually more than twice as long as the vegetative shoot leaves; lemmas (3)3.5–4.5(6) mm; WA Cascades .... *F. b.* ssp. ***brachyphylla***

26b Anthers 0.9–1.2 mm when dry, 1.2–1.8 mm when fresh or rehydrated, retained or lost after anthesis .... ***F. saximontana*** (with 2 varieties)

    28a Culms 25–50 cm, usually 3–5× the height of the vegetative shoot leaves; lower surfaces of leaf blades usually scabrous; fiber bundles 3–5, sometimes partly confluent or forming a continuous band; mts. of WA .... ***F. s.* var. *saximontana***

    28b Culms (5)8–20 cm, usually 2–3× the height of the vegetative shoot leaves; lower surfaces of leaf blades smooth or scabrous; fiber bundles 5–7, narrow; widespread in mts. of WA and OR .... ***F. s.* var. *purpusiana***

25b Mature anthers ≥ 2 mm; plants to 100 cm, rarely dwarfed except on strong serpentine substrates; lemmas (3)4–10(11) mm; lemma awns < or > half as long as lemma bodies; plants usually not alpine

    29a Some lemma awns usually > lemma bodies; apex of ovary and seed densely pubescent; mesic to dry forest, often in full or partial shade; widespread .... ***F. occidentalis***

    29b Lemma awns ≤ lemma bodies; apex of ovary and seed glabrous or pubescent; habitat various, usually in full sun; range various

        30a Mature lemmas ≥ (5.8)6 mm

            31a Leaves of vegetative shoots oval, rounded or L-shaped in cross section with fiber bundles > twice as wide as thick; lower leaf sheaths pale, not contrasting with pale veins, intact to tearing but not regularly shredding; upland grasslands

                32a Leaves very narrow, readily rolling between the fingers; high Cascades and E .... ***F. idahoensis***

                32b Leaves wider, not rolling between the fingers or rolling with angles that can be felt; western Cascades and W, and Columbia Gorge .... ***F. roemeri*** (with 2 varieties, see above, lead 13)

            31b Leaves of vegetative shoots diamond-shaped, oval, or round in cross section with fiber bundles < twice as wide as thick; lower leaf sheaths brown, contrasting with pale veins, shredding; range various

                33a Ovary apex pubescent after anthesis; lower (outer) surface of leaves densely scabrous on the veins; leaf sclerenchyma usually extending around major lateral veins from upper to lower surface of leaf; in and E of the Cascades in WA and S BC .... ***F. washingtonica***

                33b Ovary apex glabrous; lower (outer) surface of leaves glabrous to pubescent but not scabrous, or scabrous uniformly over the surface; leaf sclerenchyma not extending around major lateral veins; widespread

                    34a Plant strictly cespitose; lemmas 4.6–6 mm; awns 1–3.3 mm .... ***F. nigrescens***

                    34b Plants somewhat to strongly rhizomatous; lemmas 4.2–8 mm; awns 0.4–4.5 mm .... ***F. rubra***

        30b Mature lemmas ≤ 5.5(5.8) mm

35a Leaves smooth; lower leaf sheaths brown, ± contrasting with pale veins, shredding; leaves of vegetative shoots ± diamond shaped in cross section, forming a ± acute angle at the midrib; leaf fiber bundles < 2× as wide as thick; leaf sheaths of vegetative shoots closed about 75% their length, margins fused (best assessed on the second from top leaf of a vegetative shoot); widespread
    36a Plants strictly cespitose; lemmas 4.6–6 mm; awns 1–3.3 mm .... ***F. nigrescens***
    36b Plants somewhat to strongly rhizomatous; lemmas 4.2–8 mm; awns 0.4–4.5 mm .... ***F. rubra***
35b Leaves rough or smooth; leaf sheaths mostly whitish to light brown, sometimes pink or purplish (old overwintering sheaths sometimes dark brown), not contrasting with pale veins, generally not shredding; leaves of vegetative shoots ± oval to obovate in cross section, rounded at the midrib; leaf fiber bundles > 2× as wide as thick, sometimes confluent into a ± continuous band; leaf sheaths of vegetative shoots open to the base; margins overlapping; range various
    37a Leaf sclerenchyma forming a continuous or nearly continuous band, uniform in thickness; inner surface of folded leaf with 2–4 grooves; leaves smooth to finely scabridulous; lower leaf sheaths usually loose; used for roadside revegetation; recently documented, scattered in OR and WA .... ***F. lemanii***
    37b Leaf sclerenchyma forming 3 or more distinct fiber bundles, rarely continuous, but if so, uneven in thickness; inner surface of folded leaf with 4–6 grooves; leaves often rough; lower leaf sheaths tight; range various
        38a Leaf blades (0.2)0.3–0.5(0.6) mm wide; upper glumes 2.5–3.9 mm; lemmas 3.4–4.9(5.2) mm; leaf sclerenchyma forming 3 fiber bundles, on midrib and margins, sometimes with much smaller bundles between; mostly E of the Cascades and SW OR, expected elsewhere ...... ***F. valesiaca***
        38b Leaf blades (0.5)0.6–1.1 mm wide; upper glumes 3.4–4.5 mm; lemmas 4.2–5(6.5) mm; leaf sclerenchyma usually forming more than 3 fiber bundles, on midrib and margins and in between, sometimes continuous but uneven in thickness; ± widespread .... ***F. trachyphylla***

Glyceria, Mannagrass

Rhizomatous perennials (occasionally annuals) with leafy culms. Leaf sheaths closed at least 75% their length; ligules membranous. Inflorescences panicles or racemes; disarticulation above the glumes. Glumes 1-veined; lemmas strongly veined, the veins not converging toward the tip, awnless.

Glyceria spikelets come in 2 types, like pasta: short-fat and long-skinny. Short-fat spikelets are laterally compressed, 1–4× longer than wide, and oval in outline; long-skinny spikelets are cylindrical, > 5× longer than wide, and narrowly oblong in outline. *Puccinellia* and *Torreyochloa* can be similar, but they have open leaf sheaths. Mannagrasses grow in and near streams, lake margins, and swamps. Unlike *Puccinellia*, they do not tolerate highly alkaline soils.

1a Spikelets ovate to elliptic, 1–4× as long as wide, laterally compressed; palea keels not winged
    2a Upper glumes 3–4 mm; culms 60–250 cm, 6–12 mm in diameter; leaves sometimes variegated; rare introduction, NW WA (not treated) .... ***G. maxima***

2b Upper glumes 0.6–2.7 mm; culms 50–150(200) cm, 1.5–12 mm in diameter; leaves green
   3a Veins of 1 or both glumes extending to the tips; sheaths usually smooth; anthers 3... .**G. grandis var. grandis**
   3b Veins of glumes terminating below the tips; sheaths usually scabrous; anthers 2
      4a Spikelets (2.5)3–5 mm wide; lemma veins evident but not raised distally; paleas 1.5–1.8× as long as wide; W WA and NW OR .... **G. canadensis var. canadensis**
      4b Spikelets 1.2–2.9 mm wide; lemma veins distinctly raised throughout; paleas 1.5–3.5× as long as wide; widespread
         5a Leaf blades 2–6 mm wide; anthers 0.2–0.6 mm; culms 1.5–3.5 mm in diameter ....**G. striata**
         5b Leaf blades 6–15 mm wide; anthers 0.5–0.8 mm; culms 2.5–8 mm in diameter...**G. elata**
1b Spikelets linear to narrowly oblong, > 5× as long as wide, usually round in cross section; palea keels usually winged toward the tip
   6a Lemmas 3(5)-toothed; leaf blades 3–12 cm; primary panicle branches 1.5–9.5 cm; W of the Cascades .... **G. declinata**
   6b Lemmas entire or crenulate, not strongly toothed; leaf blades 5–30 cm; primary panicle branches 3–18 cm
      7a Lemmas 5–8 mm
         8a Anthers 0.6–1.6 mm; lemmas ≤ 5.9 mm, usually irregularly crenulate; in and W of the Cascades..**G. × occidentalis**
         8b Anthers 1.5–3 mm; lemmas 5.2–8 mm, usually entire; mostly W of the Cascades ....**G. fluitans**
      7b Lemmas 2.4–5 mm
         9a Lemmas smooth, or minutely scabrous only on the veins at 20× (rarely also between the veins, but the scabers shorter than those on the veins); mostly E of the Cascades, occasionally W.................. **G. borealis**
         9b Lemmas minutely scabrous both on and between the veins at 20×; mainly W of the Cascades
            10a Lemmas 4.5–5.9 mm, tips acute; anthers 2, 0.6–1.6 mm ....**G. × occidentalis**
             10b Lemmas 2.5–4.5 mm, tips truncate to obtuse; anthers 3, 0.3–0.9 mm .... **G. leptostachya**

Graphephorum, False Oat, Trisetum

Cespitose perennials. Leaf sheaths open; ligules membranous, sometimes minutely ciliate. Inflorescences open to spike-like panicles; disarticulation above the glumes. Spikelets with 2–5 florets; glumes longer than lower florets, lemmas awned from the back. Our *Graphephorum* species were formerly included in *Trisetum*.

1a Lemmas awnless or with awns to 2 mm; in and E of the Cascades ....**G. wolfii**
1b Lemmas with awns 7–14 mm
   2a Most panicle branches spikelet-bearing nearly to the base; panicles erect or nodding at the tips; branches ascending to somewhat spreading; lower glumes 2.8–5+ mm; ligules (1.5)3–6 mm.. **G. canescens**

2b Most panicle branches spikelet-bearing only toward the tips; panicles nodding; lower branches spreading or drooping; lower glumes 0.7–2.5(3.5) mm; ligules 1.5–3(3.4) mm..**G. cernuum**

Hierochloë, Sweetgrass

Loosely cespitose perennials. Leaf sheaths open; ligules membranous. Inflorescences open panicles. Spikelets laterally compressed, with 3 florets, the lowest 2 staminate, the distal floret bisexual; disarticulation above the glumes with the florets falling together. Glumes from slightly < to > florets. Lower lemmas awnless or awned; upper lemmas somewhat hard, shiny, inconspicuously 3–7-veined, awnless.

Hierochloë foliage contains chemical compounds called coumarins that produce a sweet smell when crushed or burned. *Hierochloë* is closely related to *Anthoxanthum* and is sometimes included in that genus.

1a Uppermost culm leaves 1–3(6) cm long; leaf blades 2.5–5.5 mm wide, green on both sides; glumes > florets; fertile lemmas 2.9–3.5 mm; lemmas acute, usually entire, rather uniformly pubescent over the back and margins; mostly in and E of the Cascades .... **H. odorata**
1b Uppermost culm leaves (1.5)3–13.5 cm long; leaf blades (3)5–15 mm wide, often glaucous on 1 side; glumes ≤ florets; fertile lemmas 3.5–4.5 mm; lemmas rounded and shallowly bilobed, puberulent over the back and papillose-ciliate along the margins; W of the Cascade Crest, mainly in OR
.... **H. occidentalis**

Holcus, Velvetgrass

Eurasian perennials; rhizomatous or cespitose. Leaf sheaths open; ligules membranous. Inflorescences congested panicles. Spikelets 2-flowered; glumes longer than florets. Lemmas shiny, lower one awnless, upper one with a short awn just below the tip.

1a Culms usually velvety-hairy on both internodes and nodes; awns < 2 mm, hook-shaped at maturity; hidden or exserted sideways from the glumes; plants cespitose; widespread..**H. lanatus**
1b Culm internodes glabrous or nearly so, nodes densely hairy; awns > 2.5 mm, ± straight or bent, not hook-shaped at maturity, exserted > 1 mm beyond the glumes; plants rhizomatous; W of the Cascades.. **H. mollis**

Hordeum, Barley

Annuals and perennials, tufted or cespitose, culms sometimes decumbent. Leaf sheaths open, ligules membranous. Inflorescences spikes, usually disarticulating along the inflorescence axis at maturity; spikelets falling in 3s. Spikelets 3 per node, the central spikelet fertile and the lateral spikelets rudimentary. Glumes very narrow, often awn-like.

1a Auricles well-developed, at least some > 1 mm
2a Lemmas of central florets ≥ 3 mm wide, usually with awns 30–180 mm, or if awnless then prominently 3-lobed near the tip; inflorescence axis not disarticulating; glume margins not ciliate; ligules 0.5–1 mm; cultivated barley ....**H. vulgare** ssp. **vulgare** (see p. 295)

  2b Lemmas of central florets ≤ 2 mm wide with awns 20–40 mm, never 3-lobed near the tip; inflorescence axis disarticulating at maturity; glume margins ciliate; ligules 1–4 mm; widespread in disturbed habitats .... *H. murinum* (with 3 subspecies, see p. 296)
1b Auricles lacking or < 1 mm
  3a Glumes, including awns, 35–85 mm; inflorescences arching to horizontal at anthesis; widespread .... *H. jubatum* ssp. *jubatum*
  3b Glumes, including awns, ≤ 26 mm; inflorescences usually ± erect
    4a Plants perennial, mostly (20)40–80(100) cm; leaf blades 2–6(9) mm wide; widespread .... *H. brachyantherum*
    4b Plants annual, usually < 50 cm; leaf blades ≤ 4.5 mm wide
      5a Lemmas of lateral spikelets with awns 2.5–8 mm; seasonally moist disturbed areas; widespread .... *H. marinum* ssp. *gussoneanum*
      5b Lemmas of lateral spikelets awnless or with awns to 1 mm; moist, often alkaline sites; rare E of the Cascades .... *H. depressum*

Leymus, Wildrye

Perennials, usually rhizomatous, sometimes cespitose; leaf sheaths open; ligules membranous; leaf blades stiff; inflorescence a spike not disarticulating at maturity, with 2–35 spikelets per node; glumes awnless; lemma tips acute, awnless or tapering into a short awn.

  The name *Leymus* is an anagram of *Elymus* and its members were formerly included in that genus. *Leymus* species are often tolerant of alkaline soils.

1a Glumes flat or rounded over the back, widest at or above midlength, flexible, the central portion scarcely thicker than the margins; coastal sand dunes .... *L. mollis* ssp. *mollis*
1b Glumes narrow, often subulate or keeled at least distally, widest below midlength, stiff, the central portion thicker than the margins; habitat various; not coastal; mostly E of the Cascades
  2a Plants cespitose (sometimes weakly rhizomatous); culms 35–350 cm
    3a Leaf blades 10–30+ mm wide, usually glaucous; inflorescences often branched, especially at the lower nodes; spikelets 5–35 per node, including side branches .... *L. condensatus*
    3b Leaf blades 3–12 mm wide, usually not glaucous; inflorescences not branched; spikelets 2–7 per node .... *L. cinereus*
  2b Plants rhizomatous; culms 18–120 cm
    4a Culms 8–12 mm thick; spikelets 3–8 per node; dunes and sandy soils; E of the Cascades in WA and N OR .... *L. racemosus*
    4b Culms 1–4 mm thick; spikelets 2 per node at middle of spike, sometimes with 1 or 3 spikelets per node near base or tip of spike
      5a Glumes glabrous or scabrous distally; lemmas usually with awns to 3 mm, glabrous to sparsely hairy; auricles to 1 mm; dry to moist, often alkaline meadows; widespread, mostly E of the Cascades .... *L. triticoides*
      5b Glumes hairy at least distally; lemmas awnless or with awns to 2 mm, densely hairy with hairs 2–3 mm long; auricles lacking or poorly developed; sandy soils and dunes; Columbia and Snake River valleys .... *L. flavescens*

Lolium, Ryegrass

Introduced annuals and perennials; cespitose; leaf sheaths open; auricles often present; ligules membranous. Inflorescence a spike with 1 spikelet per node; the spikelets oriented edgewise to the axis and sometimes sunken into it; disarticulation above the glumes; spikelets laterally flattened with 2–22 florets; glumes 1, except 2 on the terminal spikelet; lemmas awned or awnless.

1a Plants annual; lemmas ovate to ovate-elliptic, ≤ 3× as long as wide, becoming thick and hard at the base in fruit; longest glumes usually ≥ the distal florets; anthers 1.2–3.2 mm .... **L. temulentum ssp. temulentum**
1b Plants annual or perennial; lemmas narrowly oblong-ovate, > 3× as long as wide, not becoming thick and hard at the base in fruit; longest glumes usually < the distal florets; anthers 2–5 mm .... **L. perenne** (with 2 subspecies)
   2a Florets 2–10 per spikelet; leaves folded in bud; plants perennials with vegetative shoots present at flowering and fruiting; lemmas awnless or with awns to 8 mm .... **L. p. ssp. perenne**
   2b Florets 10–22 per spikelet; leaves rolled in bud; plants annuals or short-lived perennials, often lacking vegetative shoots at flowering and fruiting; lemmas with awns to 15 mm, rarely awnless .... **L. p. ssp. multiflorum**

Melica, Oniongrass, Melic

Perennials, cespitose or rhizomatous, often with bulb-like corms at the base. Leaf sheaths closed to the top (or nearly so), sometimes splitting distally; ligules membranous, sometimes with a fringe of hairs, sometimes closed in front; blades flat or involute. Inflorescences panicles, occasionally racemes; disarticulation above or below the glumes. Spikelets with 1–7 bisexual florets proximally, and 1–4 sterile florets distally with empty lemmas that enclose one another and lack paleas, forming a rudiment that is either similar in shape to (but smaller than) the fertile florets, or truncate, club-shaped, or ± round. Glumes < the lemmas, hyaline in the distal 33%, margins usually purple. Lemmas 7–11-veined, hyaline in distal 33–50%, awned or not; awns, if present, arising from the tip or from between 2 apical teeth. In species with bulb-like bases, the bases may be onion-like, thick, and rounded, or they may be more gradually tapered.

1a Lemmas conspicuously silky-hairy on the margins, the long, white hairs 3.5–5 mm, exserted between the glumes; inflorescence narrowly cylindrical, dense, and showy; escaped from ornamental plantings in Seattle .... **M. ciliata** (see p. 316)
1b Lemmas glabrous to short-hairy, the hairs < 1.5 mm, not exserted between the glumes; panicles open to narrowly cylindrical, not particularly showy
   2a Lemmas awned, awns 0.5–12 mm; culms without bulb-like bases
      3a Lemma awns 0.5–3(4) mm, often inconspicuous and/or deciduous; lemmas usually long-hairy on the margins and near the base, the hairs 0.7–1.3 mm; in and W of the Cascades .... **M. harfordii**
      3b Lemma awns 3–12 mm, conspicuous and persistent; lemmas glabrous, scabrous, or short-hairy on margins and toward base, the hairs 0.1–0.6 mm
         4a Leaf blades 2–6 mm wide; ligules usually pubescent to the top; lower panicle branches usually paired with 1–4 spikelets per branch; Cascades and Klamath Mts., S WA to CA .... **M. aristata**

    4b Leaf blades 5–13 mm wide; ligules usually glabrous except at the base; panicle branches usually solitary with 4–7 spikelets per branch; E and W of the Cascades in WA, and mts. of NE OR; rare, often reported in error .... ***M. smithii***
2b Lemmas awnless (rarely with short point to 0.4 mm); culms with or without bulb-like bases
    5a Pedicels sharply bent just below the spikelets; spikelets disarticulating below the glumes; panicles narrow, raceme-like, ± 1-sided; dry montane to alpine; SE OR .... ***M. stricta* var. *stricta***
    5b Pedicels ± straight; spikelets disarticulating above the glumes; panicles spreading or narrow, generally not 1-sided
        6a Lemma tips acuminate; lemmas usually with sparse long hairs at least on veins and margins near the base; culms usually with bulb-like bases; widespread .... ***M. subulata***
        6b Lemma tips truncate, obtuse, or acute, sometimes shallowly notched or toothed; lemmas glabrous, scabrous, or hairy; culms with or without bulb-like bases
            7a Rachilla internodes swollen when fresh, wrinkled and dull orange-brown when dry; culms 10–40(60) cm, with bulb-like bases; in and E of the Cascades, and Klamath Mts. .... ***M. fugax***
            7b Rachilla internodes not swollen when fresh, not wrinkled and orange-brown when dry; culms (30)40–200 cm, with or without bulb-like bases
                8a Inflorescence branches flexuous, spreading to reflexed; lemmas glabrous or scabrous; culms with bulb-like bases; W of the Cascades in OR, especially SW OR, rarely north to the northern Willamette Valley .... ***M. geyeri* var. *geyeri***
                8b Inflorescence branches usually appressed; lemmas glabrous, scabrous, or hairy; culms with or without bulb-like bases
                    9a Lemmas usually strongly long-hairy on the margins and near the base, the hairs 0.7–1.3 mm; culms without bulb-like bases; in and W of the Cascades .... ***M. harfordii***
                    9b Lemmas glabrous or scabrous; culms usually with bulb-like bases
                        10a Bulb-like culm bases connected to rhizomes by root-like structures 10–30 mm long, not clustered; ligules 0.1–3 mm; glumes usually < half as long as the spikelets; E of the Cascades, Olympic Peninsula, and SW OR .... ***M. spectabilis***
                        10b Bulb-like culm bases sessile on rhizomes, tending to be clustered; ligules 2–6 mm; glumes usually > half as long as the spikelets; in and E of the Cascades, and SW OR .... ***M. bulbosa***

Muhlenbergia, Muhly

Annuals or perennials, rhizomatous, cespitose, or mat-forming. Culms solid or hollow. Leaf sheaths open, ligules membranous, sometimes with minute hairs on the margin. Inflorescences open to contracted panicles. Spikelets laterally flattened, with 1(3) florets, usually disarticulating above the glumes. Glumes 1(3)-veined, the veins sometimes very faint. Lemmas 3-veined, awnless or awned from the tip. Paleas well-developed, mostly subequal to lemmas. NOTE: Nodulose culms have pale, blunt bumps that are thickenings of fibers in the culms.

1a Plants densely cespitose perennials; leaf blades 10–50 cm; inflorescences 15–60 cm, spike-like; SW OR..***M. rigens***
1b Plants rhizomatous or cespitose perennials or tufted annuals; leaf blades 0.5–20 cm; inflorescences 1–21 cm, spike-like or open
  2a Spikelets on long, slender pedicels in open, spreading panicles (1.5)4–21 cm wide; inflorescence often 33–67% the total height of the plant
    3a Plants with long scaly white rhizomes, perennial; anthers 1–1.3 mm; moist alkaline meadows E of the Cascades ....***M. asperifolia***
    3b Plants cespitose, annual or perennial; anthers 0.2–0.9 mm
      4a Glumes sparsely hairy near the tip, 0.5–0.9 mm; lemmas 0.8–1.5 mm; anthers 0.2–0.7 mm; plants annual; spikelets with 1 floret; E of the Cascades ....***M. minutissima***
      4b Glumes glabrous, 0.4–1.3 mm; lemmas 1.2–2 mm; anthers 0.6–0.9 mm; plants perennial, but often appearing annual; spikelets occasionally with 2 florets; introduced to coastal cranberry bogs .... ***M. uniflora*** (see p. 325)
  2b Spikelets subsessile or short-pedicellate in contracted, often spike-like panicles 0.1–2(3) cm wide; inflorescence generally < 33% the total height of the plant
    5a Lemmas glabrous, scabrous, or short-hairy, the hairs < 0.3 mm; glumes < lemmas; leaf blades 0.2–6.5 mm wide, flat, involute, or folded at maturity
      6a Plants perennial, rhizomatous; culms nodulose-roughened (with small bumps between the veins), at least below the nodes, sometimes short-hairy below the nodes; leaf blades 0.4–6.5 mm wide; glume tips acute, sometimes mucronate; lemmas glabrous; mainly E of the Cascades .... ***M. richardsonis***
      6b Plants annual, not rhizomatous; culms not nodulose-roughened, glabrous; leaf blades 0.6–1.6 mm wide; glume tips rounded to subacute; lemmas pubescent on margins and midvein; in and E of the Cascades, also SW OR .... ***M. filiformis***
    5b Lemmas hairy toward the base, the hairs 0.3–3.5 mm; glumes ≥ lemmas (glumes sometimes slightly < lemmas in *M. mexicana*); leaf blades 2–6 mm wide, flat at maturity
      7a Calluses and lemma bases with hairs about as long as the florets, the hairs 2–3.5 mm; SW and E OR, NE WA ....***M. andina***
      7b Calluses and lemma bases with hairs < the florets, the hairs ≤ 1.2 mm
        8a Glumes, including awns, 3–8 mm, about 1.3–2× as long as the lemmas; anthers 0.8–1.5 mm; lemma awns 0–1 mm; NE WA ....***M. glomerata***
        8b Glumes including awns 1.5–3.7(4) mm, about = or slightly < the lemmas; anthers 0.3–0.5 mm; lemma awns 0–10 mm ....***M. mexicana*** (with 2 varieties)

9a Lemma awns 3–10 mm; scattered E and W of the Cascades in OR, and SE WA .... ***M. m.* var. *filiformis***
9b Lemma awns < 3 mm or absent; rare native in E WA; introduced in coastal SW OR ....***M. m.* var. *mexicana***

Nassella, Feathergrass, Needlegrass

Cespitose perennials. Leaf sheaths open; ligules membranous. Panicles open, often nodding; disarticulation above the glumes. Spikelets with 1 floret; glumes longer than the floret, usually purple at anthesis but fading with age; florets round in cross section or somewhat laterally compressed, cylindrical to ellipsoid; calluses blunt to sharp-pointed; lemmas often finely papillose, glabrous or hairy, margins strongly overlapping their entire length at maturity (wrapping 1.2–1.5 times around the seed), lemmas narrowed near the tip into a cylindrical or cup-like crown topped by a row of hairs, surrounding the base of the awn; awns geniculate or (in *N. tenuissima*) curved throughout. Paleas up to half as long as lemmas.

1a Florets (1.5) 2.5–3 mm (excluding awn); lemma crown 0.1–0.2 mm; anthers 1.2–1.5 mm; leaf blades 0.2–1.5 mm wide; escaped ornamental; scattered W of the Cascades.. ***N. tenuissima***
1b Florets 4–11.5 mm (excluding awn); lemma crown 0.25–1.1 mm; anthers 2–5.5 mm; leaf blades 0.8–3.5 mm wide; rare in SW OR
   2a Florets 4–7 mm (excluding awn); awns 12–55mm; lemma crown 0.25–0.3 mm; anthers 2–2.5 mm .... ***N. lepida***
   2b Florets 7.5–11.5 mm (excluding awn); awns 38–100 mm; lemma crown 0.6–1.1 mm; anthers 3.5–5.5 mm .... ***N. pulchra***

Panicum, Panicgrass, Witchgrass

Cespitose annuals without overwintering basal rosettes, or perennials with short, scaly rhizomes. Leaf sheaths open; ligules a fringe of long, straight hairs, sometimes with a membranous base; sheaths and collars sometimes with papillose-based hairs. Inflorescences panicles; disarticulation below the glumes. *Panicum* and related genera have dorsiventrally compressed spikelets (see illustration on p. 9). Each spikelet has 2 glumes and 2 florets. Lower glume short; upper glume about the same length as the lower floret. Lower floret an empty, sterile lemma, which resembles the upper glume. Both glumes and the sterile lemma are membranous and have obvious veins. The upper floret is fertile with a hard, shiny, veinless lemma. To see the fertile floret pull the spikelet apart.

    Some species traditionally included in *Panicum* were recently moved to *Dichanthelium*. That genus is included in this key.

1a Pedicels < 2 mm; spikelets crowded on 1 side of panicle branches; SW OR ....***Panicum rigidulum* ssp. *rigidulum***
1b At least some pedicels > 2 mm; panicle branches not 1-sided
   2a Plants perennial, either cespitose with leaves forming basal rosettes, or forming clumps on short rhizomes
      3a Leaves 2–12 cm, of 2 forms, the basal leaves shorter and usually wider than the cauline leaves; basal rosettes well-developed, usually persistent; plants cespitose; culms 15–75 cm .... ***Dichanthelium*** (see p. 55)

    3b Leaves 10–60 cm; ± alike, not forming basal rosettes; plants with short, scaly rhizomes, often forming dense clumps; culms 40–300 cm; commonly planted ornamental escaped in N WA .... *P. virgatum*
2b Plants annual; leaves not forming basal rosettes, not rhizomatous
    4a Spikelets 4–6 mm .... *P. miliaceum* (with 2 subspecies)
        5a Mature upper florets blackish, disarticulating at maturity; culms 70–210 cm; panicles erect, ± exserted from sheath at maturity; panicle branches ascending to spreading; widespread but seldom collected ....*P. m.* ssp. *ruderale*
        5b Mature upper florets straw-colored to orange, not disarticulating; culms 20–120 cm; panicles usually nodding, not fully exserted from sheath at maturity; panicle branches ascending to appressed; widespread .... *P. m.* ssp. *miliaceum*
    4b Spikelets 2–3.5 mm
        6a Leaves with long, papillose-based hairs; stem nodes short-hairy; ultimate inflorescence branches widely spreading ....*P. capillare* ssp. *capillare*
        6b Leaves glabrous or nearly so, hairs if present not papillose-based; stem nodes glabrous; ultimate inflorescence branches ± appressed .... *P. dichotomiflorum* ssp. *dichotomiflorum*

Paspalum, Knotgrass, Dallisgrass

Native and introduced perennials; rhizomatous, sometimes stoloniferous. Leaf sheaths open, ligules membranous. Inflorescences panicles with spike-like branches; disarticulation below the glumes. Spikelets borne in 2 rows on 1 side of the inflorescence branches. Spikelets dorsiventrally flattened, as in *Panicum*, but lower glume minute or lacking; see the description of *Panicum* spikelets on p. 9.

1a Inflorescences with (2)4–7 spike-like branches; branches 4–13 cm; spikelets 2 per node, 1.7–2.5 mm wide; plants cespitose, with short rhizomes; W of the Cascades ....*P. dilatatum*
1b Inflorescences with 2(3) spike-like branches; branches 1.5–5.5 cm; spikelets solitary, 1.1–1.6 mm wide; plants strongly rhizomatous and stoloniferous; W of the Cascades, and in the Columbia and Snake River basins ....*P. distichum*

Phalaris, Canarygrass

Annuals or perennials; cespitose or rhizomatous. Leaf sheaths open; ligules membranous. Inflorescences dense panicles, sometimes spike-like with spikelets borne singly or in clusters; if in clusters, then the spikelets of 2 forms, the lower staminate or sterile and the upper bisexual or pistillate. Spikelets laterally compressed with 1–3(4) florets, the terminal floret sexual; the lower florets, if present, sterile; glumes longer than the florets, keeled, the keels often winged; sterile florets reduced to tiny knobs, linear bristle-like structures, or lanceolate lemmas to 75% as long as the bisexual floret; terminal florets bisexual (or in spikelet clusters that are sterile or staminate); lemmas awnless. The sterile florets of *Phalaris* are often misinterpreted as tufts of hairs. This genus includes 2 of our most troublesome invasive grasses, reed canarygrass (*P. arundinacea*) and Harding grass (*P. aquatica*).

1a Plants long-rhizomatous, perennial; inflorescences elongate-oblong to lanceolate, generally lobed at least at the base; glume keels not winged or with wing ≤ 0.2 mm wide; leaves 5–20 mm wide; widespread .... *P. arundinacea*
1b Plants cespitose or short-rhizomatous, perennial or annual; inflorescences cylindric to ovoid, usually smooth in outline or slightly lobed near the base; glume keels unwinged or with wing to 0.6 mm wide; leaves usually ≤ 12 mm wide
  2a Inflorescences about 10× as long as wide; glume keel wings about 0.4 mm wide; plants annual; SW OR; possibly native .... *P. angusta*
  2b Inflorescences ≤ 6× as long as wide; glume keel wings absent or to 0.6 mm wide; plants annual or perennial
    3a Glume keel wings with 1 large, triangular tooth; spikelets in clusters, each cluster with 4–7 narrow, staminate or sterile spikelets, and 1 broader fertile spikelet; plants annual; W of the Cascades .... *P. paradoxa*
    3b Glume keel wings entire to ± irregularly toothed, without 1 large triangular tooth; spikelets not in clusters, all similar; plants annual or perennial
      4a Glume keels not winged or with wings ≤ 0.2 mm wide; plants perennial, cespitose; 1 historic collection near the coast in SW OR .... *P. californica*
      4b Glume keels winged, the wings 0.2–0.6 mm wide; plants annual or perennial
        5a Plants perennial, cespitose and short-rhizomatous; inflorescences 1.5–6× as long as wide; glumes 4–7.5 × 1.2–1.5 mm; ligules 2–12 mm; culm bases bulb-like; W of the Cascades in OR, expanding northward..*P. aquatica*
        5b Plants annual, tufted; inflorescences 1–2.5× as long as wide; glumes 7–10 × 2–2.5 mm; ligules 3–6 mm; culm bases not bulb-like; widespread (often near bird feeders)............................ *P. canariensis*

Phleum, Timothy

Native and introduced cespitose perennials. Leaf sheaths open; ligules membranous. Inflorescence a tight, cylindrical, spike-like panicle, similar to *Alopecurus*. Glumes > florets, stiffly awned, the awns resembling little horns; lemmas awnless or with a weak awn arising below the tip.

1a Inflorescences 1–6 cm, usually > 1 cm wide when pressed; glume awns usually > 2 mm; upper leaf sheath inflated; culm bases not bulbous; montane to alpine, and coastal headlands..*P. alpinum*
1b Inflorescences 5–10(16) cm, < 1 cm wide when pressed; glume awns ≤ 2 mm; upper leaf sheath not inflated; culm bases bulbous; lowland to montane .... *P. pratense*

Phragmites australis, Common Reed

One of our "taller than you are" grasses, with native and introduced subspecies that are difficult to differentiate if you collect only the inflorescence. Stems stout, erect, with loose sheaths that twist in the wind to align the blades to the leeward side. Ligules membranous with a fringe of long hairs. Inflorescences large feathery panicles. Spikelets with 3 to 6 florets, rachillas with obvious, long silky hairs.

1a Lower culm internodes reddish-purple; lower glumes 3.8–7.0 mm; membranous portion of ligule on middle culm leaves (0.2)0.4–0.9 mm .... *P. a.* ssp. *americanus*
1b Lower culm internodes yellowish or yellow-brown; lower glumes 2.5–4.2(4.8) mm; membranous portion of ligule on middle culm leaves 0.1–0.4 mm .... *P. a.* ssp. *australis*

Pleuropogon, Semaphoregrass

Native rhizomatous perennials. Leaf sheaths closed; ligules membranous; leaf tips often awned or with short points. Inflorescences racemes; disarticulation above the glumes. Spikelets with multiple florets; glumes < the florets. Lemmas awned from the tip; paleas awned or toothed.

1a Mature spikelets dangling; lowest lemmas in each spikelet 8–10 mm; lower glumes 3–6 mm; palea keels awnless, but with a triangular appendage or tooth 0.2–1 mm; in and W of the Cascades .... *P. refractus*
1b Mature spikelets ascending; lowest lemmas in each spikelet 4.5–7.5 mm; lower glumes 2–3 mm; palea keels awned from the lower 33–50%, the awns 3–9 mm; rare; NE and SE OR.. *P. oregonus*

Poa, Bluegrass

Annuals or perennials; cespitose, rhizomatous, or stoloniferous; usually bisexual but sometimes dioecious or monoecious; also reproducing asexually by apomixis or by bulblets that replace flowers in the inflorescences (see section on sex in grasses p. 13); basal branching intra- or extravaginal. Leaf sheaths mostly closed to mostly open; ligules membranous; blades flat, folded, or involute, upper side with 2 longitudinal grooves, tips usually prow-shaped. Inflorescences panicles; disarticulation above the glumes. Spikelets laterally compressed, occasionally nearly round in cross section; lanceolate to ovate, occasionally forming bulblets. Glumes usually shorter than the lowest lemma, usually keeled, 1–3(5)-veined. Calluses glabrous or with a tuft of cobwebby hairs, or with a crown of straight to slightly sinuous hairs. Lemmas usually keeled, sometimes rounded over the back, 5(7–11)-veined, often with a purplish band at about midlength, margins scarious-hyaline distally, awnless.

Poa is the most diverse grass genus in the PNW and comprises about 10% of the taxa in this guide. *Poa* species are community dominants or important components of dry to moist grasslands, forests, and alpine habitats. They are economically important both as cultivated grasses for lawns, pastures, and soil stabilization, and as weeds in crops. Some species are significant invasives in native ecosystems.

Identifying *Poa* is a challenging endeavor. For example, William Cusick collected a *Poa* specimen in the high Wallowas in 1900 that appeared as a recently described species in this field guide in 2019. In his journal article describing it as a new species, *Poa wallowensis*, Robert Soreng notes that between 1900 and 2017 Cusick's collection had been assigned eight different names by eight different taxonomists, Soreng himself having changed his mind five times over twenty-five years. If you don't identify your *Poa* specimen correctly the first time or even the fifth time, don't beat yourself up. And if you collect *Poa secunda* a hundred times, thinking you have something new every time, rest assured you aren't alone. *Poa interior* has been repeatedly reported in error but not documented from WA and OR; it is often included in other keys that cover our region. (We do not, saving you from making that mistake.)

In keying *Poa*, keep in mind the following: Leaf sheath closure is measured on the uppermost culm leaf (flag leaf) and runs from the ligule to the node below; sometimes you may need to pull the next lower leaf sheath away from the culm to find the node. In the field and with fresh specimens you may be able to gently pull the blade away from the stem; the open part of the sheath may pull away also, but the closed part won't, until it tears. Running your thumbnail along the overlapped sheath margins down to the point of closure can also work if you watch carefully to see where the overlap ends. With dried specimens, insert a probe into the overlap of the margins and gently move it toward the base until it is stuck. Often there is a bit of a thickening or unevenness where the closed and open parts meet, but it takes some experience to see that. Using a dissecting microscope can help you see where the sheath is closed. Ligule length is measured on the upper 1 to 2 culm leaves unless otherwise specified. Length-width ratio should not be calculated for spikelets when florets are open. Floret pubescence is evaluated on the lower florets in a spikelet and should be checked on several florets. The length of callus hairs is measured when they are stretched out. Anther measurements should be made on well-developed, functional anthers that produce pollen, are plump before anthesis, and open at and after anthesis.

1a Culms and vegetative shoots with bulbous bases; florets usually producing tiny bulblets rather than seeds, sometimes producing normal flowers and seeds
    2a Ligules of lowest leaves mostly (0.8)1–3 mm, as long as or longer than wide, obtuse to acute, usually smooth, rarely slightly scabrous; longest blades of basal tufts mostly less than 4 cm long; panicles ± tightly contracted; widespread and common..*P. bulbosa*
    2b Ligules of lowest leaves < 1 mm, no longer than wide, truncate to obtuse, usually ± scabrous or short hairy; longest blades of basal tufts mostly 4–15 cm long; panicles ± loosely contracted; recently documented in the PNW, widespread in OR, status unknown in WA ....*P. iconia*
1b Culms and vegetative shoots lacking bulbous bases; florets producing normal flowers and seeds, not producing bulblets (rarely producing bulblets in *P. arctica*, a plant of alpine habitats)
    3a Plants annual (rarely longer lived)
        4a Callus glabrous; plants 2–20(45) cm
            5a Anthers 0.6–1.1 mm, oblong; lemmas 2.5–4 mm; lower panicle branches usually spreading to reflexed with spikelets relatively uncrowded; disturbed areas; widespread and abundant...........*P. annua*
            5b Anthers 0.1–0.5(0.6) mm, spherical to short-elliptical; lemmas 2–2.5 mm; lower panicle branches usually ascending with spikelets crowded; widespread but uncommon, likely overlooked ....*P. infirma*
        4b Callus with a tuft of cobwebby hairs (sometimes sparse); plants (10)20–80(120) cm

6a Lemmas glabrous or scabrous; rachilla internodes usually 1–1.2 mm or longer; montane to alpine; widely scattered in SE WA, NE and SW OR .... *P. bolanderi*

6b Lemmas puberulent; rachilla internodes ≤ 1 mm; lowland to montane; in and W of the Cascades, and Columbia Gorge .... *P. howellii*

3b Plants perennial

    7a Lemmas 7.5–11 mm; plants strongly rhizomatous; coastal sands .... *P. macrantha*

    7b Lemmas 2–8 mm, if > 7 mm, then plants cespitose and not of coastal sands

        8a Culms and nodes flattened, not rolling between the fingers; sheaths smooth, not scabrous; plants strongly rhizomatous; panicle branch length 0.5–3 cm; calluses usually cobwebby; widespread in disturbed habitats .... *P. compressa*

        8b Culms and nodes round in cross section or slightly compressed, generally rolling easily between the fingers; sheaths scabrous or not; plants rhizomatous or not; panicle branch length various; calluses cobwebby or not

            9a Spikelets subterete, (3.8)4–5× as long as wide; lemmas ± rounded over the back or weakly keeled, the keel usually not extending to the base; lemmas glabrous or hairy, if hairy, the areas between the veins with hairs about as long as those on the veins; calluses usually glabrous, sometimes with hairs 0.1–0.5 mm; plants densely cespitose; mesic to dry, lowlands to alpine, not coastal except in the San Juan Islands, WA .... *P. secunda* (with 2 subspecies, see p. 386)

            9b Spikelets laterally compressed, distinctly flattened, generally 1.5–3.8(4)× as long as wide; lemmas distinctly keeled to the base, glabrous or hairy; calluses glabrous or hairy, the hairs, when present, often long and cobwebby; plants rhizomatous, stoloniferous, or cespitose; habitat and range various

                10a Callus lacking cobwebby hairs

                    11a Plants rhizomatous .... Group 1, p. 80

                    11b Plants usually densely cespitose, sometimes loosely so .... Group 2, p. 82

                10b Callus with cobwebby hairs

                    12a Plants of moist lowland forest W of the Cascades, elevation < 4,000 feet; panicles (6)12–30 cm, usually sparsely flowered, nodding or with drooping branches; plants loosely cespitose to rhizomatous .... Group 3, p. 85

                    12b Differing in some way from the above combination of habitat, range, and characters

                        13a Plants rhizomatous or stoloniferous .... Group 4, p. 85

                      13b Plants cespitose .... Group 5, p. 87

Group 1 Calluses lacking cobwebby hairs; plants rhizomatous

1a Blades of uppermost culm leaves absent or vestigial, if present, usually < 1 cm, ± thick and firm; ligules (1.5)1.8–18 mm, decurrent; lemmas long-hairy on keel and marginal veins for about half their length; sagebrush steppe to dry alpine

grasslands; SE OR (NOTE: consider all characters—many *Poa* individuals with short flag leaf blades are misidentified as *P. fendleriana*)
....***P. fendleriana* ssp. *longiligula***

1b Blades of upper culm leaves well-developed, usually > 1 cm; lemmas usually sparsely pubescent on the keel and marginal veins for < half their length, sometimes glabrous; ligules 0.2–6 mm; habitat and range various

    2a Panicles open to loosely contracted, 5–18 cm long, branches 1.7–8 cm; sheaths of lower leaves retrorsely hairy or retrorsely scabrous on or near the collar, sometimes obscurely so (check several shoots at 25× to confirm)

        3a Collars of lower leaves hairy, at least near the sheath margins; collar hairs distinctly > than those elsewhere on the sheaths; anthers usually 2.5–4 mm (sometimes sterile and 0.1–0.2 mm); plants gynomonoecious (having both pistillate and bisexual flowers on the same plant); leaf sheaths closed 65–90% their length; wet rocks and cliffs, and mesic forest; Columbia Gorge, and in and W of the Cascades in N OR and SW WA
....***P. nervosa***

        3b Collars of lower leaves glabrous or hairy; collar margin hairs, if present, ≤ those elsewhere on the sheaths; anthers vestigial and 0.1–0.2 mm, or aborted late in development and up to 2 mm, rarely normal; plants pistillate (rarely some individuals with bisexual florets); leaf sheaths closed 33–75% their length; mesic to ± dry forest, subalpine meadows; mostly E of the Cascade Crest, and Klamath Mts., SW OR ....***P. wheeleri***

    2b Panicles ± dense, 1–10(12) cm long, branches 0.9–3.2(4) cm; leaf sheaths (including collars) glabrous and generally not scabrous, even on the lower leaves (check collars on several leaves)

        4a Plants of coastal sands; lemmas 2.5–4(4.5) mm .... ***P. confinis***

        4b Plants of montane forest openings to alpine meadows; lemmas 4–7 mm

            5a Plants of alpine slopes; Steens Mt., SE OR; culms 8–33 cm
....***P. mansfieldii***

            5b Plants montane to alpine; Olympic Mts., Cascades, and Mt. Ashland, OR; culms (10)25–50 cm

                6a Plants male; anthers functional, 1.8–3.7 mm; leaf margins smooth, not scabrous; montane forest openings in the W Cascades, OR
....***P. chambersii***

                6b Plants female; anthers vestigial, 0.1–0.2 mm; leaf margins smooth or scabrous; Olympic Mts., Cascades of OR and WA, and Mt. Ashland, OR

                    7a Leaf blade margins smooth; plants short-rhizomatous, stems often clustered, montane forest openings in the W Cascades, OR
....***P. chambersii***

                    7b Leaf blade margins minutely scabrous; plants usually densely cespitose but sometimes with basal shoots branching extravaginally, and resembling rhizomes; subalpine to alpine meadows, scree, and thickets; Olympic Mts., Cascades of OR and WA, and Mt. Ashland, OR .... ***P. cusickii* ssp. *purpurascens***

Group 2 Calluses lacking cobwebby hairs; plants densely to loosely cespitose

1a Spikelets broadly ovoid, 1.5–2.5× as long as wide, compact; rachilla internodes 0.5–0.8 mm; lemmas with prominent silky hairs on the keel and marginal veins and sparse silky hairs between the veins; calluses glabrous (infrequently with a few short dorsal hairs); blades thick, short, usually flat; flowers bisexual; Cascades in N WA and Wallowa Mts. in NE OR .... *P. alpina* ssp. *alpina*

1b Plants without the above combination of characters; spikelets ovoid to lanceoloid, usually > 2.5× as long as wide; rachilla internodes usually > 1 mm, at least in upper florets; lemmas glabrous or hairy; calluses glabrous or variously hairy; leaves flat, folded, or involute, flowers bisexual or unisexual

  2a Blades of uppermost culm leaves absent or vestigial, if present, usually < 1 cm, ± thick and firm; ligules (1.5)1.8–18 mm, decurrent; lemmas long-hairy on keel and marginal veins for about half their length; sagebrush steppe to dry alpine grasslands; SE OR (NOTE: consider all characters—many *Poa* individuals with short flag leaf blades are misidentified as *P. fendleriana*) .... *P. fendleriana* ssp. *longiligula*

  2b Blades of upper culm leaves well-developed, usually > 1 cm long; ligules 0.2–6 mm; lemmas short- or long-hairy on the keel and marginal veins for < or > half their length, sometimes glabrous; habitat and range various

    3a Panicles very densely ovoid with ± smooth outlines; longest panicle branches 0.5–1.5 cm; coastal bluffs; SW OR .... *P. unilateralis* ssp. *unilateralis*

    3b Panicles contracted to open, elongate, with somewhat irregular outlines; longest panicle branches (0.5)1–15 cm; habitat and range various

      4a Glumes broadly hyaline, nearly transparent except along the veins (pull glume away from lemma to check), shiny; plants dioecious, either staminate or pistillate; subalpine and alpine rocky slopes; Klamath Mts., SW OR .... *P. pringlei*

      4b Glumes hyaline only on the margins, sometimes ± broadly so, but the hyaline portion not extending to the veins, somewhat shiny or dull; plants bisexual or all pistillate; habitat and range various

        5a Culm leaf blades all short, flat, and firm, (1)1.5–3 mm wide, prominently veined, the margins usually thickened and whitish; flag leaf blades 0.2–1.8 cm, rarely absent; spikelets 1–4 per branch; subalpine and alpine serpentine slopes; Wenatchee Mts., WA .... *P. curtifolia*

        5b Culm leaf blades often longer, flat, folded, or involute, thin, soft to ± firm, 0.4–3(5) mm wide, generally not prominently veined, the margins not thickened and whitish; flag leaf blades 0.8–10(17) cm; spikelets (1)2–20(60+) per branch; habitat and range various

          6a Panicles open to loosely contracted at flowering, the primary and secondary branches spreading

            7a Leaf sheaths closed 40–80% their length; anthers (1)2–4 mm; vernally wet habitats; Columbia Gorge and E of Cascades in OR and S WA .... *P. leibergii*

            7b Leaf sheaths closed 10–33% their length; anthers 1.2–2 mm; habitat and range various

8a Upper glumes 2.5–3.8 mm; ligules 0.3–1(2.8) mm, rounded to truncate, often minutely ciliate; leaf blades 0.5–1.5 mm wide; moist slopes and streamsides; Columbia Gorge and N Willamette Valley; hybrids of *P. secunda* and *P. nervosa* .... ***P. × multnomae***

8b Upper glumes 3.5–6.5 mm; ligules (0.5)1.5–5 mm, obtuse to acuminate, not minutely ciliate; leaf blades 0.5–4(5) mm wide; habitat various; widespread

9a Spikelets laterally compressed, distinctly flattened, generally 3–3.8 × as long as wide; lemmas distinctly keeled to the base, the keels and marginal veins hairy, areas between the veins glabrous or with shorter hairs than on keels and veins; calluses usually with hairs 0.2–2 mm; coastal bluffs, and montane to low alpine forests, meadows, and cliffs; sometimes very similar to forms of *P. secunda* ssp. *secunda* .... ***P. stenantha* var. *stenantha***

9b Spikelets subterete, (3.8)4–5× as long as wide; lemmas ± rounded over the back or weakly keeled, the keel usually not extending to the base, glabrous or hairy, if hairy, the areas between the veins with hairs about as long as those on the veins; calluses usually glabrous, sometimes with hairs 0.1–0.5 mm; mesic to dry, lowlands to alpine, not coastal except in the San Juan Islands, WA .... ***P. secunda*** (with 2 subspecies, see p. 386)

6b Panicles contracted, ± dense at flowering, branches generally not spreading

10a Habitat cliffs and grasslands at the immediate coast; calluses with hairs 0.1–0.2 mm; lemmas short-hairy on keels and marginal veins for the lower 33–50% their length, area between the veins glabrous or with very short hairs near the base; NW OR to SW WA .... ***P. unilateralis* ssp. *pachypholis***

10b Habitat lowlands to alpine, grasslands and shrub steppe to slopes and ridges, not coastal except in NW WA; calluses glabrous or with hairs 0.1–0.5 mm; lemmas glabrous or the keels and marginal veins scabrous to short-hairy, area between the veins glabrous or with hairs as long as those on keels and marginal veins

11a Spikelets subterete, (3.8)4–5× as long as wide; lemmas ± rounded over the back or weakly keeled, the keel usually not extending to the base, glabrous or hairy, if hairy, the areas between the veins with hairs about as long as those on the veins; calluses usually glabrous, sometimes with hairs 0.1–0.5 mm; anthers 1.5—3 mm; mesic to dry, lowlands to alpine, not coastal except in the San Juan Islands, WA .... ***P. secunda*** (with 2 subspecies, see p. 386)

11b Spikelets laterally compressed, distinctly flattened, 2–3.6× as long as wide; lemmas distinctly keeled to the base, hairy or glabrous; anthers 0.2–3.5 mm; habitat and range various

12a Plants 10–60(70) cm; leaf sheaths closed 25–75% their length; upper sagebrush zone to alpine (with 4 subspecies) .... ***P. cusickii***

13a Cauline leaf blades usually < 1.5 mm wide; panicle branches moderately to densely scabrous; all vegetative shoots intravaginal; lemmas glabrous; calluses glabrous; functional anthers present or absent

14a Longest panicle branches usually 1.7–4(5) cm, stout to slender, with 2–15 spikelets; functional anthers sometimes present; riparian meadows and swales in upper sagebrush zone to alpine slopes and ridges; widespread E of the Cascades ....*P. c.* ssp. *cusickii*

14b Longest panicle branches 0.5–2 cm, stout, with 2–5 spikelets; functional anthers generally absent; alpine grasslands and ridges; mts. E of the Cascades.. *P. c.* ssp. *pallida*

13b Some cauline leaf blades 1.5–3.5 mm wide; panicle branches smooth to moderately scabrous; sometimes some vegetative shoots extravaginal; lemmas glabrous or sparsely pubescent on the keel and marginal veins; calluses sometimes with a few short hairs; functional anthers absent

15a Lemmas of all florets in each spikelet glabrous; calluses glabrous; panicles densely contracted, branches smooth or with very sparse, long scabers; subalpine to lower alpine, openings and meadows; Olympic Mts. and in and E of the Cascades .... *P. c.* ssp. *epilis*

15b Lemmas of lower florets in each spikelet usually sparsely pubescent at least on the lower marginal veins, lemmas of upper florets glabrous; calluses sometimes with sparse, short hairs; panicles densely to loosely contracted, branches usually sparsely scabrous; subalpine to alpine meadows and open forest; Olympic Mts., Cascades of OR and WA, and Mt. Ashland, OR ....*P. c.* ssp. *purpurascens*

12b Plants 1–15(25) cm; leaf sheaths closed 10–25(33)% their length; habitat alpine

16a Nearly all shoots bearing panicles at flowering time; lemmas short-hairy on keel and veins, with shorter, sometimes crisped hairs in areas between the veins; panicle branches strongly angled and densely scabrous along the ridges; Olympic Mts., N Cascades, Wallowa Mts., and Steens Mt. ....*P. glauca* ssp. *rupicola*

16b Most shoots vegetative at flowering time; lemmas glabrous (lemma keels and marginal veins rarely with sparse hairs in *P. lettermanii* and *P. cusickii*); panicle branches terete to weakly angled, smooth to sparsely scabrous, or if densely scabrous, scabers not lined up on ridges

17a Lemmas 2.5–4 mm; anthers 0.2–0.8 mm, well-developed or aborted late in development; Cascades from central OR to BC..................*P. lettermanii*

17b Lemmas (3)4–7 mm; anthers functional, 0.8–3 mm, or aborted early in development and 0.1–0.2 mm; widespread

18a Leaf sheaths closed 15–25(33)% their length; spikelets 4.2–7 mm; upper glumes nearly as long as or longer than lower lemmas; lemmas smooth, glabrous; anthers 0.8–1.2(1.7) mm, usually well developed; panicle branches usually smooth, sometimes sparsely scabrous; Olympic Mts., Cascades, and Wallowa Mts. .... *P. suksdorfii*

18b Leaf sheaths closed ≥ 25–75% their length; spikelets (3)4–5 mm; upper glumes distinctly shorter than (< 75% as long as) lower lemmas; lemmas smooth or scabrous, veins glabrous or puberulent proximally; anthers often aborted, 0.1–0.2 mm, less often 2–3.5 mm and fertile or aborted late in development; panicle branches smooth to sparsely or densely scabrous; widespread........... *P. cusickii* (see key to subspecies at lead 13a)

Group 3 Calluses with cobwebby hairs; plants of moist, lowland forest west of the Cascades, elevation < 4000 feet

1a Lemma keels and marginal veins glabrous; leaf sheaths closed > 90% their length; lemmas acuminate; panicle usually nodding and narrow .... *P. marcida*

1b Lemma keels and marginal veins short- to long-hairy; leaf sheaths closed 10–75% their length; lemmas acute or occasionally obtuse; panicle nodding to erect, usually broader

 2a Leaf sheaths closed 10–20% their length; lemma tips usually partially bronze-colored

  3a Ligules (1)1.5–6 mm; cobwebby hairs of callus ≥ (50)67% as long as the lemma; culms retrorsely scabrous or hairy below lower nodes; moist meadows, shorelines, forests, disturbed areas............................. *P. palustris*

  3b Ligules 0.2–0.8(1) mm; cobwebby hairs of callus < 50% as long as the lemma; culms smooth, glabrous below lower nodes; moist forests, streamsides, meadows..*P. nemoralis*

 2b Leaf sheaths closed 25–75% their length; lemma tips not bronze-colored, usually green and white or hyaline, sometimes purple

  4a Culms retrorsely scabrous at least below the nodes; sheaths densely, retrorsely scabrous; sheaths closed 50–75% their length; anthers 0.5–1.1 mm; panicle branches (5.5)8–12(15) cm with 2–6(13) spikelets borne near the tips; moist riparian terraces in late-successional conifer forest; W OR (Coos County and north) and SW WA, extending N to AK .... *P. laxiflora*

  4b Culms smooth; sheaths ± smooth, infrequently ± hairy; sheaths closed 25–50% their length; anthers 1.2–2 mm; panicle branches (1)2–9 cm with 4–30(50) spikelets fairly crowded in the distal half; habitat various; widespread .... *P. pratensis*

Group 4 Calluses with cobwebby hairs; plants rhizomatous or stoloniferous

1a Cobwebby hairs of callus < 25% length of lemma when straightened; substrate not strongly ultramafic

 2a Panicles open, 8–15 cm with branches 2.5–8 cm, loosely ascending to spreading; sheaths of lower leaves retrorsely hairy or retrorsely scabrous on or near the collar, sometimes obscurely so (check several shoots at 25× to confirm); low elevation moist forest Columbia Gorge, and in and W of the Cascades in N OR and SW WA ....*P. nervosa*

 2b Panicles contracted, 2–10(12) cm, with branches 0.9–3.2(4) cm, strongly ascending (sometimes somewhat spreading in *P. chambersii*); leaf sheaths and collars glabrous, not scabrous; Olympic Mts., Cascades, and Mt. Ashland, OR

  3a Plants male; anthers functional, 1.8—3.7 mm; leaf margins smooth, not scabrous; montane forest openings in the W Cascades, OR.. *P. chambersii*

  3b Plants female; anthers vestigial, 0.1—0.2 mm; leaf margins smooth or scabrous; Olympic Mts., Cascades of OR and WA, and Mt. Ashland, OR

   4a Leaf margins smooth; plants short-rhizomatous, stems often clustered; montane forest openings in the W Cascades, OR .... *P. chambersii*

4b Leaf margins minutely scabrous; plants usually densely cespitose but sometimes with basal shoots branching extravaginally, and resembling rhizomes; subalpine to alpine meadows, scree, and thickets; Olympic Mts., Cascades of OR and WA, and Mt. Ashland, OR.... ***P. cusickii*** **ssp.** ***purpurascens***

1b Cobwebby hairs of callus ≥ 33% length of lemma; substrate strongly ultramafic or not

    5a Lemmas 4–7 mm; substrate strongly ultramafic; Klamath Mts., SW OR

        6a Panicles loosely contracted; lemma keels scabrous; ligules of cauline leaves 1–2 mm, obtuse to truncate; blades firm, involute; forest openings on serpentine substrates .... ***P. piperi***

        6b Panicles open; lemma keels silky-hairy; ligules of cauline leaves 2–8 mm, acute to acuminate; blades softer, mostly flat; rocky gabbro or peridotite soils in montane conifer forest.. ***P. rhizomata***

    5b Lemmas 2–4.3(6) mm; substrate various but not strongly ultramafic; range various

        7a Sheaths densely scabrous; panicles erect; plants cespitose, sometimes weakly stoloniferous; lower glumes sickle-shaped, 1-veined; ligules of upper culm leaves 3–10 mm, acute to acuminate; lowlands to mid-montane, moist deciduous forest, riparian areas, disturbed grasslands; widespread.. ***P. trivialis*** **ssp.** ***trivialis***

        7b Sheaths smooth or nearly so; panicles erect or nodding; plants rhizomatous (sometimes obscurely so) or stoloniferous, sometimes tufted along rhizomes; lower glumes usually not sickle-shaped, 1–3-veined; ligules of upper culm leaves 0.5–4(6) mm, truncate to acuminate; lowlands to alpine; range various

            8a Plants coastal; habitat sand dunes and forests with sandy soils

                9a Cauline leaf blades 0.5–1(1.5) mm wide, involute; rhizomes typically vertical; lemmas glabrous, or keel and marginal veins sparsely puberulent on lower portion; calluses with diffuse cobwebby hairs; spikelets gynodioecious............................. ***P. confinis***

                9b Cauline leaf blades 0.4–4.5 mm wide, flat, folded, or involute; rhizomes usually horizontal; lemmas with prominent silky hairs on keel and marginal veins; calluses with cobwebby hairs originating from discrete tufts on the callus back and sometimes at the base of the marginal veins; spikelets with bisexual flowers .... ***P. pratensis***

            8b Plants of various habitats but not coastal sands

                10a Plants of alpine habitats

                    11a Paleas soft-hairy between the keels, the keels short- to long-villous for most of their length (rarely nearly glabrous or scabrous); ligules 1.5–2.5(3) mm; reported in error from Mt. Rainier; present in BC but not OR or WA....................... ***P. arctica***

                    11b Paleas glabrous between the keels, the keels sometimes puberulent; ligules 0.5–2(3.1) mm; range not well documented in our area (*P. pratensis* ssp. *alpigena*) .... ***P. pratensis***

                10b Plants not of alpine habitats

12a Lemmas 2–3 mm, tips usually partially bronze-colored; leaf sheaths closed 10–20% their length; plants stoloniferous; most shoots producing a panicle at flowering time (sometimes additional vegetative shoots form late in the season, changing the proportion) .... *P. palustris*
12b Lemmas 3–4.3(6) mm, tips usually green, purple, or white, not bronze-colored; leaf sheaths closed 25–50% their length; plants rhizomatous; most shoots vegetative at flowering time .... *P. pratensis*

Group 5 Calluses with cobwebby hairs; plants densely to loosely cespitose

1a Plants of low to mid-elevation habitats
    2a Panicles 4–8 cm; lemmas 4–6(7) mm; low to mid-elevation forest openings and Jeffrey pine savanna on ultramafic soils; SW OR .... *P. piperi*
    2b Panicles 7–30(41) cm; lemmas 2–4 mm; low elevation moist forest, meadows, shorelines, disturbed areas, not on ultramafic soils; widespread
        3a Leaf sheaths usually densely scabrous, closed 33–50% their length; lower glumes sickle-shaped, 1-veined; many shoots vegetative at flowering time .... *P. trivialis* ssp. *trivialis*
        3b Leaf sheaths smooth or sparsely scabrous, closed 10–20% their length; lower glumes generally not sickle-shaped, 3-veined; most shoots producing a panicle at flowering time (sometimes additional vegetative shoots form late in the season, changing the proportion)
            4a Ligules (1)1.5–6 mm; cobwebby hairs of callus ≥ (50)67% as long as the lemma; culms retrorsely scabrous or hairy below lower nodes; moist meadows, shorelines, forests, disturbed areas .... *P. palustris*
            4b Ligules 0.2–0.8(1) mm; cobwebby hairs of callus < 50% as long as the lemma; culms smooth, glabrous below lower nodes; moist forests, streamsides, meadows .... *P. nemoralis*
1b Plants of subalpine to alpine habitats
    5a Panicles condensed to somewhat open, erect; panicle branches 0.5–4(5) cm, steeply ascending; leaf sheaths closed 10–75% their length; anthers 0.6–3 mm or aborted and tiny or aborted late in development and therefore up to 3 mm long but poorly developed
        6a Lemmas 4–7 mm; spikelets 7–10 mm; anthers aborted late in development and poorly developed; Olympic Mts., Cascades of OR and WA, and Mt. Ashland, OR .... *P. cusickii* ssp. *purpurascens*
        6b Lemmas 2.5–4 mm; spikelets 3–7(9) mm; anthers 0.6–2.5 mm, well-developed, rarely aborted
            7a Panicle branches angled, moderately to densely scabrous on the angles; anthers (1)1.2–2.5 mm; leaf sheaths closed 10–20% their length; most shoots bearing panicles at flowering time (sometimes additional vegetative shoots form late in the season; Wallowa Mts., OR, Olympic Mts., and N Cascades, WA .... *P. glauca* ssp. *glauca*
            7b Panicle branches terete to few-angled, smooth to sparsely scabrous on angles; anthers 0.6–1 mm; leaf sheaths closed 20–75% their length; most shoots vegetative, not producing a panicle at flowering time; Wallowa Mts., Steens Mt., and central Cascades, OR ...... *P. wallowensis*

5b Panicles very open, often sparse, usually nodding, sometimes ± erect; panicle branches (2)3–8 cm, widely spreading to reflexed (sometimes ascending in *P. paucispicula*); leaf sheaths closed 25–67% their length; anthers 0.2–1.1 mm
   8a Glume keels scabrous; lemma keels hairy ≤ 50% their length, lateral veins glabrous; panicle branches usually moderately scabrous (sometimes sparsely so); subalpine to alpine wet meadows, margins of lakes and streams; mostly in and E of the Cascades, also in Olympic Mts., WA .... ***P. leptocoma***
   8b Glume keels smooth or nearly so; lemma keels hairy 50–75% their length, lateral veins glabrous or sparsely short-hairy; panicle branches usually smooth, rarely some branches sparsely scabrous; mesic to dry, sometimes seasonally moist, alpine meadows and ridges
      9a Lower panicle branches usually reflexed, with (3)6–18 spikelets per branch; lateral veins of lemmas usually sparsely short-hairy; subalpine and alpine meadows and talus; Wallowa Mts., OR .... ***P. reflexa***
      9b Lower panicle branches spreading to ascending, with 1–3(5) spikelets per branch; lateral veins of lemmas glabrous; mesic rocky alpine slopes; Cascades and Olympic Mts., WA .... ***P. paucispicula***

Podagrostis, Bentgrass

Cespitose or rhizomatous perennials; leaves mostly basal; sheaths open; ligules membranous, subacute, obtuse, or truncate. Inflorescences panicles; disarticulation above the glumes. Spikelets weakly laterally compressed with 1 floret; rachilla prolonged beyond the base of the floret, sometimes absent from the spikelets in the lower part of a panicle; calluses glabrous or with hairs ≤ 0.5 mm; lemmas ≤ glumes, (3)5-veined, awned or awnless; paleas > 50% as long as lemmas.

   Podagrostis species were long treated as members of *Agrostis* but differ in having paleas > half as long as lemmas, rachillas that are prolonged beyond the base of the lemmas, and firmer, opaque lemmas with (3)5 relatively distinct veins.

1a Glumes 2.3–4.3 mm, usually ± equal; rachilla prolonged 0.5–1.9 mm beyond base of floret; anthers 0.8–1.3 mm; NW WA...................................... ***P. aequivalvis***
1b Glumes 1.6–2.3 mm, equal, or lower glumes longer; rachilla absent or prolonged 0.1–0.5 mm beyond base of floret; anthers 0.4–1 mm; widespread in mts.
   2a Plants often much < 15 cm, not rhizomatous; leaves usually folded, 0.3–0.8(1.5) mm wide; panicles 1–6 cm × 0.2–1.5 cm, branches appressed; rachilla absent or prolonged 0.1 mm beyond base of floret .... ***P. humilis***
   2b Plants usually > 15 cm, short-rhizomatous; leaves flat, 1–3 mm wide; panicles (3)5–10 cm × 0.5–2.5 cm, branches sometimes ascending; rachilla prolonged 0.1–0.5 mm beyond base of floret............................. ***P. thurberiana***

Polypogon, Rabbitsfoot Grass, Beardgrass

Annuals or cespitose perennials. Leaf sheaths open; ligules membranous, erose, ciliate. Inflorescences dense, often spike-like panicles. Spikelets borne on a stipe that is connected to a pedicel or directly to the panicle branch; disarticulation below the stipe. Spikelets laterally compressed with 1 floret. Glumes nearly = in length, > florets, often 2-lobed, usually awned from the tip or from between the lobes. Lemmas 1–3(5)-veined, often awned.

Polypogon species resemble *Agrostis* but differ in having glumes usually awned, spikelets with stipes, and disarticulation below the stipe. *Polypogon australis*, an introduced perennial with glume lobes short or lacking, was collected from ballast in Portland in the early twentieth century but has not been seen since. It is established in California.

1a Glumes with awns 4–12 mm; range various
    2a Glume lobes 0.1–0.2 mm, ≤ 10% as long as the glume body; common, widespread ....*P. monspeliensis*
    2b Glume lobes 0.3–1.2 mm, > 15% as long as the glume body; SW OR, Columbia Basin; probably overlooked and more widespread .....*P. maritimus*
1b Glumes awnless or with awns to 3.2 mm; W of the Cascades and Columbia Gorge
    3a Glumes awnless ....*P. viridis*
    3b Glumes awned, the awns 0.6–3.2 mm
        4a Plants annual; glumes 1.8–2.4 mm, lobes 0.1–0.2 mm, awns 0.6–3 mm ....*P. fugax*
        4b Plants perennial, occasionally flowering in the first year; glumes 2–3 mm, lobes < 0.1 mm, awns 1.5–3.2 mm .... *P. interruptus*

Puccinellia, Alkali Grass

Annuals and perennials; cespitose, sometimes stoloniferous. Leaf sheaths open; ligules membranous. Inflorescences panicles; disarticulation below the glumes. Spikelets with 2–7 florets; lemmas awnless.

Puccinellia species grow in alkaline and saline soils both inland and at the coast. Similar to *Glyceria*, which has closed sheaths and grows in freshwater habitats. See also *Torreyochloa*, which has open sheaths but also grows in freshwater habitats. *Puccinellia* might be confused with *Poa*, but the distal margin of the lemma is minutely jagged in most *Puccinellia* species.

1a Plants annual, 2–25 cm; panicles contracted, linear, branches erect; anthers 0.2–0.5 mm; seasonally moist alkaline flats along the Columbia River; NE OR; rare introduction, to be expected elsewhere, distributed by migratory waterfowl ....*P. simplex*
1b Plants perennial, 5–100 cm; panicles often more open, compact to diffuse, branches spreading to erect; anthers 0.4–2.6 mm
    2a Lemmas 1.5–2(2.2) mm; anthers 0.4–0.8 mm; lowest panicle branches horizontal to descending at maturity; mostly E of the Cascades, also coastal in NW WA ....*P. distans*
    2b Lemmas (2)2.2–5 mm; anthers 0.5–2.6 mm; lowest panicle branches erect to descending at maturity
        3a Range coastal
            4a Upper glumes 3–4.5 mm; lower glumes 2–3.4 mm; lemmas leathery, 3–5 mm; anthers 1.5–2.6 mm; introduced to NW WA .... *P. maritima*
            4b Upper glumes 0.9–3 mm; lower glumes 0.5–2(4) mm; lemmas herbaceous, not leathery, 1.5–4.5(5) mm; anthers 0.5–2 mm
                5a Plants 8–40 cm; pedicels smooth; lemma tip margins smooth or with a few scattered minute teeth ....*P. pumila*
                5b Plants 10–100 cm; pedicels scabrous; lemma tip margins densely minutely jagged

6a Lemmas (2.2)3–4.5(5) mm; lower glumes 1–1.6 mm; pedicels papillose or with slightly swollen cell surfaces distally (visible at 15×), sparsely to densely scabrous; lower panicle branches usually erect to ascending, occasionally spreading to descending; coastal habitats .... ***P. nutkaensis***
  6b Lemmas (1)2.2–3(3.5) mm; lower glumes 0.5–1.5 mm; pedicels scabrous, not papillose, cell surfaces not swollen; lower panicle branches erect to descending; mostly E of the Cascades; rarely coastal in NW OR and NW WA .... ***P. nuttalliana***
3b Range E of the Cascades
  7a Leaves concentrated near the base of the plant, involute and very narrow, 1.2–1.9 mm wide when flattened; lemma tips acute; lemma apical margins smooth to minutely jagged .... ***P. lemmonii***
  7b Leaves either concentrated near the base of the plant or distributed along the stem, involute or flat, usually wider, 1–4 mm wide when flattened; lemma tips usually obtuse, sometimes acute; lemma apical margins densely minutely jagged .... ***P. nuttalliana***

Rytidosperma, Wallabygrass

Cespitose to short-rhizomatous perennials. Leaf sheaths open; ligules a fringe of hairs. Inflorescences panicles or racemes; disarticulation above the glumes. Spikelets with 3–10 florets; glumes usually exceeding the florets. Lemmas with 2 transverse rows of tufts of hairs, the rows sometimes reduced to marginal tufts; lemma tips with two awn-like bristles ≥ the lemma body; lemma awns originating between and exceeding the bristles, twisted and bent. Australian and New Zealand genus similar to *Danthonia*. Introduced W of the Cascades, mostly near the coast.

  1a Uppermost culm leaf usually reaching or exceeding the inflorescence; upper row of lemma hairs in a ± continuous row of tufts, the hairs much exceeding the base of the awn .... ***R. biannulare***
  1b Uppermost culm leaf usually not reaching the inflorescence; upper row of lemma hairs in isolated tufts or only at the lemma margins, the hairs not or only just exceeding the base of the awn .... ***R. penicillatum***

Schedonorus, Tall Fescue

Cespitose or rhizomatous perennials; sheaths open, auricles present; inflorescences panicles; 2+ florets per spikelet; spikelets laterally compressed; lemmas with short awns. Both species are introduced and widespread.

  Schedonorus has been separated from *Festuca* because it is more closely related to *Lolium*, with which it sometimes hybridizes. Sometimes merged into *Lolium*.

  1a Auricles and flared rounded area at base of blade ciliate (sometimes just a few hairs present; hairs wear off with age); plants densely cespitose (rarely rhizomatous); lemma awns usually 0.5–1.5(4) mm; leaves 3–12 mm wide .... ***S. arundinaceus***
  1b Auricles and flared rounded area at base of blade not ciliate; plants clearly rhizomatous, often with culms loosely clustered along the rhizome; lemma awns 0–0.5 mm; leaves 3–5(7) mm wide .... ***S. pratensis***

Setaria, Foxtail, Bristlegrass, Cattail Grass

Introduced annuals. Leaf sheaths open; ligules membranous and ciliate, or a fringe of hairs. Inflorescences dense, spike-like panicles; disarticulation below the glumes; spikelets dorsiventrally flattened, subtended by 1–several bristles; 2 florets per spikelet.

1a Bristles retrorsely scabrous and as a result, inflorescences tending to stick to clothing; inflorescence branches often whorled (sometimes alternate) .... ***S. verticillata***
1b Bristles antrorsely scabrous; inflorescences not sticking to clothing; inflorescence branches alternate
    2a Upper lemmas smooth and shiny, occasionally obscurely transversely rugose, exposed at maturity; inflorescences 2–3 cm wide, usually lobed throughout at maturity; spikelets about 3 mm; cultivated grain sprouting beneath bird feeders and in disturbed areas but not persisting (not treated) .... ***S. italica***
    2b Upper lemmas ± distinctly transversely rugose, dull, exposed or hidden at maturity; inflorescences usually 0.5–1(2) cm wide, usually straight-sided, occasionally basally lobed; spikelets 1.8–3.4 mm
        3a Inflorescences arching or drooping from near base; upper surface of leaf blades usually soft-hairy; few records in W WA and NE OR .... ***S. faberi***
        3b Inflorescences erect; upper surface of leaf blades smooth, scabrous, or with papillose-based hairs near the base of the blade
            4a Upper glumes approximately as long as upper lemmas; spikelets 1.8–2.2 mm, each subtended by 1–3 bristles .... ***S. viridis*** (with 2 varieties)
                5a Culms 20–100 cm; leaf blades 4–12 mm wide; panicles 3–8(15) cm, not lobed; widespread .... ***S. v.* var. *viridis***
                5b Culms 100–250 cm; leaf blades 10–25 mm wide; panicles 10–20 cm, usually somewhat lobed at base; uncommon in N OR and SE WA .... ***S. v.* var. *major***
            4b Upper glumes 50–67% as long as upper lemmas; spikelets 2–3.4 mm, each subtended by 4–12 bristles .... ***S. pumila*** (with 2 subspecies)
                6a Spikelets 2–2.5 mm; bristles reddish to purplish; few records SW O .... ***S. p.* ssp. *pallide-fusca***
                6b Spikelets 3–3.4 mm; bristles yellow to orange or rusty; widespread .... ***S. p.* ssp. *pumila***

Sorghum, Sorghum, Johnson grass, Shattercane

Introduced annuals or strongly rhizomatous perennials. Leaf sheaths open; ligules membranous and ciliate. Inflorescences panicles; disarticulation in the panicle branches. Spikelets dorsiventrally compressed, in clusters. Spikelet clusters consist of 1 sessile, bisexual spikelet and 1–2 pedicellate, staminate spikelets. Glumes hard, shiny or leathery; lemmas awned or awnless. Large coarse weeds and crop species from tropical and warm-temperate regions.

1a Annual; spikelets not or tardily disarticulating at maturity; leaf blades 5–100 mm wide.. ***S. bicolor***
1b Perennial from deep rhizomes; spikelets disarticulating at maturity; leaf blades (8)10–20(40) mm wide.. ***S. halepense***

Spartina, Cordgrass

Rhizomatous perennials. Leaf sheaths open; ligules appearing to be a fringe of hairs, but with a short, inconspicuous, membranous basal portion. Inflorescences panicles with spike-like branches; spikelets sessile on 1 side of branches, often closely arranged like the teeth of a comb, or appressed and ± overlapping along the branch axis; disarticulation below the glumes. Spikelets laterally compressed, with 1 floret. Glumes keeled; lower glumes shorter than the florets, upper glumes usually longer than the florets; lemmas shorter than the paleas.

    It's easy to misinterpret the short, crowded branches of *Spartina* as spikelets. Cordgrasses grow in saline and alkaline habitats both at the coast and E of the Cascades. Introduced species are significant invasives in coastal habitats. Some authors now include *Spartina* within *Sporobolus*.

1a Range E of the Cascades, native
    2a Upper glumes awnless, 6–10 mm; culms 40–100 cm, 1.5–5.5 mm thick; inflorescences with 3–12 branches .... *S. gracilis*
    2b Upper glumes awned, 10–25 mm (including awn); culms 50–250 cm, 2.5–11 mm thick; inflorescences with 5–50 branches .... *S. pectinata*
1b Range coastal, introduced, often invasive
    3a Leaf blades inrolled when fresh, with strongly scabrous margins; panicle branch axes not prolonged beyond spikelets
        4a Rhizomes lacking or short; culms usually clumped; upper glumes 1-veined; panicle branches tightly appressed .... *S. densiflora*
        4b Rhizomes present, well-developed, thin and wiry; culms usually solitary, sometimes a few together on the spreading rhizomes; upper glumes with 2 veins on the same side of the keel; panicle branches appressed to strongly divergent .... *S. patens*
    3b Leaf blades flat when fresh, at least near the base, with smooth or slightly scabrous margins; panicle branch axes often prolonged beyond the spikelets
        5a Glumes mostly glabrous on the sides, sometimes with appressed hairs near base; panicles with 3–25 branches; fresh culms with unpleasant sulfurous odor; internodes fleshy; anthers 3–6 mm .... *S. alterniflora*
        5b Glumes usually appressed-hairy on the sides; panicles with 1–12 branches; fresh culms lacking unpleasant sulfurous odor; internodes firm; anthers 5–13 mm
            6a Ligules 2–3 mm; anthers (5)7–13 mm, well-developed, dehiscent at maturity .... *S. anglica*
            6b Ligules 1–1.8 mm; anthers 5–10 mm, sometimes poorly developed and indehiscent at maturity (see *S. anglica* treatment, p. 428) .... *S.* × *townsendii*

Sphenopholis, Wedgegrass

Cespitose perennials. Leaf sheaths open; ligules membranous. Inflorescences dense to open, erect to nodding panicles; disarticulation below the glumes. Spikelets with 2–3 florets; glumes about the length of the lowest floret; upper glumes distinctly wider than the lower glumes, with tips often rounded, truncate, or hooded.

1a Upper glumes hooded, 30–50% as wide as long, tips rounded to truncate;

panicles erect, often spike-like, the spikelets densely arranged; wet meadows, streamsides, sometimes on alkaline soil; E of the Cascades .... ***S. obtusata***
1b Upper glumes not hooded, 17–35% as wide as long, tips acute, rounded, or subtruncate; panicles usually nodding, not spike-like, the spikelets more loosely arranged; wet ground and shorelines; E WA .... ***S. intermedia***

Sporobolus, Dropseed

Annuals or cespitose perennials. Leaf sheaths open, usually with a tuft of hairs at the top; ligules a fringe of hairs. Inflorescences panicles, sometimes hidden in the upper leaf sheaths; disarticulation above the glumes. Spikelets rounded to laterally compressed, with 1 floret (rarely up to 3); glumes usually slightly to distinctly < the lemmas; lemmas usually 1(3)-veined, awnless. Fruit an achene with an outer layer free from the seed, which often becomes mucilaginous when wet, sometimes forcibly ejecting the seed when it dries. Recent research would place the genera *Calamovilfa*, *Crypsis*, and *Spartina* in *Sporobolus*. For this guide we have chosen not to adopt those changes.

1a Plants annual, 10–60(70) cm; panicles ≤ 5 cm × 0.2–0.5 cm, usually ± concealed within the upper leaf sheath, branches appressed; leaf blades 0.6–2 mm wide; scattered sites on both sides of the Cascades in WA
   2a Lemmas sparsely appressed-hairy; 2.1–5.4 mm, spikelets 2.3–6 mm; anthers 1.2–3.2 mm; sandy or gravelly roadsides, disturbed sites; few scattered sites in WA, to be expected elsewhere .... ***S. vaginiflorus* var. *vaginiflorus***
   2b Lemmas glabrous, 1.6–2.9 mm; spikelets 1.6–3 mm; anthers 1.1–1.6 mm; silty or sandy streamsides; E WA .... ***S. neglectus***
1b Plants perennial, (20)30–130(150) cm; panicles ≥ 5 cm × 0.4–25 cm, exserted from or concealed within terminal leaf sheath, branches appressed to spreading; leaf blades 1.5–10 mm wide; E of the Cascades
   3a Spikelets 4–6(10) mm; anthers 0.2–3.2 mm; E of the Cascades in WA
   .... ***S. compositus* var. *compositus***
   3b Spikelets 1.3–2.8 mm; anthers 0.5–1.8 mm; range various
      4a Panicles 15–25 cm wide, very open and diffuse, with primary and secondary branches spreading; upper culm leaf blades ascending; leaf sheath tops glabrous or with sparse scattered hairs; SE OR, disjunct in Okanogan County, WA .... ***S. airoides***
      4b Panicles 2–12(14) cm wide, contracted and spike-like, becoming open with age, primary branches appressed to spreading, secondary branches appressed; upper culm leaf blades nearly perpendicular to the culms; leaf sheath tops with a conspicuous tuft of white hairs; widespread E of the Cascades... ***S. cryptandrus***

Thinopyrum, Wheatgrass

Introduced cespitose or rhizomatous perennials. Leaf sheaths open, often ciliate on the margins; ligules membranous. Inflorescences spikes with 1 spikelet per node; disarticulation above the glumes, sometimes in the inflorescence axis. Glumes stiff, hard, or leathery, truncate to acute; lemmas awnless or awned from the tip.

Formerly part of *Agropyron*, *Thinopyrum* differs in its thick, stiff glumes and lemmas. Our species are tolerant of alkaline soils.

1a Plants cespitose, to 200 cm; glume tips truncate with all veins approximately equal in length and prominence .... *T. obtusiflorum*
1b Plants strongly rhizomatous, to 115 cm; glume tips usually acute to obtuse with veins unequal in length, the midvein usually more prominent and longer than the lateral veins... *T. intermedium* (with 2 subspecies)
   2a Lemmas usually hairy throughout, occasionally hairy only toward the margins; glumes usually hairy throughout, sometimes glabrous .... *T. i.* ssp. *barbulatum*
   2b Lemmas and glumes usually glabrous; lemmas occasionally with hairs on the margins .... *T. i.* ssp. *intermedium*

Torreyochloa, Mannagrass, False Mannagrass

Rhizomatous perennials. Leaf sheaths open, ligules membranous. Inflorescences panicles; disarticulation above the glumes. Spikelets with 2–8 florets; glumes shorter than the lowest florets; lemmas with (5)7–9 prominent, parallel veins, awnless.

   Torreyochloa differs from *Glyceria* in having open sheaths. Both grow in freshwater habitats. *Puccinellia* has faint rather than prominent lemma veins and usually grows in saline or alkaline habitats.

1a Inflorescences linear to narrowly elliptic in outline, 5–19× as long as wide, < 1 cm wide; panicle branches appressed (rarely slightly spreading); widest culm leaf blades 3.4–7.2 mm wide; subalpine to alpine shores and wetlands; Cascades and Klamath Mts., OR...*T. erecta*
1b Inflorescences elliptic to ovate in outline, 1.2–7.5× as long as wide, 1–14 cm wide; panicle branches spreading; widest culm leaf blades 3.6–18 mm wide; wetlands and along streams, often emergent in shallow water; widespread .... *T. pallida* var. *pauciflora*

Zizania, Wild Rice

Tall annuals, emergent in aquatic habitats. Leaf sheaths open; ligules membranous. Upper part of panicles pistillate, lower part staminate (occasionally boys and girls mingle in the middle); disarticulation beneath the spikelets. Glumes absent; pistillate lemmas awnless or awned; staminate lemmas awnless.

   Like *Leersia, Zizania* has no glumes. Culms grow several meters tall, often in shallow lakes. The dark brown, cylindrical seeds are often gathered for food (wild rice). *Zizania* is sometimes planted as food for waterfowl and other wildlife.

1a Pistillate panicle branches usually appressed at maturity, sometimes 1 or a few somewhat spreading; lemmas of pistillate spikelets stiff, hard, shiny, glabrous or with hairs only near the tip; aborted pistillate spikelets 0.6–2.6 mm wide; W OR and WA ....*Z. palustris* var. *palustris*
1b Pistillate panicle branches usually spreading to drooping at maturity; lemmas of pistillate spikelets flexible, papery, dull, or slightly shiny, with short, scattered hairs evenly distributed or slightly denser toward the tip; aborted pistillate spikelets 0.4–1 mm wide; 1 record in NW WA .... *Z. aquatica* var. *aquatica*

SPECIES ACCOUNTS

NOTE: [E] indicates an introduced or exotic taxon; [N?] indicates uncertainty as to whether a taxon is native in OR or WA; and [E, N] indicates both exotic and native subtaxa or strains of the species. Absence of a symbol indicates it is native to OR and/or WA.

Aegilops cylindrica Host
Jointed Goatgrass [E]

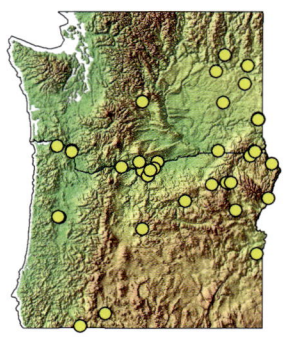

Winter annual; **culms** 14–75 cm. **Leaf sheaths** open, hyaline margins sometimes ciliate; **blades** 2–5 mm wide, with long hairs on the lower margin and collar; **ligules** ~ 0.5 mm, membranous, truncate; **auricles** ciliate. **Seedling leaves** twist clockwise, the first leaf to emerge reddish-green to brownish-green; leaf margins, sheaths, and auricles ciliate. **Spikes** cylindric, breaking into barrel-shaped joints. **Spikelets** 9–12 mm with 3–5 florets, the lower 2–3 florets fertile. **Glumes** of lower spikelets awnless or with an awn 2–5 mm; glumes of terminal spikelets 7–9 mm with awns 50–80 mm. **Lemmas** 9–10 mm, mucronate to awned, awns of the lower lemmas 1–5 mm, of terminal lemmas 40–80 mm. **Anthers** 3–3.5 mm. **Habitat**: Dry, disturbed sites, roadsides, cultivated fields. Native to the Mediterranean and Central Asia. **Comments**: Jointed goatgrass is a serious weed in winter wheat, which emerges with a whitish-green first leaf without ciliate margins. Jointed goatgrass has been spread to revegetation sites in wheat straw.

Aegilops triuncialis L. var. *triuncialis*
Barbed Goatgrass [E]

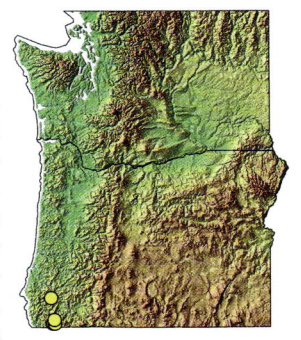

Winter annual, usually with several tillers; **culms** 17–60 cm, geniculate to semiprostrate at the base. **Leaf sheaths** open with hyaline margins, lower ones often ciliate; **blades** 2–3 mm wide; **ligules** ~0.5 mm, membranous, truncate, with ciliate hairs at the throat and on the clasping auricles. **Spikes** ellipsoid with 2–7 fertile spikelets below 2–3 rudimentary spikelets. Disarticulation is at the base of the spikelets. **Lower spikelet glumes** have 2–3 awns 15–60 mm, sometimes with 1 awn reduced to a stout tooth. **Upper spikelet glumes** have 3 awns 25–80 mm, or 1 awn and 2 lateral teeth. **Lemmas** 7–11 mm with 2–3 teeth. **Anthers** 4–4.5 mm. **Habitat**: Disturbed sites, roadsides, wheat fields, and relatively undisturbed grasslands on ultramafic substrates. Native to the Mediterranean and Central Asia. **Comments**: Barbed goatgrass differs from jointed goatgrass in bearing more awns and having a more ellipsoid spike. This grass threatens serpentine **habitats**. It often grows with medusahead and, like it, forms a thatch that decomposes slowly. Unpalatable to livestock, even goats.

Agropyron cristatum (L.) Gaertn.
Crested Wheatgrass [E] (yellow dots)

Cespitose perennial; occasionally rhizomatous; **culms** 25–110 cm. **Blades** 1.5–6 mm wide, **ligules** to 1.5 mm, **auricles** usually present. **Spikes** broadly lanceolate to ovate. **Spikelets** diverging from rachis at 35–90°, 7–16 mm long, with 3–6(8) florets. **Glumes** 3–6 mm, usually with awns 1.5–3 mm. **Lemmas** 5–9 mm with awns 1–6 mm. **Anthers** 3–5 mm. **Habitat**: Common in juniper and sagebrush steppe, roadsides, disturbed areas. **Comments**: Seeded E of the Cascades for forage; competes with native species.

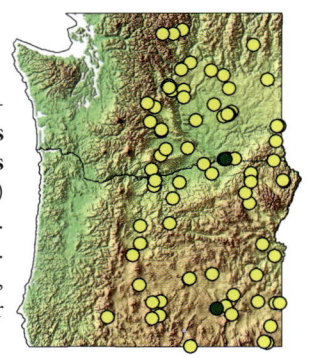

Agropyron fragile (Roth) P. Candargy
Siberian Wheatgrass [E] (green dots)

Similar to *A. cristatum* but **spikes** linear or narrowly lanceolate; **spikelets** diverging from rachis at < 30–35°; **lemmas** awnless or with a short point to 0.5 mm; **anthers** 4–5 mm. **Comments**: More drought tolerant than crested wheatgrass; documented only twice in the PNW, in SE OR and SE WA.

Agrostis blasdalei Hitchc.
Blasdale's Bentgrass

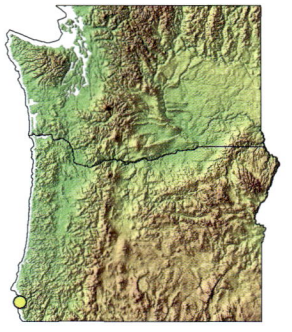

Small, densely tufted perennial; **culms** 6–30 cm; decumbent to erect. **Leaf blades** < 1 mm wide, tightly inrolled; **ligules** 0.7–2.3 mm, scabridulous on the back. **Panicles** narrow, spike-like, enclosed at the base by the sheath of the flag leaf; scabrous branches strongly appressed. **Glumes** 1.8–4 mm, scabrous to smooth; **calluses** glabrous; **lemmas** 1.5–2.5 mm, veins obscure or extending as 0.2 mm teeth at the tip, sometimes with awns to 1.2 mm; paleas to 0.3 mm. **Anthers** 0.7–2 mm. **Habitat**: Coastal cliffs, dunes, seasonally wet swales in sandy soil. **Comments**: Until recently discovered in S OR, this rare CA endemic was limited to a few coastal CA counties, where it is considered imperiled. Hybridizes with *A. densiflora*.

Agrostis canina L.
Velvet Bentgrass [E]

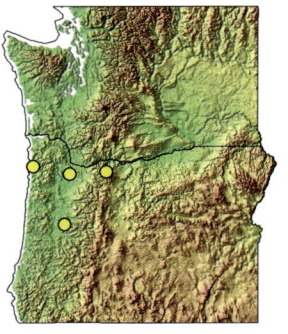

Stoloniferous perennial, appearing loosely cespitose; **culms** 15–75 cm. **Leaf blades** 1–3 mm wide, flat; **ligules** 1–4 mm. **Panicles** contracted to ± open; branches erect to spreading, spikelet-bearing in the outer 33%. **Glumes** 1.7–3 mm; **calluses** with hairs to 1 mm; **lemmas** 1–2(2.3) mm, minutely pubescent at base, otherwise glabrous, awned from near base; awns to 5 mm, sometimes absent or minute; **paleas** 0–0.2 mm, < 40% as long as lemmas; **anthers** 1–1.5 mm. **Habitat**: Wetlands, moist open forest, disturbed sites. Native to Eurasia. **Comments**: Short to nonexistent paleas distinguish *A. canina* from our other introduced bentgrasses. It may be confused with *A. gigantea* and *A. stolonifera*, which have panicle branches bearing spikelets to near the base, and with rare *A. howellii*, which has longer lemmas and is restricted to the Columbia Gorge.

Agrostis capillaris L.
Colonial Bentgrass, Browntop [E]

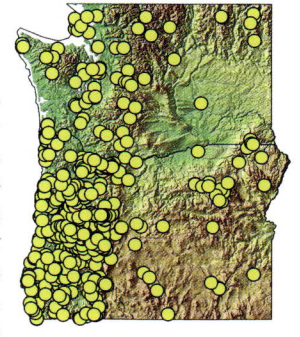

Rhizomatous or stoloniferous perennial; **culms** 10–75 cm, often clustered. **Leaf blades** 1–5 mm wide, flat; **ligules** 0.3–2 mm, shorter than wide on upper leaves. **Panicles** open, branches spreading, spikelet-bearing on outer 50%. **Glumes** 1.7–3 mm; **calluses** glabrous or with sparse hairs to 0.1 mm; **lemmas** 1.2–2.5 mm, usually awnless or awned from near the middle, awns to 2 mm, bent or straight; **paleas** ≥ 50% as long as lemmas. **Anthers** 0.8–1.3 mm. **Habitat**: Roadsides, meadows, sand dunes, disturbed areas, lawns, generally in mesic, well-drained soils. Native to Europe. **Comments**: Colonial bentgrass is one of five introduced perennial bentgrasses that are distinguished based on ligules, paleas, awns, growth habit, and habitat. Common in our area, it is invasive in mesic grasslands. It is often a component in lawn grass seed mixes.

Agrostis castellana Boiss. & Reuter
Highland or Dryland Bentgrass [E]

Rhizomatous perennial; **culms** 3–80 cm. **Leaf blades** 1–3 mm wide, flat; **ligules** 0.5–3 mm, shorter than wide on upper leaves. **Panicles** open to ± contracted; branches spreading to ascending; **spikelets** ± clustered on outer 33% of branches. **Glumes** 2–3 mm; **calluses** with many hairs 0.3(0.6) mm; **lemmas** 1.3–1.9 mm, sometimes pubescent, often awned from lower 33%, awns to 5 mm, usually present on lemmas of the terminal spikelets of panicle branches, shorter or lacking on lemmas lower on the branches; **paleas** ≥ 50% as long as lemmas. **Anthers** 1–1.5 mm. **Habitat**: Disturbed sites, lawns, golf courses. Native to S Europe. **Comments**: Lemmas of spikelets near tips of panicle branches often have awns to 5 mm, while lemmas of spikelets lower on panicle branches are awnless or have very short awns. Similar *A. capillaris* has less hairy calluses, and lemmas without awns or with shorter awns. *Agrostis canina*, *A. gigantea*, and *A. stolonifera* have ligules longer than wide. Without close inspection of details, you are not likely to distinguish this from the other weedy *Agrostis* species. It is probably more widespread than reported.

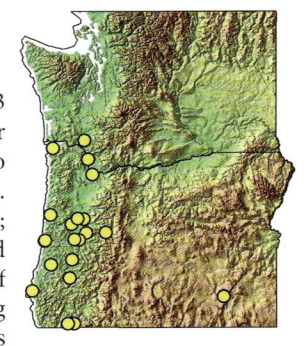

Agrostis densiflora Vasey
California Bentgrass

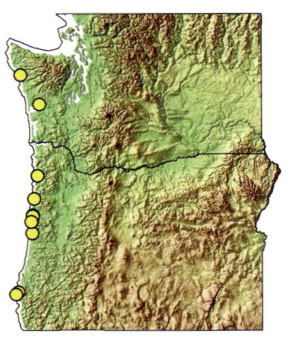

Cespitose perennial, sometimes rhizomatous when buried by sand; **culms** 9–85 cm. **Leaf blades** 2–10 mm wide, flat; **ligules** 1–3(7.5) mm. **Panicles** narrow, dense, the base often included in uppermost sheath; branches appressed, mostly hidden by spikelets. **Glumes** 2–3.3 mm, finely scabrous over the back; **calluses** with dense hairs to 0.3 mm; **lemmas** 1.5–2.1 mm, awnless or awned from above midlength, awns to 3.5 mm; **paleas** (0.3)0.5–0.7 mm, ~ 30% of lemma length. **Anthers** 0.3–0.7 mm. **Habitat**: Sandy soils on coastal dunes and bluffs. Uncommon. **Comments**: Similar *A. exarata* has glumes more or less glabrous over the back with finely scabrous keels; it usually has shorter paleas and longer ligules than *A. densiflora*, though there is overlap in both characters. The coastal form of *A. pallens*, a plant of sand dunes and coastal cliffs, is strongly rhizomatous, with longer anthers, narrower leaves, and panicles that are exserted from upper leaf sheaths.

Agrostis exarata Trin.
Spike Bentgrass

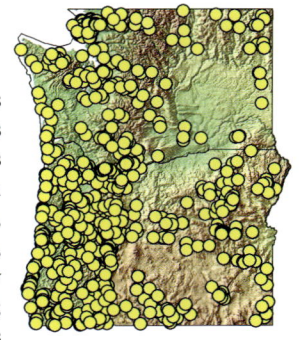

Cespitose perennial, rarely short-rhizomatous, sometimes flowering the first year; **culms** 8–100(150) cm. **Leaf blades** 2–7 mm wide, flat; **ligules** (1)2.5–4(11) mm. **Panicles** dense, spike-like, usually exserted from upper sheath, often interrupted near base, branches ascending to appressed, spreading at anthesis and closing again after pollen release, spikelet-bearing to near base. **Glumes** 1.5–3.5 mm, usually ± glabrous over the back, except finely scabrous on the keel; **calluses** with sparse to abundant hairs to 0.3 mm; **lemmas** 1.2–2.2 mm, awnless or awned from above midlength, awns 0–3.5(6) mm; **paleas** absent or to 0.5 mm. **Anthers** 0.3–0.6 mm. **Habitat**: Wet meadows, shorelines, ditches, disturbed areas. **Comments**: Spike bentgrass is a common, widespread grass of wet, often disturbed habitats. It is distinguished by dense, spike-like panicles and tiny anthers. See similar *A. densiflora* and *A. microphylla*.

Agrostis gigantea Roth
Redtop [E]

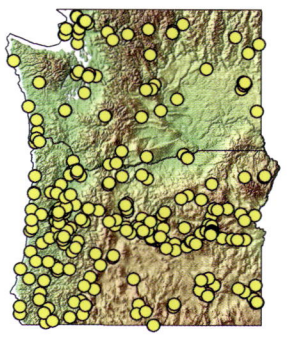

Strongly rhizomatous perennial; **culms** 20–120 cm, often clustered on rhizome. **Leaf blades** 3–8 mm wide, flat; **ligules** 1–7 mm, longer than wide on upper leaves. **Panicles** 8–30 cm, usually open, reddish, sometimes contracted and pale green to tan; branches usually spreading, spikelet-bearing to base. **Glumes** 1.7–3.2 mm; **calluses** glabrous or with sparse hairs to 0.5 mm; **lemmas** 1.5–2.2 mm, usually awnless, rarely with straight awn 0.4–1.5(3) mm, awn arising from near tip to near base of lemma; **paleas** 40–60% as long as lemmas. **Anthers** 1–1.4 mm. **Habitat**: Mesic meadows, roadsides, disturbed areas. Native to S Europe. **Comments**: This rhizomatous bentgrass has a large, open, reddish panicle and ligules that are longer than wide. It is widespread, abundant, and invasive in mesic grasslands. The reddish panicles are distinctive enough during flowering to identify from a speeding automobile; later in the season they fade to tan or straw-colored. Often confused with *A. stolonifera,* which is stoloniferous, has narrower, paler panicles, and usually grows in wetter habitats. See similar *A. canina.*

Agrostis hallii Vasey
Hall's Bentgrass

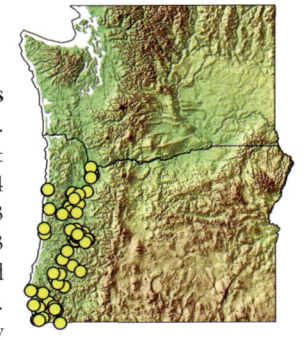

Strongly rhizomatous, slightly glaucous perennial; **culms** 17–100 cm. **Leaf blades** 2–5 mm wide, flat; **ligules** 2.3–7 mm. **Panicles** narrow to widely open; branches ascending to ± appressed, mostly branching above the middle. **Glumes** 2.5–4 mm; **calluses** with dense, long hairs, 0.8–2 mm; **lemmas** 2–3 mm, awnless. **Paleas** absent or to 0.2 mm. **Anthers** 1.5–2.3 mm. **Habitat**: Open oak or mixed forest, oak savanna, upland meadows. **Comments**: The long callus hairs distinguish *A. hallii* from our other bentgrasses. In the field, the slightly glaucous foliage helps it stand out from other grasses. Unlike in the introduced bentgrasses, the palea of *A. hallii* is nonexistent or minute. Its leaves are stiffer and broader than other bentgrasses, giving it a coarser look. See also *A. pallens*.

Agrostis hendersonii Hitchc.
Henderson's Bentgrass

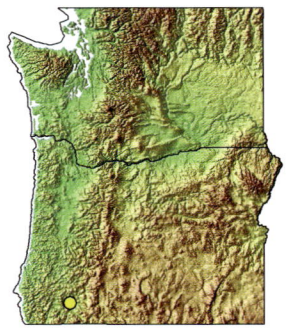

Annual; **culms** 6–70 cm. **Leaf blades** 1–4.5 mm wide, flat or ± involute; **ligules** 0.5–5 mm, acute to obtuse, erose-lacerate, usually scabridulous. **Panicles** dense, spike-like, cylindrical, sometimes interrupted near base, appearing bristly with awns; branches ascending to appressed. **Glumes** 3.5–5 mm, narrowly acuminate to awn-tipped, scabrous on midvein and often scabridulous on back; **calluses** with abundant hairs to 0.7 mm; **lemmas** 2.5–4 mm, tips with 2 teeth to 1.5 mm, awned from near midlength, awns (5)8–10 mm; **paleas** absent. Anthers 0.4–0.7 mm. **Habitat**: Clay or rocky soil around vernal pools; on serpentine in CA. **Comments**: Henderson's bentgrass was first collected in 1930 in Jackson County, OR, but has not been found in OR since. It is listed as imperiled in CA. *Agrostis microphylla* has shorter lemmas and awns, and also grows on serpentine and vernally wet sites.

Agrostis howellii Scrib.
Howell's Bentgrass

Cespitose perennial; **culms** 40–80 cm, often hanging down on steep slopes and cliffs. **Leaves** mostly cauline; **blades** 3–5 mm wide, flat; **ligules** (0.9)2.7–5 mm. **Panicles** open, diffuse, branches wavy, spreading, spikelet-bearing on outer 50%. **Glumes** 2.3–3.5 mm; **calluses** with abundant hairs to 0.3 mm; **lemmas** (2.2)2.5–3 mm, tips with 2 teeth to 1.5 mm, awned from lower 33%, awns 4–6 mm, bent; **paleas** absent or minute. **Anthers** 1–1.3 mm. **Habitat**: Moist rocks and cliffs, waterfall spray zones, moist, shady forest. **Comments**: Howell's bentgrass is a rare endemic of the Columbia River Gorge in OR and WA. Sometimes confused with introduced *A. canina* and native *A. mertensii*. See also newly described *A. swalalahos*.

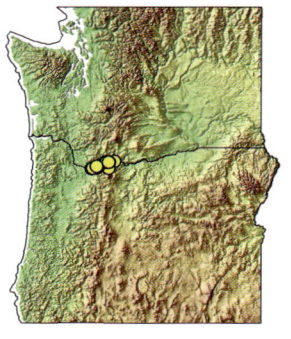

Agrostis idahoensis Nash
Idaho Redtop

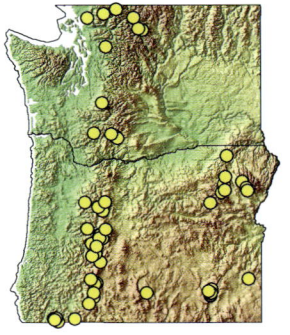

Cespitose perennial; **culms** 8–40 cm. **Leaves** mostly basal; **blades** 0.5–2 mm wide, flat to involute; **ligules** (0.7)1–3.8 mm. **Panicles** diffuse, reddish to purple; branches ± ascending; **spikelets** often solitary at branch tips. **Glumes** 1.5–2.5 mm; **calluses** with sparse hairs to 0.3 mm; **lemmas** 1.2–2.2 mm, awnless; **paleas** absent or to 0.2 mm. **Anthers** 0.3–0.6 mm. **Habitat**: Standing water of fens, seeps, and wet meadows, montane to alpine. **Comments**: Idaho redtop is a delicate, cespitose grass with narrow, mainly basal leaves and an open, dark red to purple inflorescence. It can be confused with dwarfed *A. scabra,* which has longer lower panicle branches and grows in drier habitats. *A. oregonensis* grows in the same habitat but has wider leaves. Be careful to measure fully developed anthers. Persistent anthers late in the season may be aborted and atypically small.

Agrostis mertensii Trin.
Northern Bentgrass

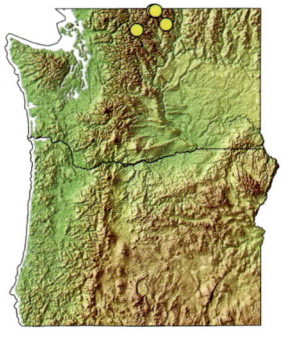

Cespitose perennial; **culms** (5)10–40 cm. **Leaves** mostly basal; **blades** 0.5–3 mm wide, usually flat, occasionally involute or folded; **ligules** 0.7–3.3 mm. **Panicles** (2)3–10 mm, 1–7 mm wide, open, dark brown or purplish, branches spreading, spikelet-bearing on outer 50%. **Glumes** 2–4 mm; **calluses** with sparse hairs to 0.4 mm; **lemmas** 1.6–2.6 mm, awned from just below midlength, awns (2)3–4.4 mm, bent; **paleas** absent or to 0.1 mm. **Anthers** 0.5–0.8 mm. **Habitat**: Rocky areas, gravel bars, edges of wet sites, cliffs, subalpine to alpine in our area.
Comments: This circumboreal bentgrass enters our area in the mountains of N WA. Sometimes confused with dwarfed *A. scabra,* which has panicles that disarticulate to become tumbleweeds. See also *A. idahoensis.*

Agrostis microphylla Steud.
Small-leaf Bentgrass

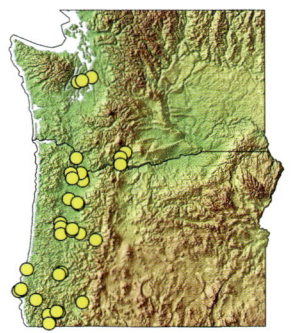

Annual; **culms** 8–45 cm. **Leaf blades** 0.7–2.5 mm wide, flat, becoming involute; **ligules** 1.4–4.5 mm. **Panicles** dense, spike-like, cylindrical, appearing bristly with awns; branches ascending to appressed. **Glumes** 2.5–5 mm, narrowly acuminate to awn-tipped, scabrous on midvein and often scabridulous on back; **calluses** with dense hairs to 0.5(1) mm; **lemmas** 1.5–2.3 mm, tips with 2 teeth to 0.5 mm, awned from midlength or above, awns 2.3–5(8) mm; **paleas** absent or to 0.2 mm. **Anthers** 0.4–0.5 mm. **Habitat**: Thin, rocky, vernally moist soils, vernal pools; also ultramafic substrates in SW OR. **Comments**: Uncommon *A. microphylla* typically grows in thin, rocky soils that are wet in spring and become bone-dry by midsummer. It can be locally common on ultramafic soils that are not necessarily shallow but have the same moisture regime. Its narrowly tapered glumes, longer awns, narrower leaves, and annual life cycle distinguish it from *A. exarata*. *Agrostis hendersoni* grows in similar habitats but has longer lemmas and awns.

Agrostis oregonensis Vasey
Oregon Redtop

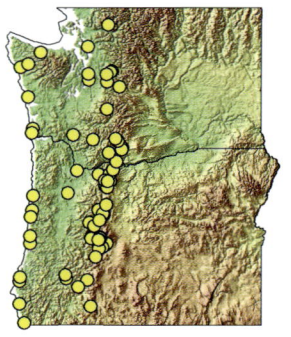

Cespitose perennial; **culms** 12–75 cm. **Leaf blades** (1)2–5 mm wide, flat; **ligules** 1.2–6.3 mm. **Panicles** open, dark red to purple, branches ascending, primary branches dividing above midlength, spikelet-bearing in outer 50%. **Glumes** 2–3.6 mm; **calluses** with sparse hairs to 0.2 mm; **lemmas** 1.5–2.5 mm, usually awnless but sometimes with awned and awnless lemmas on the same plant; lemma awns to 2 mm, arising at about midlength; **paleas** absent or to 0.2 mm. **Anthers** 0.5–1.2 mm. **Habitat**: Wet meadows, shorelines, sphagnum bogs, coastal to montane. **Comments**: *Agrostis oregonensis* has an open, dark reddish panicle. It resembles immature *A. scabra*, which has a more diffuse panicle and narrower leaves, and grows in drier habitats. *Agrostis idahoensis* grows in similar habitats but has narrower leaves and smaller anthers. There have been no coastal collections of *A. oregonensis* in OR since the 1940s.

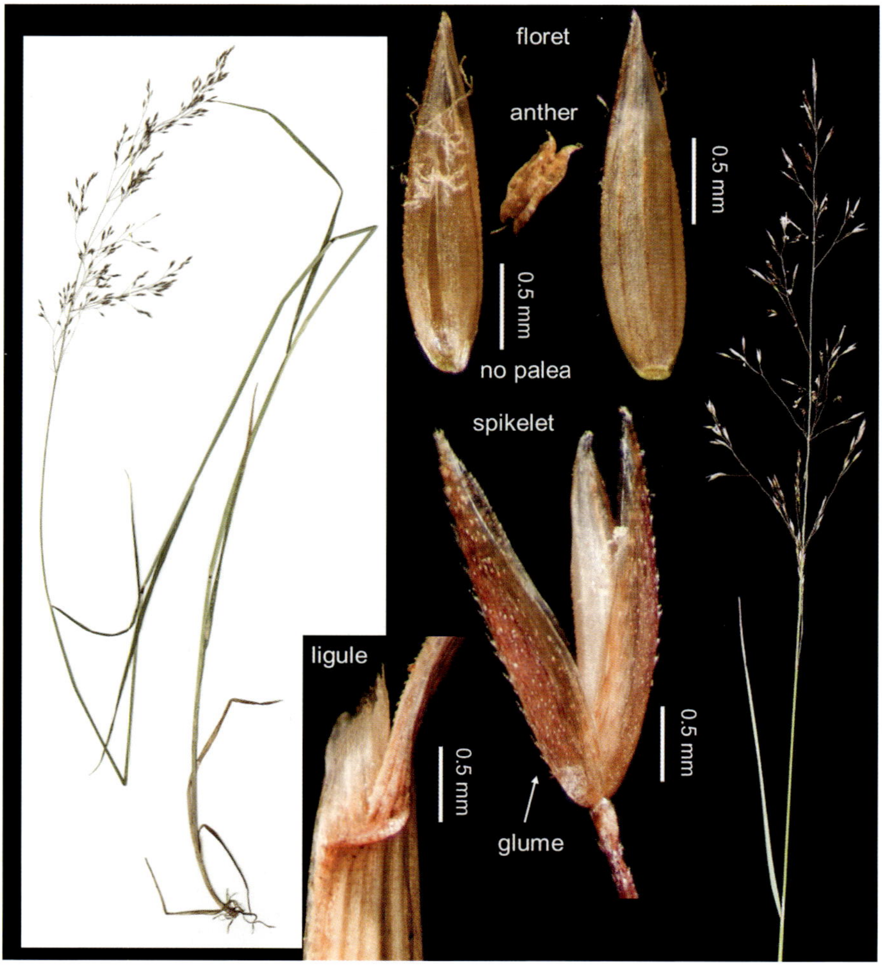

Agrostis pallens Trin.
Thin, Seashore, or Dune Bentgrass

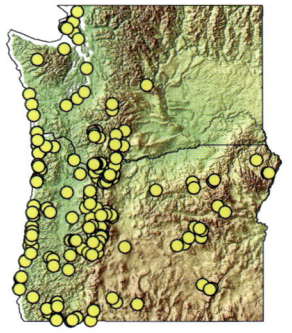

Strongly rhizomatous perennial; **culms** 10–70 cm. **Leaf blades** 1–6 mm wide, flat, becoming involute; **ligules** 1–6 mm. **Panicles** condensed to open, pale to reddish-purple; branches ascending, spikelet-bearing to the base. **Glumes** 2–3.5 mm; **calluses** with sparse hairs to 0.3(1) mm; **lemmas** 1.5–2.5 mm, smooth, scabridulous or warty, awnless or awned usually from below the middle, awns 0–2.5 mm; **paleas** absent or to 0.2 mm. **Anthers** 0.7–1.8 mm. **Habitat**: Sand dunes, coastal cliffs, meadows, and open woods, from sea level to alpine.

Comments: Coastal *A. pallens* has dense, spike-like panicles and may be confused with *A. densiflora* and *A. exarata,* which both have smaller anthers. In inland habitats *A. pallens* panicles are less dense, with short, spreading branches; it forms discrete patches (from rhizomes) with upright leaves and culms. Similar *A. hallii* has longer callus hairs and anthers and coarser foliage.

Agrostis scabra Willd.
Ticklegrass

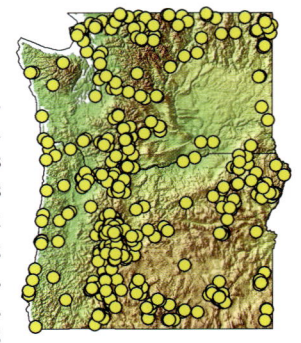

Cespitose perennial (rarely annual); **culms** (7.5)15–90 cm. **Leaves** mostly basal; **blades** to 2 mm wide; **ligules** 0.7–5 mm. **Panicle** very open, diffuse, delicate, often nearly as wide as long, branches capillary, widely spreading, with spikelets mostly at the ends; mature panicle detaching at base to form a tumbleweed. **Spikelets** dark purple; **glumes** 1.8–3.4 mm; **calluses** with sparse hairs to 0.2 mm; **lemmas** 1.4–2 mm, awnless or awned from below midlength, awns 0.2–3 mm, bent or straight; **paleas** absent or to 0.2 mm. **Anthers** 0.4–0.8 mm. **Habitat**: Moist to dry grasslands, deflation plains, shrub steppe, forest, roadsides, disturbed areas, from sea level to alpine. **Comments**: Widespread and common, ticklegrass has a diffuse, delicate inflorescence that appears oversized relative to the small tuft of basal leaves. At maturity, the panicle breaks off at the base and rolls in the wind, dispersing seeds like a tumbleweed. It is easy to misinterpret the lifespan of this species, guessing it is an annual, given the small tuft of leaves, shallow roots, and the tendency to grow in disturbed locations. Immature and dwarfed plants can be confusing. See *A. idahoensis*, *A. mertensii*, and *A. oregonensis*.

Agrostis stolonifera L.
Creeping Bentgrass [E]

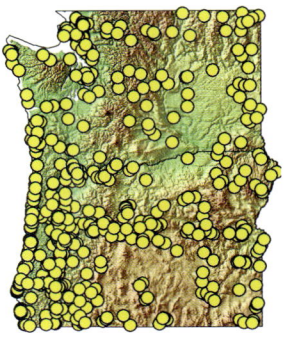

Stoloniferous perennial; stolons rooting at nodes, sometimes forming dense mats; **culms** (8)15–60 cm. **Leaf blades** 2–6 mm wide, flat; **ligules** 0.7–7 mm, longer than wide. **Panicles** 3–20 cm, contracted, pale green to tan, branches ascending to appressed, spreading at anthesis and closing again after pollen release, spikelet-bearing to the base. **Glumes** 1.6–3 mm; **calluses** with sparse hairs to 0.5 mm; **lemmas** 1.4–2 mm, usually awnless, or with awn to 1 mm arising near tip; **palea** ≥ 50% lemma length. **Anthers** 0.9–1.4 mm. **Habitat**: Moist meadows, marshes, shorelines, ditches, ocean beaches, and disturbed, moist, open areas. Native to Eurasia. **Comments**: Creeping bentgrass is widespread and common in disturbed wetlands. The more common *A. gigantea* is often misidentified as *A. stolonifera*. It is a larger, rhizomatous plant of somewhat drier habitats, with longer, more open, reddish inflorescences.

Agrostis swalalahos Otting
Saddle Mountain Bentgrass

Short-rhizomatous, turf-forming perennial; **culms** 20–75 cm. **Leaves** basal and cauline; **blades** 0.8–3 mm wide, flat; **ligules** to 2(3) mm on lower leaves, to 3.5(4) mm on uppermost leaves. **Panicles** 10–22 cm, 1–8 cm wide, open, ovate to lanceolate, branches spreading, spikelet-bearing in distal 33–50%. **Glumes** (2.5)3–5(5.7) mm; **calluses** with hairs 0.2–0.8 mm; **lemmas** 2.3–3.1 mm, awned from the lower 30–50(55)%, awns (3)3.6–5 mm, bent; **paleas** absent or minute. **Anthers** (1.2)1.5–2 mm. **Habitat**: Grassy balds and cliffs, and roadsides in mountains, within 20 km of the coastline, elevation 460–975 m. **Comments**: *Agrostis swalalahos* is a recently described narrow endemic restricted to the N OR Coast Range. It resembles *A. howellii* of the Columbia Gorge but has the awn inserted higher on the lemma back, and longer anthers. *Agrostis pallens* is more strongly rhizomatous, and its lemmas are usually awnless, sometimes with a shorter, more delicate awn.

Agrostis variabilis Rydb.
Mountain Bentgrass

Cespitose or rarely short-rhizomatous perennial; **culms** 5–30 cm. **Leaf blades** 0.5–2 mm wide, flat, folded, or involute; **ligules** 0.7–2.8 mm. **Panicles** 0.3–1.2(2) cm wide, narrow, dense, red to purple; branches erect to ascending, spikelet-bearing to base. **Glumes** 1.8–2.5 mm; **calluses** with sparse to abundant hairs to 0.2 mm; **lemmas** 1.5–2 mm, usually awnless, rarely awned from above midlength, awns to 1(2.8) mm, usually not reaching lemma tips; **paleas** absent or to 0.4 mm. **Anthers** 0.4–0.7(1) mm. **Habitat**: Wet meadows, forests, and talus, subalpine to alpine. **Comments**: This short, delicate bentgrass has a tuft of fine leaves and a dark, narrow inflorescence. It can be confused with *Podagrostis humilis,* and with immature *A. idahoensis* before its panicle branches spread. *Podagrostis humilis* has a much longer palea, ≥ 65% as long as the lemma; *A. idahoensis* has spikelets borne near the tips of panicle branches.

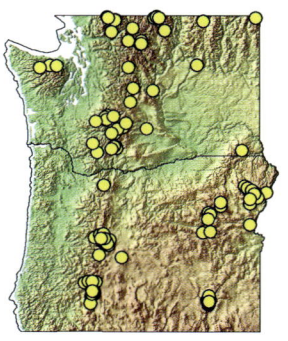

Aira caryophyllea L.
Silver Hairgrass [E]

Delicate tufted annual; **culms** 4.5–55 cm. **Leaf blades** 0.3–2.5 mm wide, involute; **ligules** 1.2–8 mm, membranous. **Panicles** 1.2–13.5 × 1.5–6 cm, open; pedicels 1–2× as long as spikelets; spikelets in 2s and 3s at branch tips; disarticulation above glumes. **Spikelets** 2–3.5 mm with 2 florets; **lemmas** 2–2.6 mm, hidden by glumes, awned from below middle, awns 2.1–3 mm, straight or bent, exserted beyond glumes. **Anthers** 0.2–0.5 mm. **Habitat**: Disturbed grasslands, rocky areas, roadsides, often on very shallow soil over rock. Native to the Mediterranean region. **Comments**: This common annual grass has airy, open panicles and 2 awns projecting from each spikelet. At maturity, the florets fall separately, so sometimes only 1 awn is visible peeking out of the glumes. See *A. elegans*.

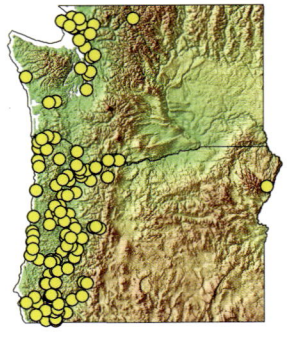

Aira elegans Willd. ex Roem. & Schult.
Elegant Hairgrass [E]

Delicate tufted annual; **culms** 4.5–55 cm. **Leaf blades** 0.3–2.5 mm wide, involute; **ligules** 1.2–8 mm, membranous. **Panicles** 1.2–13.5 × 1.5–13 cm, open; pedicels 1–8× as long as spikelets; spikelets single at branch tips; disarticulation above glumes. **Spikelets** 1.7–2.5 mm with 2 florets; **lemmas** 1.3–2.1 mm, hidden by glumes, lower lemma awnless, upper lemma awned from below middle, awns 2.1–3.9 mm, straight or bent, exserted beyond glumes. **Anthers** 0.2–0.4 mm. **Habitat**: Disturbed areas, often on shallow soil over rock or pavement. Native to Eurasia, North Africa. **Comments**: *Aira elegans* is similar to *A. caryophyllea* and is sometimes treated as a variety of that species. *Aira elegans* has larger, more open panicles, longer pedicels, and only 1 awned lemma in each spikelet, along with an unawned lemma. It is much less common than *A. caryophyllea*.

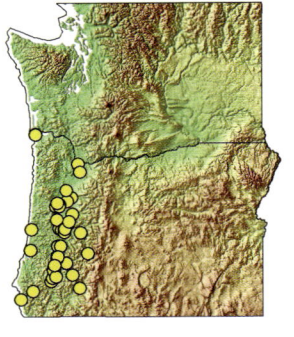

Aira praecox L.
Spike Hairgrass [E]

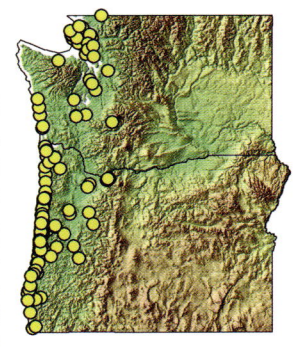

Tufted annual; **culms** 4–15(36) cm. **Leaf blades** 0.3–2 mm wide, involute; **ligules** 1.4–5.3 mm, membranous. **Panicles** narrow, spike-like; pedicels shorter than spikelets; disarticulation above the glumes. **Spikelets** 2.8–3.8 mm with 2 florets, **lemmas** 2.4–3.3 mm, hidden by glumes, awned from below middle, awns 3–4.5 mm, bent, exserted beyond glumes. **Anthers** 0.2–0.4 mm. **Habitat**: Sand dunes, rock ledges, shallow soil, disturbed areas, roadsides; most common at the coast, but occasionally found high in the mountains. Native to Europe. **Comments**: *Aira praecox* is a small annual grass with a silvery, spike-like panicle. Larger individuals can resemble *Anthoxanthum aristatum*, a more robust plant with larger spikelets (5–9 mm) that have 2 sterile lemmas below a hard, fertile lemma.

Alopecurus aequalis Sobol var. *aequalis*
Shortawn Foxtail

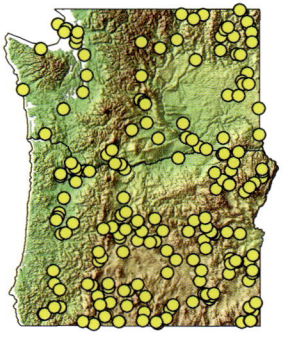

Cespitose perennial; **culms** 9–75 cm, often decumbent, rooting at lower nodes. **Upper leaf sheaths** not inflated; **blades** 1–5(8) mm wide, flat; **ligules** 2–6.5 mm. **Panicles** 1–9 × 0.3–0.6 cm, neat and trim because lemma awns are hidden or only slightly exserted from spikelets. **Glumes** 1.8–3 mm, fused near the base, hairy on the sides, keels ciliate; **lemmas** 1.5–2.5(3.5) mm, awns 0.7–3 mm, straight, usually < lemmas or exceeding them by < 1 mm. **Anthers** 0.5–0.9 mm. **Habitat**: Vernal pools, springs, wet meadows, forest openings, disturbed wet sites. **Comments**: *Alopecurus aequalis* has smooth, soft-silky panicles that have a trim appearance because the short lemma awns are hidden in the spikelets or just barely stick out. Panicles of other small foxtails appear shaggier because of their longer lemma awns.

Alopecurus arundinaceus Poir.
Black Foxtail, Garrison Grass, Creeping Meadow Foxtail [E]

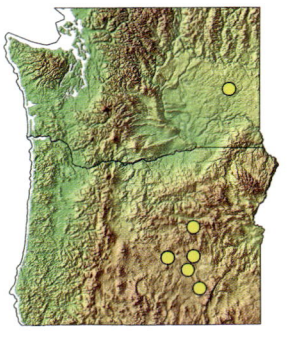

Rhizomatous perennial; **culms** 30–180 cm. **Upper leaf sheaths** ± inflated; **blades** 3–12 mm wide, flat; **ligules** 1.3–5 mm. **Panicles** 3–10 × 0.7–1.3 cm, dark brown to blackish at maturity. **Glumes** 3.6–5 mm, fused near base, keels ciliate, tips outcurved, green, often turning lead-gray at maturity; **lemmas** 3.1–4.5 mm, tips usually obliquely truncate, awns 1.5–7.5 mm, bent, exceeding lemmas by 0–3 mm. **Anthers** 2.2–3.5 mm. **Habitat**: Wet or seasonally saturated meadows, pastures, ditches; tolerates moderately alkaline soils. Native to Eurasia. **Comments**: In July, the mature black inflorescences of black foxtail stand out in eastside ditches and wet meadows. *Alopecurus arundinaceus* tends to be more robust than similar *A. pratensis*, which has acute lemmas and straw-colored glumes with ± straight tips. *Alopecurus pratensis* is often present with *A. arundinaceus*. The cultivar "Garrison," creeping meadow foxtail, was developed for hay and forage and is planted by ranchers. *Alopecurus arundinaceus* is probably underreported due to its similarity to *A. pratensis*.

Alopecurus carolinianus Walter
Tufted Foxtail [E]

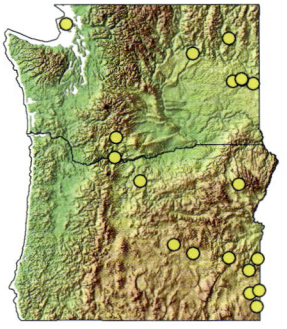

Tufted annual; **culms** 5–50 cm, erect or decumbent, not rooting at lower nodes. **Upper leaf sheaths** not or slightly inflated; **blades** 0.9–3 mm wide, flat; **ligules** 2.8–4.5 mm. **Panicles** 1–7 × 0.3–0.6 cm, appearing soft-bristly due to long, exserted awns. **Glumes** 2.2–3.3 mm, fused at base, keels ciliate; **lemmas** 1.9–2.7 mm, awns 3–6.5 mm, bent, exceeding lemmas by 1.5–4 mm. **Anthers** 0.3–1 mm. **Habitat**: Vernal pools, ponds, ditches, wet meadows. Native to E North America. **Comments**: The shaggy panicles, long, exserted awns, and more delicate, annual growth form distinguish *A. carolinianus* from *A. aequalis*. This grass is also mistaken for *A. saccatus,* which has conspicuously inflated upper leaf sheaths and larger spikelets.

Alopecurus geniculatus L.
Water Foxtail [N?]

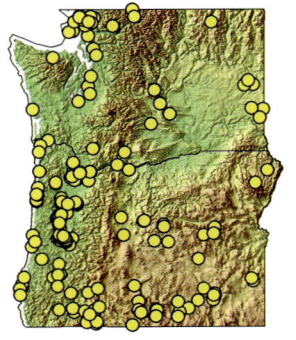

Cespitose perennial; **culms** (5)10–60 cm, erect or decumbent, rooting at lower nodes. **Upper leaf sheaths** somewhat inflated; **blades** 1–4 mm wide, flat; **ligules** 2–5 mm. **Panicles** 1.5–7 × 0.4–0.8 cm, appearing soft-bristly due to long, exserted awns. **Glumes** 1.9–3.5 mm, fused at base, keels ciliate; **lemmas** 2.5–3 mm, awns 3.5–6 mm, bent, exceeding lemmas by (1.2)2–4 mm. **Anthers** (0.9)1.4–2.2 mm. **Habitat**: Vernal pools, wet prairies, disturbed wet areas, low to middle elevations. **Comments**: *Alopecurus geniculatus* grows in wet, often disturbed habitats that dry out by late summer. *Alopecurus aequalis* has shorter awns and anthers; *A. saccatus* has strongly inflated upper leaf sheaths, and *A. carolinianus* has shorter anthers. Water foxtail is common, especially W of the Cascades. There is uncertainty about whether it is native in the PNW.

Alopecurus myosuroides Huds.
Blackgrass, Slender Meadow Foxtail [E]

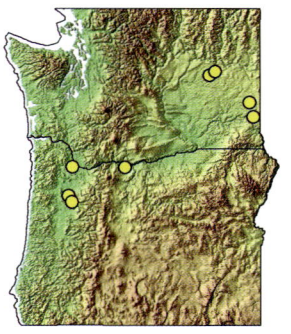

Tufted annual; **culms** (10)40–85 cm, erect. **Upper leaf sheaths** slightly inflated; **blades** 3.5–6 mm wide, flat; **ligules** 2–6 mm. **Panicles** 4–12 × 0.3–0.7 cm, narrow, usually tapering toward tip. **Glumes** 4.5–7.5 mm, lower 30–50% fused, sides glabrous, keels short-ciliate; **lemmas** 4–7 mm, awns to 12 mm, bent, exceeding lemmas by 3–6 mm. **Anthers** 2.4–4.1 mm. **Habitat**: Grain and grass seed fields. Native to Eurasia. **Comments**: *Alopecurus myosuroides* is a relatively tall, erect weed of grain crops, especially winter wheat. It is rare in the PNW and seldom spreads outside of cultivated grain fields. The large spikelets combined with its annual growth habit are unique among PNW *Alopecurus*. Noxious weed in WA. (It isn't wanted in OR, either.)

Alopecurus pratensis L.
Meadow Foxtail [E]

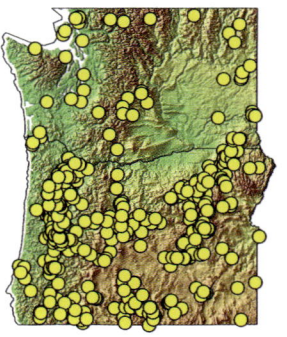

Strongly rhizomatous perennial; **culms** 30–110 cm. **Upper leaf sheaths** not inflated; **blades** 1.9–8 mm wide, flat; **ligules** 1.5–3 mm. **Panicles** 3.5–9 × 0.6–1 cm. **Glumes** 4–6 mm, lower 25% fused, keels ciliate, tips not outcurved; **lemmas** 4–6 mm, tips acute, straw-colored at maturity, awns 5–10.5 mm, bent, exceeding lemmas by (1)2.2–5.5 mm. **Anthers** 2–4 mm. **Habitat**: Wet or mesic meadows, pastures, ditches. Native to Eurasia and N Africa. **Comments**: Meadow foxtail is a common, robust grass of wet meadows that flowers in winter or early spring. An important early spring forage in irrigated pastures, it is often confused with *Phleum pratense*, a later-flowering species that grows in drier habitats and has stiff glumes with 2 short awns, giving the spikelets a 2-horned look. Vegetative *Schedonorus arundinaceus* resembles *A. pratensis* but has very short ligules, ciliate auricles, and more leathery leaves. See also *A. arundinaceus*.

Alopecurus saccatus Vasey
Pacific Meadow Foxtail

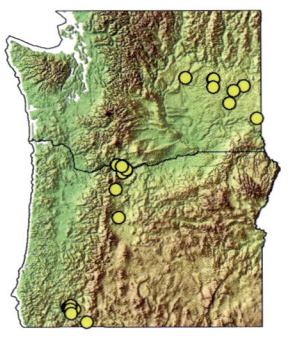

Annual; **culms** 12–45 cm; erect or decumbent, not rooting at lower nodes. **Upper leaf sheaths** conspicuously inflated; **blades** 1.2–4 mm wide, flat; **ligules** 1.5–5.5 mm, obtuse. **Panicles** 1.5–6.5 × 0.5–1.3 cm, appearing soft-bristly due to long, exserted awns. **Glumes** 3–5 mm, fused at base, sides pubescent, keels ciliate; **lemmas** 3–5 mm; awns 6–10 mm, bent, exceeding lemmas by 3–6 mm. **Anthers** 0.7–1.8 mm. **Habitat**: Vernal pools in low-elevation, interior valleys, not coastal. **Comments**: The upper leaf sheaths of *A. saccatus* are conspicuously inflated (if you're not sure if it's inflated, it probably isn't). Also, the panicles are proportionately wider than in similar species. Similar *A. aequalis* and *A. geniculatus* have shorter glumes and awns, and less inflated sheaths, and they root at lower culm nodes. Also see *A. carolinianus*. Possibly rare.

Anthoxanthum aristatum Boiss. ssp. *aristatum*
Annual Vernalgrass [E]

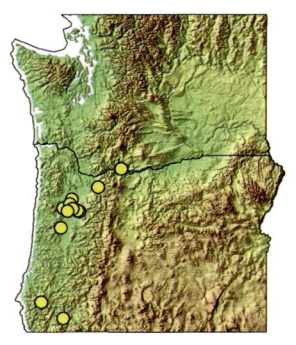

Tufted annual; **culms** 5–60 cm. **Leaf sheaths** open; **blades** 1–5 mm wide, long-ciliate at the base; **ligules** 1–3 mm, membranous. **Panicles** 1–4 cm, congested, spike-like. **Spikelets** 4–9 mm, with 3 florets; lowest 2 florets staminate, upper floret fertile. **Glumes** unequal, exceeding the florets; lower glumes 3–5 mm, upper glumes 5–7.5 mm; **lower 2 lemmas** ~ 3 mm, hairy, awned from below midlength, awn of lowest lemma 3.5–5 mm, awn of second lemma 3.5–10 mm, exserted from glumes 2–3 mm. **Anthers** 2.8–4.1 mm. **Habitat**: Disturbed areas with thin soils, dry meadows, roadside gravels, vernally wet sites. Native to Europe. **Comments**: This uncommon annual resembles perennial *A. odoratum* but has longer awns and narrower leaves. Vernalgrasses and our native sweetgrasses (*Hierochloë*) are sometimes lumped together into *Anthoxanthum*. Both genera contain compounds called coumarins that produce a pleasant scent when plants are crushed or burned. When metabolized by certain fungi, coumarin becomes dicoumarol, which causes vitamin K deficiency and poor blood clotting. Moldy hay containing these grasses can be toxic to livestock.

Anthoxanthum odoratum L.
Sweet Vernalgrass [E]

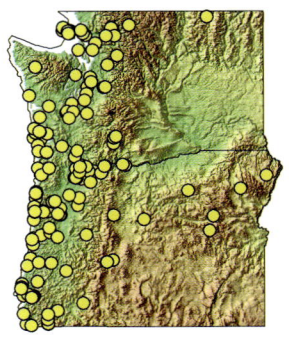

Cespitose perennial; **culms** (10)25–60(100) cm. **Leaf sheaths** open; **blades** 3–10 mm wide, long-hairy near base; **ligules** 2–7 mm, membranous; **auricles** 0.5–1 mm, long-hairy. **Panicles** 3–14 cm, dense, spike-like. **Spikelets** 6–10 mm, with 3 florets; lowest 2 florets staminate, upper floret fertile. **Glumes** unequal, exceeding florets; **lower 2 lemmas** 3–4 mm, hairy, awn of lowest lemma ~ 3 mm, arising at or above midlength, awn of second lemma 7–10 mm, arising near base, = or slightly > upper glume; upper lemma 1–2.5 mm, awnless. **Anthers** (2.9)3.5–4.8(5.5) mm. **Habitat**: Disturbed, usually moist grasslands, pastures, roadsides, mostly W of the Cascades. Native to S Europe. **Comments**: This common grass flowers in early spring. With its hairy leaf bases and auricles, vegetative *A. odoratum* is sometimes confused with *Danthonia californica*, which has tufts of hairs at the top of the leaf sheath.

Apera interrupta (L.) P. Beauv.
Interrupted Windgrass [E]

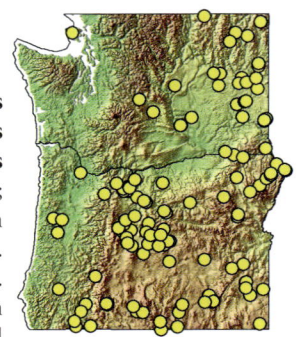

Tufted annual; **culms** 10–50 cm. **Leaf sheaths** open; **blades** 0.3–4 mm wide, flat or rolled, **ligules** 1.5–5 mm. **Panicles** narrow, branches erect, spikelet-bearing to base. **Spikelets** 2–2.8 mm, with 1 floret; **rachilla** prolonged 0.2–0.6 mm; **upper glumes** < to > florets; **lemmas** 1.5–2.5 mm, firmer than glumes, awns 4–10(16) mm; **paleas** ≥ 75% as long as lemmas. **Anthers** 0.3–0.6 mm. **Habitat**: Roadsides, cultivated fields. Native to Europe. **Comments**: *Apera* differs from *Agrostis* in having firmer lemmas, longer paleas, prolonged rachillas, and awns > twice as long as lemmas.

Apera spica-venti (L.) P. Beauv.
Common Windgrass [E] (not shown)

Collected a few times in OR prior to 1960, this species differs from *A. interrupta* by having wider, open **panicles**, **spikelets** near the branch tips, and **anthers** 1–2 mm long.

Aristida oligantha Michx.
Prairie Threeawn

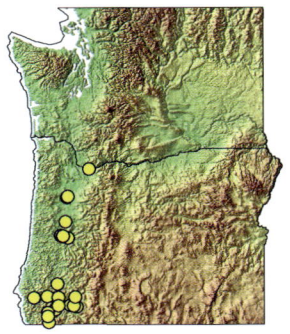

Tufted annual; **culms** 25–55 cm, much branched above base. **Leaf sheaths** open; **blades** 0.5–1.5 mm, flat or involute; **ligules** < 0.5 mm, membranous with a fringe of short hairs. **Inflorescences** usually racemes, **spikelets** divergent, with 1 floret. **Lower glumes** (9)12–22(28) mm, 3–7-veined, with delicate awn 1–13 mm; **upper glumes** 1-veined. **Lemmas** 6–9 mm, mottled with purple, awned from tip with 3 spreading awns (8)12–65(70) mm. **Anthers** usually 1 and < 0.5 mm, rarely 3 and 3–4 mm. **Habitat**: Disturbed areas, along railroads or roadsides, dry fields, usually in sandy soils, sometimes on serpentine. **Comments**: Threeawns are utterly distinctive with their 3 spreading lemma awns. *Aristida oligantha* is an uncommon warm-season annual that appears to be in decline in W OR. Its awns and minimal leaves make it poor forage for livestock. It is sometimes used for erosion control in sandy soils in the Great Plains, where it is also known as oldfield threeawn because it often recolonizes abandoned farmland. Six-weeks threeawn (*A. adscensionis*, not pictured), with shorter awns and lower glumes than *A. oligantha*, was collected once in OR in 1956.

Aristida purpurea Nutt. var. *longiseta* (Steud.) Vasey
Red Threeawn

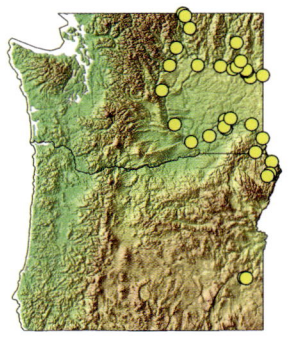

Cespitose perennial; **culms** 10–40 cm. **Leaf sheaths** open; **blades** 1–1.5 mm wide, rolled; **ligules** < 0.5 mm, membranous, with a fringe of hairs longer than the membranous part. **Inflorescences** panicles or racemes, **spikelets** ascending, with 1 floret. **Glumes** 1-veined, awnless or with awns to 2 mm; lower glumes 8–13 mm. **Lemmas** 12–16 mm, generally not mottled, awned from tip with 3 spreading awns 40–100(140) mm. **Anthers** 1, 0.7–2 mm. **Habitat**: Ponderosa pine with grass understory, grassy rocky slopes and benches, warm canyons; shows a strong preference for well-drained, sandy soils. **Comments**: Red threeawn differs from *A. oligantha* in its very long lemma awns and 1-veined lower glumes. The pointed callus and long awns aid in seed dispersal and deter grazers. This grass is scarce in OR (except in the Snake River Canyon) but more common in WA. Red threeawn is one of our native warm-season (C4) grasses.

Arrhenatherum elatius (L.) P. Beauv. ex J. Presl & C. Presl ssp. *elatius*
Tall Oatgrass [E] (yellow dots)

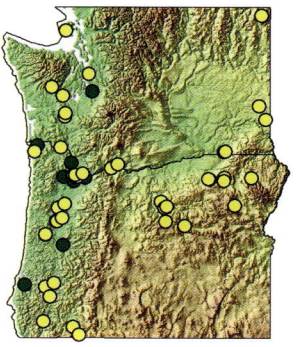

Cespitose perennial, sometimes rhizomatous; **culms** 50–140(180) cm; nodes glabrous, basal internodes 2–4 mm thick. **Leaf blades** 3–8(10) mm wide, flat; **ligules** 1–3 mm, membranous, short-ciliate. **Panicles** open, branches ascending to spreading. **Spikelets** 7–11 mm, laterally compressed, with 2 florets, the lower staminate, the upper pistillate or bisexual. **Lower glumes** 4–7 mm, **upper glumes** 7–10 mm. **Lemmas** (4)7–10 mm, bifid; lower lemma awned from below middle, awn 10–20 mm, twisted, bent; upper lemma awnless or with awn to 5 mm, arising just below tip. **Anthers** 3.6–5(6) mm. **Habitat**: Upland prairies, savannas, open forest, roadsides. Native to Europe. **Comments**: Introduced as a forage species, tall oatgrass is a serious weed of upland prairies and other grasslands, where it forms a dense overstory that competes with shorter native species.

Arrhenatherum elatius ssp. *bulbosum* (Willd.) Schübl. & G. Martens
Tall Oatgrass [E] (green dots)

Nodes densely hairy; **basal internodes** 5–10 mm thick, bulbous. Less common than ssp. *elatius*.

Arundo donax L.
Giant Reed [E]

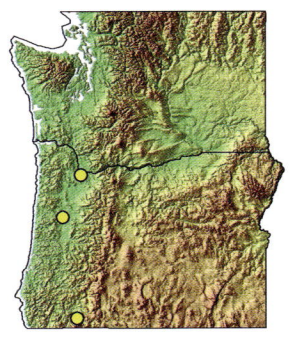

Rhizomatous perennial forming enormous tussocks or thickets; **culms** 2–10 m. **Leaf sheaths** open; **blades** 2–7 cm wide, 2-ranked, basal part with light to dark brown band and wavy margins; **ligules** 0.4–1 mm, membranous, short-ciliate. **Panicles** 30–60 cm, plumose, silvery to purplish. **Spikelets** 10–15 mm, laterally compressed, with 2–4 florets; **glumes** as long as spikelets; **lemmas** 8–12 mm, pilose, hairs 4–9 mm, tips bifid, midvein ending in a delicate awn. **Anthers** 2–3 mm. **Habitat**: Ditch margins, stream shores, wet areas. Native to Europe. **Comments**: Gigantic *A. donax* is recognizable in vegetative form by its sheer size; thickets may be mistaken for stands of good-sized willows. In CA it is highly invasive along waterways and ditches, and it has recently been documented at a few sites in W OR. It has been proposed as a biofuel crop in the Columbia Basin, igniting intense debate because of the potential impacts on riparian areas and native plant communities.

Avena barbata Pott ex Link
Slender Wild Oats [E]

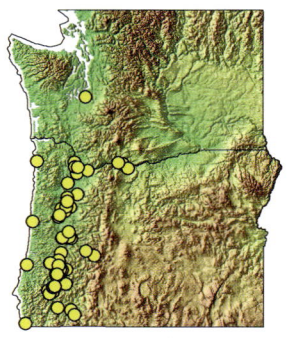

Weedy annual; **culms** 60–80(150) cm. **Leaf blades** 2–20 mm wide, flat to involute; **ligules** 1–6 mm, membranous. **Panicles** 15–35(50) × 6–12 cm. **Spikelets** 21–30 mm with 2–3 florets; **glumes** 15–30 mm, 7–9-veined; **calluses** hairy; **lemmas** 15–26 mm, tips with 2 elongated, bristle-like teeth, the teeth 2–4 mm; lemma awns 30–45 mm, arising from about midlength. **Anthers** 2.5–4 mm. **Habitat**: Disturbed sites, degraded upland grasslands, roadsides; especially on well-drained soils. Native to Mediterranean region and Central Asia. **Comments**: *Avena barbata* is characterized by large dangling spikelets and lemmas with long, narrow, bristle-like teeth. This grass is a serious invasive of upland grasslands in W OR and CA. Slender wild oats is often confused with *A. fatua*. Hold the floret up to backlighting and look for the long teeth on the lemma tip.

Avena fatua L.
Wild Oats [E]

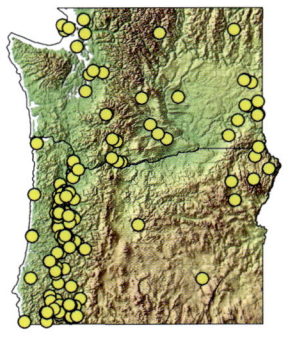

Weedy annual; **culms** 8–160 cm. **Leaf blades** 3–15(25) mm wide, usually flat; **ligules** 4–6 mm, membranous. **Panicles** 7–40 × 6–12 cm. **Spikelets** 18–32 mm with 2–3(5) florets; **glumes** 18–32 mm, 9–11-veined; **calluses** hairy; **lemmas** 14–22 mm, tips with 2 short teeth 0.3–1.5 mm, but lacking long bristle-like teeth; lemma awns 23–42 mm, arising from the middle third of lemma back. **Anthers** ± 3 mm. **Habitat**: Upland grasslands, crop fields, roadsides. Native to Europe and Central Asia. **Comments**: *Avena fatua* lacks the long, bristle-like lemma teeth of *A. barbata*. Like *A. barbata* it is a serious weed of upland grasslands as well as grain crops. *Avena fatua* and *A. sativa* hybridize. Common and widespread, wild oats often appear when straw has been used as a mulch for erosion control.

Avena sativa L.
Oats, Cultivated Oats [E]

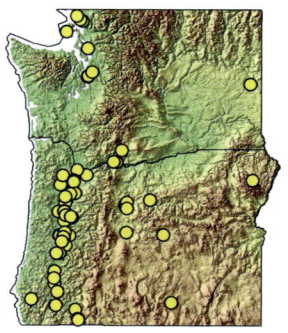

Annual; **culms** 35–180 cm. **Leaf blades** 3–14(25) mm wide, usually flat; **ligules** 4–6 mm, membranous. **Panicles** (6)15–40 × 5–15 cm. **Spikelets** (18)25–32 mm with 1–2(7) florets; **glumes** (18)20–32 mm, 9–11-veined; **calluses** glabrous; **lemmas** 14–18 mm, tips erose to toothed, longest teeth 0.2–0.5 mm, lacking long bristle-like teeth; **lemmas** usually awnless, awns 15–30 mm occasionally present. **Anthers** (1.7)3–4.3 mm. **Habitat**: Along railroads, roadsides, fallow farm fields. Native to Eurasia. **Comments**: *Avena sativa* is widely cultivated in cool temperate regions worldwide. It escapes from cultivation but seldom persists. Ancestral *A. fatua* also has acute lemma teeth, but it has hairy calluses, and awned second lemmas. The florets of *A. sativa* do not disarticulate but remain in the glumes long past maturity, facilitating harvest. Its panicles are denser and its culms are usually shorter and stouter than in other *Avena* species.

Beckmannia syzigachne (Steud.) Fernald
American Sloughgrass

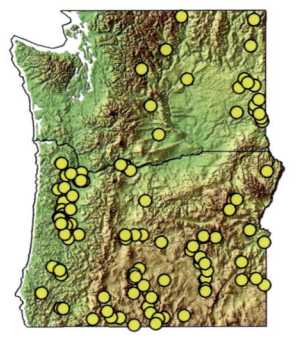

Coarse, tufted annual; **culms** 20–120 cm. **Leaf sheaths** open; **blades** 4–10(20) mm wide, flat; **ligules** 5–11 mm, membranous. **Panicles** dense, narrow, base occasionally partially hidden in uppermost leaf sheath; branches 1-sided, ± erect, with spikelets borne in crowded rows; disarticulation below glumes. **Spikelets** 2–3 mm, round to ovate in side view, with 1 floret. **Glumes** semicircular, with a small point at the tip, 3-veined, winged, the wings inflated, and the veins prominently raised at maturity. **Lemmas** 2.4–3 mm, entirely hidden by the glumes, awnless, tip sometimes with short point. **Anthers** 0.5–1.5 mm. **Habitat**: Freshwater wetlands, marshes, ditches, ponds, shorelines, often in standing water. **Comments**: American sloughgrass is distinguished by panicle branches with crowded rows of nearly circular spikelets. The spikelets fall as a unit and are adapted to floating in water. American sloughgrass provides valuable food and cover for waterfowl and small mammals. Some *Echinochloa* species bear a vague resemblance to *Beckmannia* but have dorsiventrally compressed spikelets and lack ligules.

Blepharidachne kingii (S. Watson) Hack.
King's Eyelash Grass

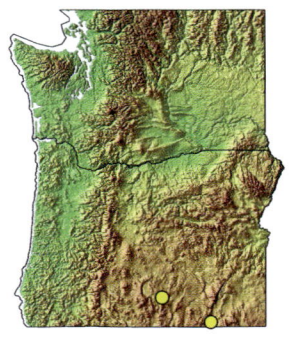

Small cespitose perennials (rarely annuals); **culms** 3–8(14) cm. **Leaf sheaths** often with a tuft of hairs at the throat, the upper two sheaths often partially enclosing the inflorescence; **blades** 0.7–3 cm, to 1 mm wide, folded or inrolled, often arching, deciduous, leaving the sheath behind; **ligules** of hairs to 0.5 mm, or absent. **Panicles** 10–25 mm, head-like, conspicuously hairy. **Spikelets** 6–9 mm, with 4 florets that are hard to recognize as such, the lower two sterile, the third fertile, the upper one vestigial. **Lemmas** fringed with long hairs, with 2 lateral lobes. **Lower lemmas** 3.4–5.8 mm with lobes 2.2–3 mm, awnless, with paleas plumose, linear. **Fertile floret lemmas** awned from between 2 lateral lobes, the **awn** 3–5 mm, the lobes 0.5–1.5 mm. **Anthers** ~1.5 mm. **Caryopses** ~2 mm. **Habitat**: arid areas of the Great Basin. **Comments**: This cute, hairy, tiny bunchgrass was recently reported in the Pueblo Mts. and near Lake Abert in SE OR. Eyelash grass is easy to overlook but disinctive when noticed. The only other grasses with deciduous leaf blades in our region are *Crypsis vaginiflora*, a low-growing annual of shoreline habitats, and *Molinia caerulea*, a tall ornamental that rarely escapes.

Brachypodium sylvaticum (Huds.) P. Beauv.
Falsebrome [E]

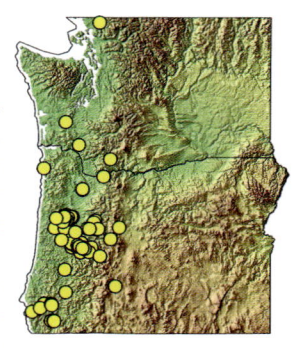

Loosely cespitose perennial; **culms** 30–120 cm. **Leaf sheaths** open, with silvery, spreading hairs; **blades** 4–12(15) mm wide, yellowish-green, flat, with spreading hairs; **ligules** 1–3(6) mm, membranous. **Racemes** spike-like, nodding; **spikelets** 17–30(50) mm, with 6–16(22) florets, nearly sessile or on pedicels to 2 mm. **Glumes** hairy; lower glumes 6–9 mm, 5–7-veined; upper glumes 8–11 mm, 7–9-veined; **lemmas** 6–12(13.5) mm, hairy, with 7 distinct veins, awned from tip, awns 7–15 mm, straight or curved. **Anthers** 3–4(5.5) mm. **Habitat**: Usually in shade, sometimes full sun, dry to moist forests, savannas, prairies, riparian zones, seasonally wet areas, roadsides. Native to Eurasia and N Africa. **Comments**: Falsebrome has tufts of yellow-green leaves and unbranched inflorescences. Its silvery spreading hairs are distinctive when viewed straight down the sheath. It can be confused with *Bromus vulgaris,* which has darker green leaves, closed sheaths, and branched inflorescences. Falsebrome is a serious invasive in forests W of the Cascades. In severely impacted areas falsebrome forms virtual monocultures in the understory, even outcompeting poison oak. Recently found on east slopes of the Cascades in Oregon.

Briza maxima L.
Rattlesnake Grass, Big Quaking Grass [E]

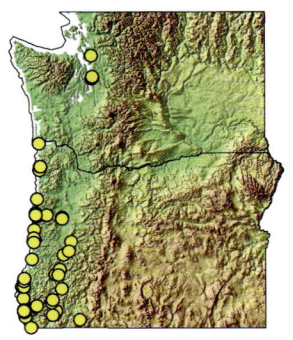

Annual; **culms** 20–80 cm. **Leaf blades** 2–8 mm wide; **ligules** 3–7 mm, membranous, sides decurrent. **Panicles** 3.5–10 × 1–5 cm. **Spikelets** 10–19(27) mm with 4–12(15) florets. **Glumes** 5–6.5 mm; **lemmas** 7–9 mm, glabrous basally, appressed-hairy on upper portion, unawned. **Anthers** 1.2–1.5 mm. **Habitat**: Dry, disturbed, often sandy soils, dunes, disturbed grasslands, cutbanks, roadsides. Native to the Mediterranean region. **Comments**: *Briza maxima* has attractive, dangling spikelets and is sometimes used in ornamental arrangements. It is often confused with *Bromus briziformis,* which differs in its closed leaf sheaths, longer spikelets, short-awned lemmas, and range E of the Cascades. *Briza maxima* is common at the coast and is increasing inland, particularly in SW OR, where it is invading disturbed grasslands and appearing along highways. It was recently found in the S Willamette Valley.

Briza minor L.
Little Quaking Grass [E]

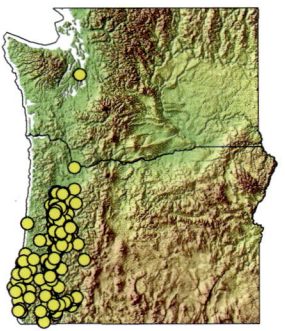

Annual; **culms** 7.5–50(80) cm. **Leaf blades** 1–8 mm wide; **ligules** 4–13 mm, membranous, sides decurrent. **Panicles** (2)4–14(18) × 3–11 cm. **Spikelets** 2–4(7) mm with 4–7(13) florets. **Glumes** 2–3.5 mm; **lemmas** 1.6–2 mm, glabrous or minutely scurfy, unawned. **Anthers** 0.4–0.5 mm. **Habitat**: Disturbed grasslands, seasonal wetlands, open woodlands, roadsides. Native to Europe. **Comments**: The tiny, triangular spikelets of *B. minor* dangle like little jewels on delicate branches in an open, ± globose panicle. *Aira caryophyllea* is equally delicate and can grow in the same habitat but has glumes that do not spread, 2 florets per spikelet, and awned lemmas.

Briza media L.
Perennial Quaking Grass [E] (not pictured)

Differs from *B. minor* in its perennial, short-rhizomatous habit, **ligules** about 0.5 mm, membranous, usually not decurrent, **lemmas** 3–4 mm, and **anthers** 1.3–2 mm. **Habitat**: Mesic meadows. Native to Europe. **Comments**: *Briza media* is sometimes grown as an ornamental. It was recently collected at 2 locations in Grant County, OR, where it has naturalized in a mesic meadow and a disturbed riparian area.

Bromus arenarius Labill.
Australian Brome [E]

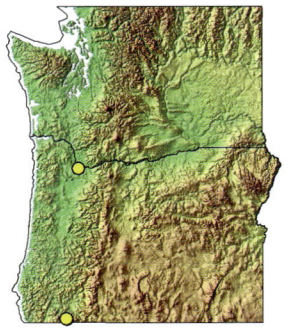

Annual; **culms** 20–40 cm. **Leaf sheaths** densely retrorsely pilose; **blades** 3–6 mm wide, hairy on both surfaces; **ligules** 1.5–2.5 mm, hairy or glabrous. **Panicles** open, nodding; branches spreading or ascending, sinuous, at least some branches > their spikelets. **Spikelets** 10–20 mm, lanceolate, with 5–9(11) florets. **Lower glumes** 7–10 mm, **upper glumes** 8–12 mm. **Lemmas** 9–11(13) mm, firm, rounded on the back, densely hairy, veins not thickened and raised, margins narrow, rounded, tips acute, bidentate, with teeth < 1 mm; awns 10–16 mm, arising ≥ 1.5 mm below lemma tips, straight to slightly spreading. **Anthers** 1.5–3 mm. **Habitat**: Dry disturbed grasslands. Native to Australia. **Comments**: Uncommon or overlooked, *B. arenarius* was recently collected in dry, montane meadows in the Upper Applegate drainage in SW OR. Previously it was only known as a ballast waif at Portland in the early 1900s. It is widespread in CA. *Bromus commutatus*, *B. japonicus*, and *B. squarrosus* have shorter glumes. *Bromus tectorum* spikelets are more slender, with longer, acuminate lemma teeth.

Bromus briziformis Fisch. & C.A. Mey.
Rattlesnake Brome [E]

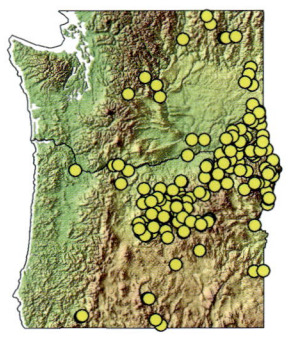

Annual; **culms** 20–62 cm. **Leaf sheaths** densely pilose, blades 2–4 mm wide, hairy on both surfaces; **ligules** 0.5–2 mm, hairy. **Panicles** open, nodding, branches sometimes > their spikelets. **Spikelets** 15–30 mm, ovate, laterally compressed, with 7–15 florets. **Glumes** meet at ~ 90° angle. **Lemmas** 9–10 mm, ± inflated, rounded over the back, margins abruptly angled near the middle, tips acute to obtuse, bidentate with teeth < 1 mm, awnless or with a short awn to 1 mm arising < 1.5 mm below the tip. **Anthers** 0.7–1 mm. **Habitat**: Disturbed areas, dry grasslands, ponderosa pine woodlands. Native to SW Asia and Europe. **Comments**: The mature spikelets of *B. briziformis* not only look like rattlesnake rattles, but sound like them too. It can be unnerving to walk through a stand of rattlesnake brome as it dries in the summer sun. Similar *Briza maxima* has open sheaths and glumes that meet at nearly a 180° angle.

Bromus catharticus Vahl var. *catharticus*
Rescue Grass [E]

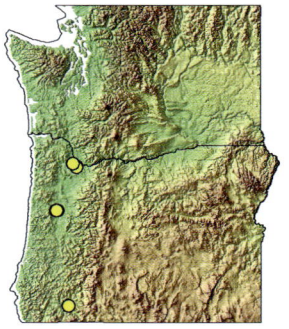

Loosely tufted annual, biennial, or short-lived perennial; **culms** 30–120 cm. **Leaf sheaths** densely hairy, sometimes only at the throat; **blades** 3–10 mm wide, glabrous or hairy on both surfaces; **ligules** 1–4 mm, glabrous or hairy, lacerate. **Panicles** open, erect to nodding, branches spreading to ascending. **Spikelets** 17–30 mm, strongly flattened; **lemmas** 11–20 mm, strongly keeled, V-shaped in cross section, glabrous or scabrous, veins often raised; awns lacking or present to 4 mm. **Anthers** 2–4 mm, 0.5 mm in self-pollinating florets. **Habitat**: Disturbed open areas, roadsides, pastures. Uncommon. Native to South America. **Comments**: The strongly flattened spikelets and awnless or short-awned lemmas are distinctive. *Bromus sitchensis* has longer lemma awns. Rescue grass was introduced as a forage species and is found occasionally in W OR in lawns and disturbed places on or near farms. It may be more widespread than reported.

Bromus ciliatus L.
Fringed Brome

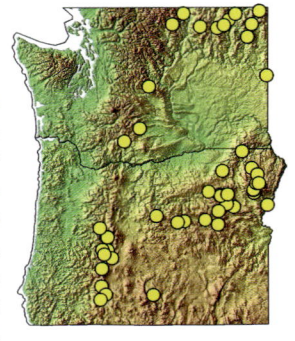

Loosely cespitose perennial; **culms** 45–120(150) cm, 2–4 mm thick. **Basal leaf sheaths** retrorsely hairy, sometimes glabrous; **upper sheaths** glabrous except hairy at the throat; **blades** 4–10 mm wide, glabrous or hairy on both surfaces; **ligules** 0.4–1.4 mm, usually glabrous. **Panicles** open, nodding, branches ascending, spreading, or drooping. **Spikelets** 15–25 mm; **glumes** glabrous, the lower 1(3)-veined, the upper 3-veined; **lemmas** 9.5–14 mm, rounded over the back, the margins hairy but the backs glabrous; awns 2–4(5) mm. **Anthers** 1.5–2 mm. **Habitat**: Moist sites: mountain meadows, stream banks, woodlands. **Comments**: Note the conspicuously hairy lemma margins. *Bromus laevipes* has similar lemmas, but it has 5-veined upper glumes, longer ligules and anthers, and it grows in drier habitats.

Bromus commutatus Schrad.
Meadow Brome, Hairy Chess [E]

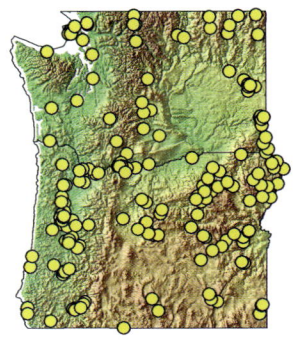

Annual; **culms** 20–90(120) cm. **Lower leaf sheaths** with dense, stiff, often retrorse hairs, **upper sheaths** pubescent or glabrous; **blades** 2–4 mm wide, hairy on both surfaces; **ligules** 1–2.5 mm, glabrous or hairy, ciliolate. **Panicles** erect or ascending, open (rarely reduced to a single spikelet), branches usually > their spikelets. **Spikelets** 10–20(30) mm, oblong-lanceolate, with 4–9 florets, **rachilla internodes** 1.5–3 mm. **Lower glumes** 5–7 mm, **upper glumes** 6–9 mm. **Lemmas** 8–11.5 mm, 1+ mm > paleas, firm, rounded, and usually glabrous over the back, veins usually not or only slightly thickened and raised, margins bluntly angled above middle, often inrolled, tips acute to obtuse, bidentate, the teeth < 1 mm; awns 3–10 mm, straight, arising < 1.5 mm below lemma tips. **Anthers** 0.7–2(3) mm. **Habitat**: Disturbed places, fields, roadsides, grain crops. Common, widespread. Native to Europe. **Comments**: Plants with short panicle branches or a single spikelet resemble *B. hordeaceus*, which has usually hairy lemmas with a papery texture and thickened, raised veins. *Bromus secalinus* lemmas have strongly inrolled, evenly rounded margins; it usually grows in moister habitats.

Bromus diandrus Roth
Ripgut Brome [E]

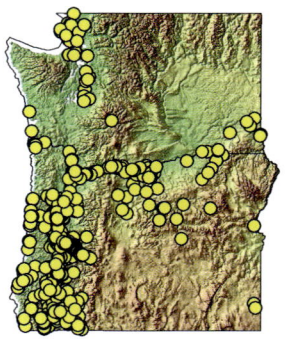

Annual; **culms** 20–90 cm. **Leaf sheaths** softly pilose, hairs retrorse or spreading; **blades** 1–9 mm wide, hairy on both surfaces; **ligules** 2–3 mm, glabrous, lacerate. **Panicles** open, usually nodding, sometimes erect, branches stiff, ascending to spreading, with 1–2 spikelets per branch. **Spikelets** 25–70 mm, often reddish at maturity; **lower glumes** 15–20 mm, 1(3)-veined; **upper glumes** 25–33 mm, 3(5)-veined; **lemmas** 20–30(35) mm, rounded on the back, scabrous, tips acuminate, bifid, with slender teeth 3–5 mm; awns 30–65 mm, stiff, straight, scabrous. **Anthers** 0.5–1 mm. **Habitat**: Disturbed open areas, degraded grasslands, roadsides. Native to Europe. **Comments**: *Bromus sterilis* is smaller in all its parts. (See *B. sterilis* for side-by-side comparison of *B. diandrus*, *B. sterilis*, and *B. tectorum*.) Ripgut brome has sharp florets with long stiff awns that penetrate animals' digestive tracts, ears, paws, noses, and gums, causing infection and even death.

Bromus hordeaceus L.
Soft Brome, Soft Chess [E]

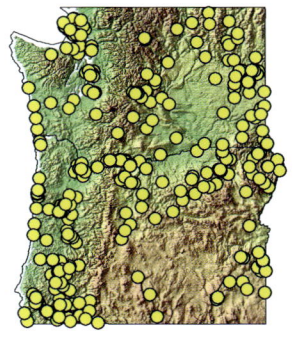

Annual; **culms** 2–70 cm. **Lower leaf sheaths** densely hairy, **upper sheaths** hairy or glabrous; **blades** 1–4 mm wide, hairy above, hairy or glabrous below; **ligules** 1–1.5 mm, hairy. **Panicles** erect, ± dense, sometimes reduced to a single spikelet, most branches < their spikelets. **Spikelets** (11)14–20(23) mm, lanceolate, with 5–10 florets. **Lower glumes** 5–7 mm, **upper glumes** 6.5–8 mm. **Lemmas** 6.5–11 mm, papery, rounded on the back, veins thickened and raised at maturity, usually hairy, margins bluntly to abruptly angled, not inrolled, tips rounded to acute, bidentate, with teeth < 1 mm; awns 6–8 mm, usually arising < 1.5 mm below lemma tips, straight to spreading at maturity. **Anthers** 0.6–1.5(2) mm. **Habitat**: Disturbed areas, roadsides, degraded grasslands. Native to S Europe and N Africa. **Comments**: One of our most common weedy annual bromes. The papery texture and raised veins of the lemmas are distinctive but subtle characters, especially when immature. *Bromus commutatus* has lemmas that are firmer in texture at maturity, but trying to distinguish immature plants may be futile. Under poor growing conditions both species may have the inflorescence reduced to a single spikelet.

Bromus inermis Leyss.
Smooth Brome [E]

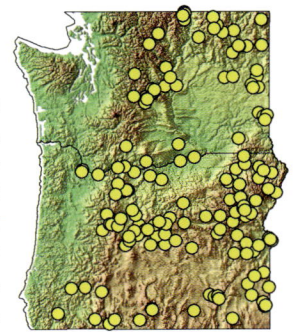

Strongly rhizomatous perennial, forming dense patches; **culms** 50–130 cm. **Leaf sheaths** glabrous or rarely hairy, **blades** 5–15 mm wide, glabrous or rarely hairy; **ligules** to 3 mm. **Panicles** erect, branches spreading. **Spikelets** 20–40 mm; **glumes** glabrous, the lower (1)3-veined, the upper 5-veined; **lemmas** rounded over the back, glabrous (rarely hairy near the base), awnless or with awns 1–2(4) mm. **Anthers** 3.5–6 mm. **Habitat**: Mesic meadows, open forest, roadsides, ditches. Native to Eurasia. **Comments**: *Bromus inermis* is easily identified from a distance by its dense stands and pink-brown panicles. Originally introduced for erosion control and forage, it is invasive in mesic habitats E of the Cascades. Once established, its thick mat of rhizomes and persistent thatch layer block the establishment of other plants. Because it is difficult to control, small unwanted patches should be eliminated before they spread.

Bromus japonicus Thunb.
Japanese Brome [E]

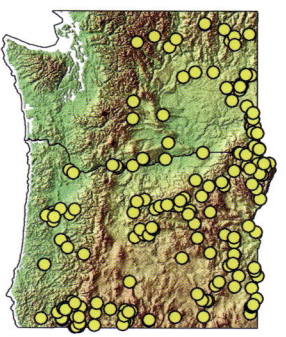

Annual; **culms** 20–70 cm. **Leaf sheaths** densely hairy, upper sheaths sometimes glabrous; **blades** 2–4 mm wide, usually hairy on both surfaces; **ligules** 1–2.2 mm, hairy. **Panicles** open, nodding, most branches > their spikelets; spikelets often > 1 per branch. **Spikelets** 20–40 mm, lanceolate, with 6–12 florets. **Lower glumes** 4.5–7 mm, **upper glumes** 5–8 mm. **Lemmas** 7–9 mm, firm, rounded on the back, glabrous, scabrous distally, veins not thickened and raised, margins hyaline, 0.3–0.6 mm wide, rounded to slightly angled above the middle, tips rounded to acute, bidentate, with teeth < 1 mm; awns 8–13 mm, arising ≥ 1.5 mm below lemma tips, strongly divergent at maturity. **Anthers** 0.7–1.5 mm. **Habitat**: Dry disturbed areas. Native to Europe and Asia. **Comments**: *Bromus japonicus* has relatively long spikelets and divergent awns that arise ≥ 1.5 mm below the lemma tips. *Bromus squarrosus* has more florets per spikelet, broader, more strongly angled lemma margins, and usually only 1 spikelet per panicle branch. See also *B. arenarius* and *B. commutatus*.

Bromus laevipes Shear
Chinook Brome

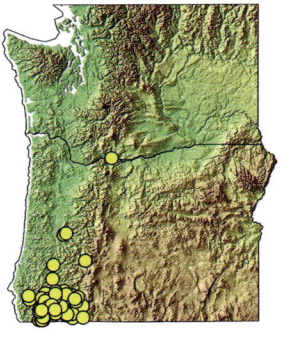

Loosely cespitose perennial; **culms** 50–150 cm. **Leaf sheaths** glabrous or lightly pubescent near the throat; **blades** 4–10 mm wide, glabrous; **ligules** 2–4.2 mm. **Panicles** nodding, the branches flexuous, ascending, spreading, or drooping. **Spikelets** 23–35 mm; **glumes** glabrous, margins often bronze-tinged, the lower (1)3-veined, the upper 5-veined; **lemmas** 12–16 mm with rounded backs sparsely hairy to scabrous, sides densely hairy; awns 4–6 mm. **Anthers** 3.5–5 mm. **Habitat**: Oak, madrone, and conifer woodlands and brushy slopes. **Comments**: *Bromus laevipes* grows in the foothills of the S Willamette, Umpqua, and Rogue Valleys, and W into the Coast Range of SW OR; it was collected more than 100 years ago in Hood River County, OR, and Klickitat County, WA. Its 5-veined upper glumes are unusual among bromes that have rounded lemma backs. See *B. ciliatus*.

Bromus orcuttianus Vasey
Orcutt's Brome

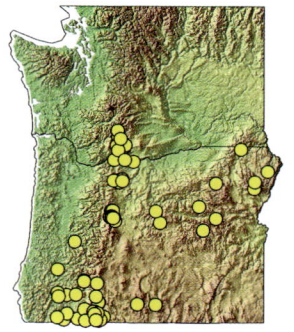

Loosely cespitose perennial; **culms** 90–150 cm, with 2–4 nodes. **Leaf sheaths** hairy (lower sheaths occasionally glabrous); **blades** 3–12 mm wide, **ligules** 1–3 mm, usually glabrous, occasionally hairy. **Panicles** erect, open, the branches ± stiff, appressed to somewhat spreading. **Spikelets** 20–40 mm; **glumes** usually glabrous, the lower 1(3)-veined, the upper 3(5)-veined; **lemmas** 9–16 mm, rounded over the back, backs and sides hairy to scabrous; awns (4)5–8 mm. **Anthers** 3–5 mm. **Habitat**: Dry montane conifer forests, especially ponderosa pine; openings, sometimes in moister sites. **Comments**: *Bromus orcuttianus* typically has an erect panicle that is more open than that of *B. suksdorfii*. *Bromus orcuttianus* has stiffer panicle branches than *B. laevipes*, which has upper glumes with 5 veins and gracefully drooping panicles. Orcutt's brome may be declining due to grazing.

Bromus pacificus Shear
Pacific Brome

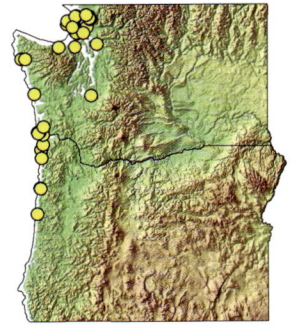

Cespitose perennial; **culms** 60–170 cm, with (5)6–8 nodes. **Leaf sheaths** hairy; **blades** 6–16 mm wide, upper surface pilose, lower surface glabrous; **ligules** 2–4 mm, glabrous. **Panicles** open, nodding, branches widely spreading or drooping. **Spikelets** 20–30 mm; **glumes** hairy, the lower 1(3)-veined, the upper 3-veined; **lemmas** 10–12 mm, backs rounded to slightly keeled, long soft-hairy especially toward the base, and with the hairs denser on sides than back; awns 3.5–7 mm. **Anthers** 2–4 mm. **Habitat**: Moist thickets, forest edges, mainly along the coast. **Comments**: *Bromus pacificus* is a tall brome of coastal thickets, with widely spreading panicle branches. It is similar to *B. sitchensis*, which has glabrous glumes and more distinctly keeled lemmas. Other similar, but usually smaller, species include *B. ciliatus, B. laevipes,* and *B. vulgaris*.

Bromus rubens L.
Red Brome, Foxtail Brome [E]

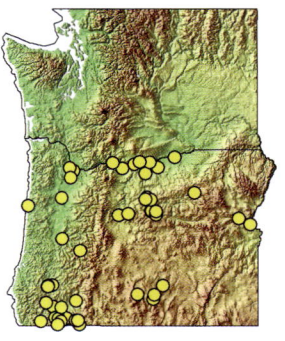

Annual; **culms** 10–40 cm. **Leaf sheaths** soft-hairy; **blades** 1–5 mm wide, hairy on both surfaces; **ligules** 1–3(4) mm, pubescent, lacerate. **Panicles** very dense, erect, about as long as wide, branches 0.1–1 cm, not readily visible, ascending, shorter than spikelets, with 1–2 spikelets per branch. **Spikelets** 18–25 mm, often reddish at maturity; **lower glumes** 5–8 mm, 1(3)-veined; **upper glumes** 8–12 mm, 3(5)-veined; **lemmas** 10–15 mm, rounded on the back, hairy, tips acuminate, bifid, with slender teeth 1–4 mm; awns 8–20 mm, straight. **Anthers** 0.5–1 mm. **Habitat**: Weedy, disturbed sites, roadsides, dry, degraded shrub steppe. Native to S Europe. **Comments**: The dense, erect, brushlike inflorescences of red brome are distinctive. It is widespread in OR but known only from extreme S WA. It is expected to turn up farther N. *Bromus madritensis*, with a somewhat looser panicle, is present in CA and likely is coming soon to the PNW.

Bromus secalinus L.
Rye Brome [E]

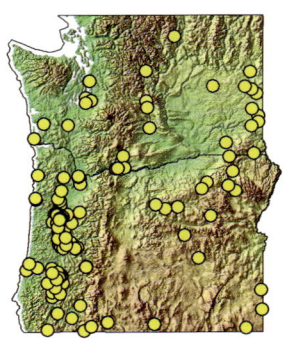

Annual; **culms** 20–80(120) cm. **Leaf sheaths** glabrous or loosely hairy, the hairs deciduous as the plant matures; **blades** 1–12 mm wide, hairy or glabrous below, hairy above; **ligules** 2–3 mm, glabrous. **Panicles** open, nodding, most branches > their spikelets. **Spikelets** 10–20 mm, ovoid-lanceolate or ovate, with 4–9(10) florets; **rachilla internodes** 1–1.6 mm, these and floret bases visible at maturity. **Lower glumes** 4–6 mm, **upper glumes** 6–7 mm. **Lemmas** 6.5–8(10) mm, firm, rounded on the back, usually glabrous, veins not thickened and raised, margins evenly rounded, inrolled at maturity, tips acute to obtuse, bidentate, with teeth < 1 mm; awns (0)3–6(9.5) mm, arising < 1.5 mm below lemma tips, straight or flexuous. Mature seeds thick, inrolled, U-shaped in cross section. **Anthers** 1–2 mm. **Habitat**: Moist meadows, ditches, disturbed areas. Native to Europe. **Comments**: Note the exposed floret bases in mature spikelets and the strongly inrolled lemmas enfolding the thick, U-shaped seeds. Rye brome is our most common weedy annual brome in moist disturbed habitats.

Bromus sitchensis Trin. var. *aleutensis* (Trin. ex Griseb.) Hultén
Aleutian Brome

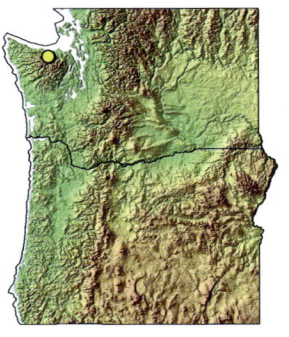

Loosely cespitose perennial; **culms** 40–130 cm, 3–7 mm thick. **Leaf sheaths** sparsely to moderately densely hairy; **blades** 6–15 mm wide, usually hairy on both surfaces, sometimes glabrous, veins mostly < one-third as broad as the area between them; **ligules** 3.5–5 mm, usually glabrous. **Panicles** erect, open to somewhat contracted, lower branches ≤ 10 cm, 1–2 per node, stiffly ascending; with 1–3 spikelets borne on the distal half. **Spikelets** 25–40 mm, strongly laterally compressed; **glumes** glabrous or hairy, the lower 3–5-veined, the upper 7(9)-veined; **lemmas** 12–17 mm, strongly keeled, at least distally, usually hairy; awns (3)5–10 mm. **Anthers** 2.2–4.2 mm. **Habitat**: Meadows, stream banks, roadsides. **Comments**: Aleutian brome ranges from the Olympic Peninsula along the coast to SE AK with a few inland records in Canada and N ID. Similar var. *sitchensis* has longer panicle branches and ligules. Also compare with *B. sitchensis* var. *marginatus*.

Bromus sitchensis Trin. var. *carinatus* (Hook. & Arn.) R.E. Brainerd & Otting
California Brome

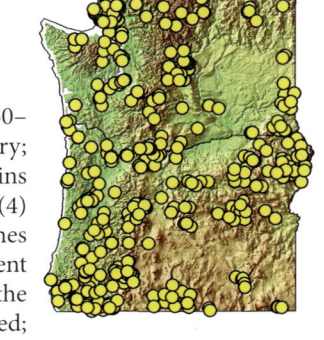

Loosely cespitose biennial or short-lived perennial; **culms** 50–100 cm, ≤ 3 mm thick. **Leaf sheaths** glabrous or retrorsely hairy; **blades** 3–5 mm wide, usually hairy on both surfaces, veins mostly ≥ half as broad as the area between them; **ligules** 1–3(4) mm, membranous. **Panicles** erect, open, lax, lower branches usually < 10 cm, 2–4 per node, ascending to strongly divergent or reflexed, with 1–4 spikelets sometimes borne to near the base. **Spikelets** 20–40(53) mm, strongly laterally compressed; **glumes** glabrous or hairy, the lower 3(5)-veined, the upper 5(7)-veined; **lemmas** 12–16 mm, strongly keeled at least distally, usually short-hairy; awns 8–17 mm. **Anthers** 1–5 mm. **Habitat**: Grasslands, oak woodlands, forest openings. **Comments**: Similar to *B. sitchensis* var. *marginatus,* which has shorter awns and a more contracted panicle. California brome is more common W of the Cascades and tends to occur at lower elevations. It intergrades with Sitka brome in W OR and WA, where some plants with intermediate form can be difficult to assign to one variety or the other.

Bromus sitchensis Trin. var. *marginatus* (Nees ex Steud.) B. Boivin
Mountain Brome

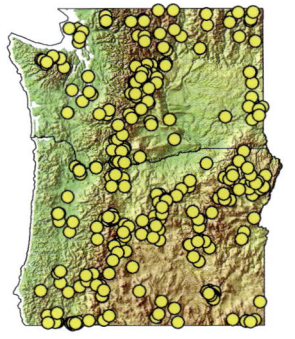

Loosely cespitose perennial; **culms** 45–120(180) cm, ≤ 3 mm thick. **Leaf sheaths** sparsely retrorsely hairy, sometimes glabrous; **blades** 1–12 mm wide, glabrous or hairy, veins mostly ≥ half as broad as the area between them; **ligules** 1–3.5 mm. **Panicles** erect, usually somewhat contracted, lower branches usually < 10 cm, 1–4 per node, erect or ascending, with 1–4 spikelets sometimes borne to near the base. **Spikelets** 20–40(53) mm, strongly laterally compressed; **glumes** glabrous or hairy, the lower 3–7(9)-veined, the upper 5–9(11)-veined; **lemmas** 10–14(17) mm, strongly keeled at least distally, usually hairy; awns 4–7 mm. **Anthers** 1–6 mm. **Habitat**: Grasslands, shrublands, and open forest; mostly montane. **Comments**: *Bromus sitchensis* var. *marginatus* has shorter awns and more-erect panicle branches than *B. s.* var. *carinatus*. Mountain brome occurs mostly in montane areas in and E of the Cascades.

Bromus sitchensis var. *maritimus*
(Piper) Otting & R.E. Brainerd
Maritime Brome

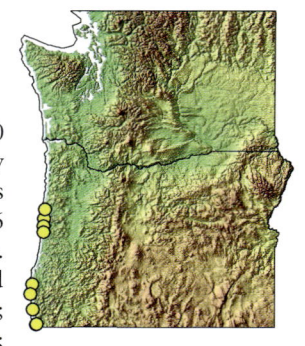

Loosely cespitose to rhizomatous perennial; **culms** 20–70 cm, ≤ 3 mm thick. **Leaf sheaths** glabrous, sometimes slightly short-hairy distally; **blades** 6–8 mm wide, glabrous, veins mostly ≥ half as broad as the area between them; **ligules** 1–6 mm. **Panicles** erect, dense, lower branches < 10 cm, erect. **Spikelets** 20–40 mm, strongly laterally compressed, crowded and overlapping, usually longer than branches and pedicels; **glumes** hairy, the lower (3)5(7)-veined, the upper 7(9)-veined; **lemmas** 12–14(17) mm, strongly keeled at least distally, hairy; awns (2)4–7 mm. **Anthers** 2–4 mm. **Habitat**: Coastal sands and bluffs from OR to S CA. **Comments**: *Bromus sitchensis* var. *maritimus* is distinguished by very dense panicles and coastal habitat. Wind and salt spray in coastal environments may induce *B. sitchensis* var. *sitchensis* and *B. s.* var. *carinatus* to have abnormally dense inflorescences, superficially resembling *B. s.* var. *maritimus*. Maritime brome is becoming rare due to habitat loss from dune stabilization and coastal development.

Bromus sitchensis Trin. var. *sitchensis*
Sitka Brome

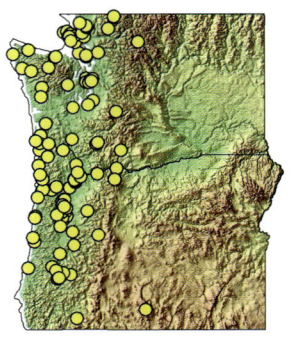

Loosely cespitose perennial; **culms** 120–180 cm, 3–7 mm thick. **Leaf sheaths** glabrous or sparsely hairy; **blades** 5–20 mm wide, hairy above or on both surfaces, veins mostly < one-third as broad as the area between them; **ligules** 3–8 mm. **Panicles** often drooping, open, lower branches to 20 cm, spreading to drooping, 2–4(6) per node, with 1–3 spikelets borne on the outer half. **Spikelets** 18–38 mm, strongly laterally compressed; **glumes** glabrous, the lower 3–5-veined, the upper 5–7-veined; **lemmas** 12–15 mm, strongly keeled at least distally, usually glabrous, sometimes short-hairy; awns 5–10 mm. **Anthers** (3)4–6 mm. **Habitat**: Moist sites, forest edges, riparian corridors, oak woodlands, disturbed areas. **Comments**: In our area Sitka brome occurs mainly in W OR and WA. Intermediates between Sitka brome and California brome are commonly encountered where the 2 varieties occur together. Recent work has joined a complex of 6 intergrading taxa as varieties of *B. sitchensis*, a single, variable species. Five of the varieties occur in OR and WA.

Bromus squarrosus L.
Squarrose Brome, Corn Brome [E]

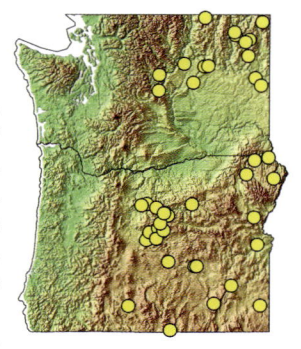

Annual; **culms** 20–60 cm. **Lower leaf sheaths** densely hairy, **upper sheaths** glabrous; **blades** 4–6 mm wide, densely hairy on both surfaces; **ligules** 1–1.5 mm, hairy. **Panicles** open, nodding, 1-sided, most branches > their spikelets; usually with only 1(2) spikelet per branch. **Spikelets** 15–70 mm, lanceolate to elliptic, with 8–30 florets, **rachilla internodes** 1.2–1.4 mm. **Lower glumes** 4.5–7 mm, **upper glumes** 6–8 mm. **Lemmas** 8–11 mm, papery, rounded on the back, glabrous to short-hairy, veins not thickened and raised, margins hyaline, 0.6–0.9 mm wide, strongly angled above the middle, tips rounded to acute, bidentate, with teeth < 1 mm; awns 8–10 mm, arising ≥ 1.5 mm below lemma tips, straight when young, divaricate at maturity. **Anthers** 0.7–1.5 mm. **Habitat**: Dry disturbed areas, roadsides. Native to central Russia and S Europe. **Comments**: Squarrose brome usually has 1 long drooping spikelet (sometimes 2) on each panicle branch, broad, abruptly angled lemma margins, and awns that arise ≥ 1.5 mm below the lemma tips. See similar *B. japonicus*.

Bromus sterilis L.
Poverty, Barren, or Sterile Brome [E]

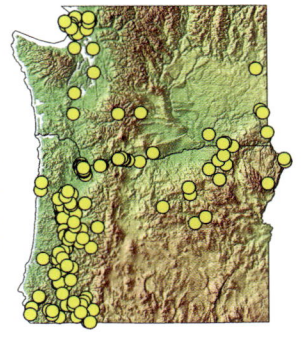

Annual; **culms** 35–100 cm. **Leaf sheaths** densely hairy; **blades** 1–6 mm wide, hairy on both surfaces; **ligules** 2–2.5 mm, glabrous, lacerate. **Panicles** open, drooping, branches spreading, ascending, or drooping, rarely with > 3 spikelets. **Spikelets** 20–35 mm, reddish to dark purple at maturity; **lower glumes** 8–10 mm, 1(3)-veined; **upper glumes** 12–15 mm, 3(5)-veined; **lemmas** 14–20 mm, rounded on the back, pubescent, tips acuminate, bifid, with slender teeth 1–3 mm; awns 15–30 mm, straight, scabrous. **Anthers** 1–1.4 mm.
Habitat: Dry disturbed areas, degraded grasslands, roadsides. Native to Europe.
Comments: Invasive in upland grasslands and savannas. Widespread in our area. Intermediate between *B. diandrus* and *B. tectorum* in the size of its glumes, lemmas, and awns. (See photo below for side-by-side comparison of panicles and spikelets.)

Bromus suksdorfii Vasey
Suksdorf's Brome

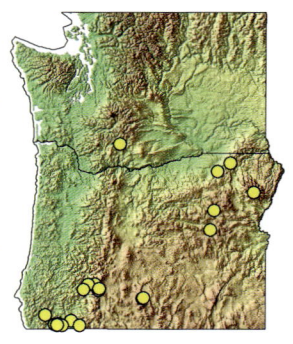

Loosely cespitose perennial; **culms** 50–100 cm. **Leaf sheaths** glabrous; **blades** 4–8(14) mm wide; **ligules** to 1 mm, glabrous. **Panicles** erect, contracted, the branches erect or ascending. **Spikelets** 15–30 mm; **glumes** glabrous or sparsely hairy, the lower 1(3)-veined, the upper 3-veined. **Lemmas** 12–15 mm, rounded over the back, backs and sides hairy to nearly glabrous; awns 2–5 mm. **Anthers** 2–3.5 mm. **Habitat**: Open montane slopes and subalpine forests, sometimes near springs or seeps. **Comments**: *Bromus suksdorfii* has a dense panicle with erect branches. *Bromus orcuttianus* and *B. laevipes* have more-open panicles; *B. inermis* is rhizomatous and has shorter awns. *B. sitchensis* var. *marginatus* also has contracted panicles, but its lemmas are distinctly keeled.

Bromus tectorum L.
Cheatgrass, Downy Brome [E]

Annual; **culms** 5–90 cm. **Leaf sheaths** densely soft-hairy; **blades** 1–6 mm wide, soft-hairy on both surfaces; **ligules** 2–3 mm, glabrous, lacerate. **Panicles** open, nodding to drooping, branches drooping, 1-sided, > spikelets, usually at least 1 branch with 4–8 spikelets. **Spikelets** 10–20 mm; lower glumes 4–9 mm, 1-veined; **upper glumes** 7–13 mm, 3(5)-veined; **lemmas** 9–12 mm, rounded on the back, glabrous or pubescent, tips acuminate, bifid, with slender teeth 0.8–2(3) mm; awns 10–18 mm, straight. **Anthers** 0.5–1 mm. **Habitat**: Overgrazed grasslands and shrub steppe, disturbed areas, roadsides. Native to Europe. **Comments**: Cheatgrass occurs in every state in the United States, but it is most abundant in the Great Basin. It outcompetes native bunchgrasses and forbs and shortens the fire return interval, preventing reestablishment of native species. Among the labels applied to it are weed, noxious weed, regulated weed, prohibited plant, invasive plant, and agricultural, nursery, and orchard pest. Or as we like to say, a bad weed.

Bromus vulgaris (Hook.) Shear
Columbia Brome

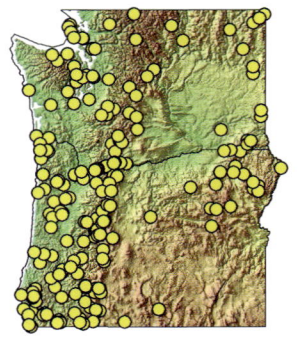

Loosely cespitose perennial; **culms** 60–120 cm. **Leaf sheaths** glabrous or with spreading hairs; **blades** 5–14 mm wide; **ligules** 2–6 mm, glabrous. **Panicles** nodding, the branches ascending to appressed; **glumes** glabrous or hairy, lower 1(3)-veined, upper 3-veined; **lemmas** 8–15 mm, rounded over the back, backs and sides hairy or glabrous; awns (4)6–12 mm. **Anthers** 2–4 mm. **Habitat**: Lowlands to subalpine mesic forests and edges, usually in shade or partial shade. **Comments**: This common and widespread forest grass has gracefully arching culms and drooping panicles. Invasive *Brachypodium sylvaticum* (falsebrome) has similar nodding inflorescences, but each spikelet is attached directly to the inflorescence axis by a short, unbranched stalk. Falsebrome also has open sheaths and more yellowish-green foliage. See also *Bromus ciliatus, B. laevipes, B. orcuttianus, B. pacificus,* and *B. suksdorfii.*

Calamagrostis arenaria (L.) Roth ssp. *arenaria*
European Beachgrass [E] (yellow dots)

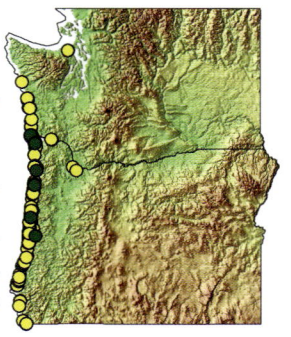

Strongly rhizomatous perennial with clustered shoots; **culms** 50–130 cm. **Leaf sheaths** open; **blades** 0.5–2.5 mm wide, inrolled, 6 mm wide when flat; **ligules** 10–35 mm, acute, bifid or lacerate. **Panicles** dense, spike-like. **Spikelets** 12–14 mm, with 1 floret. **Glumes** 10–14 mm; **callus hairs** 2–6 mm; **lemmas** 8–12 mm, awns lacking or to 0.5 mm. **Anthers** 4–7 mm. **Habitat**: Coastal sand dunes, beaches. Native to Europe and N Africa. **Comments**: Introduced from Europe to stabilize dunes, European beachgrass is abundant along the PNW coast. It prevents formation of moving dunes and associated deflation plains, important habitats for many native plant and wildlife species. The dense, spike-like inflorescences resemble those of native *Leymus mollis*, which has wider leaves and inflorescences that are true spikes. In the PNW a hybrid of *C. arenaria* and *C. breviligulata* appears to be more aggressive than either of its parent species.

Calamagrostis breviligulata (Fernald) Saarela ssp. *breviligulata*
American Beachgrass [E] (green dots)

Similar to European beachgrass but has short, truncate ligules 1–3(4.6) mm. It tends to have a more open growth form. Native to E North America.

Calamagrostis breweri Thurb.
Brewer's Reedgrass

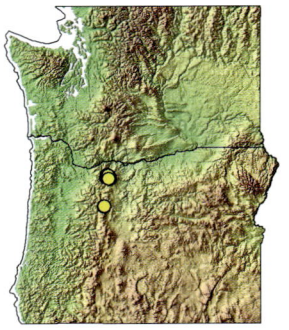

Cespitose perennial, but sometimes with rhizomes to 5 cm; **culms** (12)15–30(55) cm. **Leaves** mostly basal, **blades** 0.9–1.7 mm wide when flat, 0.4–0.6 mm when rolled; **ligules** 1.7–4 mm. **Panicles** open, usually dark purple, spikelets borne near ends of branches. **Glumes** 3–5 mm, longer than florets. **Callus hairs** 0.3–1.2 mm, 20–50% as long as lemmas, sparse. **Lemmas** 2.5–4 mm; awns 3.5–5 mm; attached to lower third of lemma, exserted > 1.5 mm beyond glumes. **Anthers** 1.3–2.6 mm. **Habitat**: Moist subalpine to alpine meadows, lake margins, stream banks. **Comments**: Brewer's reedgrass is a diminutive high-mountain bunchgrass, similar in appearance to an *Agrostis*. Similar-looking *Podagrostis* species are awnless or have shorter awns not exserted from the glumes. Brewer's reedgrass is a rare species that occurs in several disjunct populations ranging from Mt. Jefferson to Mt. Hood in OR, and a few high peaks in the Sierra Nevada and the Klamath region in CA.

Calamagrostis canadensis (Michx.) P. Beauv.
Bluejoint Reedgrass

Strongly rhizomatous perennial, often glaucous; **culms** 65–112 cm with obvious dark nodes. **Leaf blades** 2–8(11) mm wide. **Panicles** ± open and nodding at maturity, purplish. **Glumes** longer than florets. **Callus hairs** (1.5)2–3.5(4.5) mm, 50–150% as long as lemmas, abundant. **Lemmas** 2–3.1(4) mm, awns 0.9–3.1 mm, arising from lower 20–67% of lemma, usually not exserted, delicate and often difficult to distinguish from callus hairs. **Anthers** (0.8)1.2–1.6(2.6) mm. **Habitat**: Stream and lake margins, wet meadows, forest openings. **Comments**: Bluejoint reedgrass is our most common and widespread reedgrass. It may have been more abundant along rivers prior to introduction of reed canarygrass and velvetgrass. Our 2 varieties intergrade morphologically and overlap geographically.

Calamagrostis canadensis var. *canadensis*

Glumes 2.5–3.5(4) mm, acute (rarely acuminate), smooth or scabrous, often scabrous only on keels, prickles on keels straight.

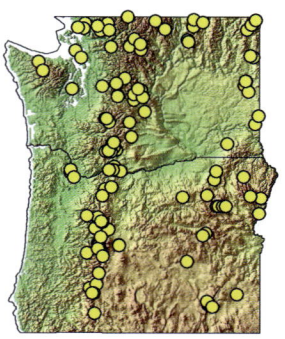

Calamagrostis canadensis var. *langsdorffii* (Link) Inman

Glumes (3.5)4–4.5(5.2) mm, acuminate, scabrous over entire surface, prickles on keels short and hooked.

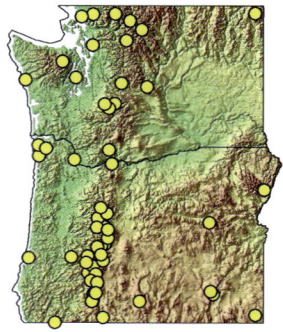

Calamagrostis howellii Vasey
Howell's Reedgrass

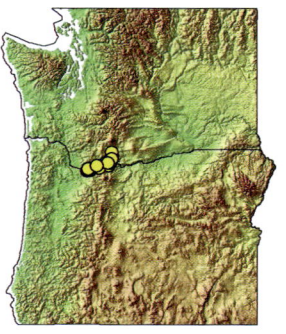

Densely cespitose perennial; often hanging down from steep, rocky slopes; **culms** (25)35–45(60) cm. **Leaf blades** 1–2.5(3) mm wide, flat to involute; **ligules** (2.5)3.5–6 mm. **Panicle** open, with spikelets borne on distal half of widely spreading branches. **Glumes** (5.5)6–8 mm, longer than florets. **Callus hairs** 2–3(4.5) mm, 40–60(67)% as long as lemmas. **Lemmas** 4.5–5 mm; awns (10)13–16 mm, arising from lower 20–40% of lemma, exserted 7–11 mm beyond glumes, strongly bent. **Anthers** (2)2.5–3(4) mm. **Habitat**: Cliffs, bluffs, dry rocky slopes. **Comments**: *Calamagrostis howellii* is a densely cespitose grass with long awns, long tapered glumes, and open panicles. It is a rare endemic that grows on both sides of the Columbia River Gorge. *Agrostis howellii* is another Columbia Gorge endemic with open panicles and relatively long awns, but it has shorter glumes, awns, and callus hairs than *C. howellii*. See also *Agrostis swalalahos*.

Calamagrostis koelerioides Vasey
Pine Reedgrass, Fire Reedgrass

Densely cespitose perennial, rhizomes 2–6 cm; **culms** (26)60–85(120) cm. **Leaf blades** 2.5–4.5 mm wide, flat, upper surface glabrous or sparsely hairy; **collars** glabrous, not thickened, veins evident; **ligules** 2–4.5 mm. **Panicles** dense, narrow, spike-like, branches spikelet-bearing to the base. **Glumes** 4–6(7) mm, longer than florets. **Callus hairs** 1.5–2 mm, 33–40% as long as lemmas, sparse. **Lemmas** 3.5–5(6) mm; awns 4–5.5 mm, arising from lower 10–20% of lemma, exserted from glumes < 1.5 mm or not exserted, stout, bent, easily distinguished from callus hairs. **Anthers** 2–3.5 mm. **Habitat**: Montane meadows, open forests, serpentine grasslands. **Comments**: The dense panicles of *C. koelerioides* resemble those of *Koeleria macrantha*, which has 2 florets per spikelet, and *Agrostis exarata*, which has shorter paleas and shorter callus hairs. *Calamagrostis koelerioides* hybridizes with coastal *C. nutkaensis*, which has wider leaves, longer callus hairs, and shorter awns and anthers.

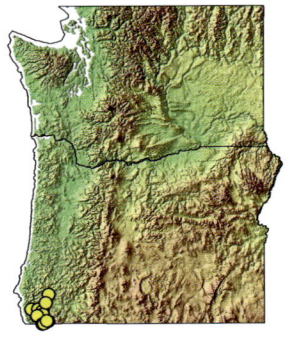

Calamagrostis nutkaensis (J. Presl) Steud.
Pacific Reedgrass

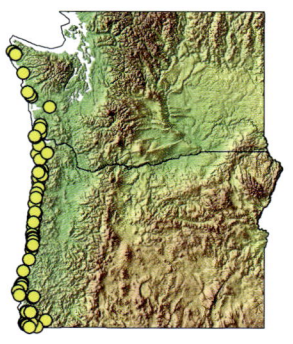

Densely cespitose perennial, rhizomes < 3 cm; **culms** (42)55–105(150) cm. **Leaf blades** (2)4–10(20) mm wide, flat, tough; **collars** thickened, veins obscure or lacking; **ligules** 1–4 mm. **Panicles** contracted to somewhat loose, erect to slightly nodding; spikelets borne on outer half of branches. **Glumes** 4–6.5(8) mm, longer than florets. **Callus hairs** (1)2–3 mm, (20)50–67% as long as lemmas, sparse. **Lemmas** (3)4–4.5(5) mm; awns 1–3 mm, attached on lower 33–50% of lemma, not exserted from glumes, straight or slightly bent. **Anthers** (1)2.4–2.6(3.3) mm. **Habitat**: Coastal wetlands, beaches, dunes, bluffs; also, inland bogs and streams at higher elevations in SW OR. **Comments**: *Calamagrostis nutkaensis* forms large, dense clumps of tough, thick leaves with usually purplish panicles. It intergrades and hybridizes with *C. koelerioides* in SW OR where some plants cannot be identified clearly to either species.

Calamagrostis purpurascens R. Br.
Purple Reedgrass

Densely cespitose perennial, rhizomes 1–4 cm; **culms** (10)30–80 cm. **Leaf blades** 2–5 mm wide, flat or involute, densely short-hairy on upper surface; **ligules** (1.5)2–4(9) mm. **Panicles** contracted, often interrupted near base, red- or purple-tinged; branches spikelet-bearing to base. **Glumes** (4.5)5.5–6.5(8) mm, longer than florets. **Callus hairs** 0.9–1.5(2.4) mm, 20–40(67)% as long as lemmas, sparse. **Lemmas** (3.5)4–4.5(5) mm; awns (4.5)6–7(9) mm, attached to lower 33% of lemma, bent, usually exserted from glumes > 1.5 mm, sometimes less or not exserted, easily distinguished from callus hairs. **Anthers** (1.3)1.7–2.9 mm. **Habitat**: Rocky alpine ridges and summits, well- to moderately drained soils. **Comments**: Densely short-hairy upper leaf surfaces separate purple reedgrass from all other PNW *Calamagrostis*. This is a subtle character that requires close scrutiny of the leaf surface under a hand lens or microscope. Frequently reported in error, *C. purpurascens* is often confused with *C. tacomensis*, which has glabrous or sparsely hairy upper leaf surfaces.

Calamagrostis rubescens Buckley
Pinegrass

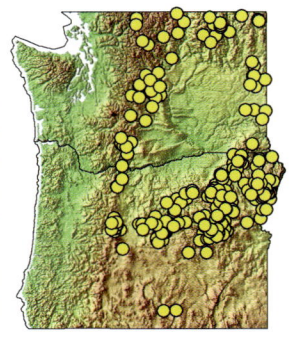

Rhizomatous perennial, sometimes loosely tufted, rhizomes 15+ cm; **culms** 60–100 cm, bases reddish. **Leaf blades** (1)2–5(8) mm wide, flat; **collars** usually pubescent; **ligules** 3–5 mm. **Panicles** contracted to somewhat open, branches spikelet-bearing to base. **Glumes** (3)4–4.5(5.5) mm, longer than florets. **Callus hairs** 0.5–1.5(2.5) mm, 20–50(60)% as long as lemmas, sparse. **Lemmas** 2.5–4 mm; awns 2.8–3.5(4.5) mm, attached to lower 20% of lemma, strongly bent, exserted < 1.5 mm beyond glumes. **Anthers** 1.2–2.6 mm. **Habitat**: Dry montane conifer forests, aspen groves. **Comments**: Pinegrass can be identified by the curly, spreading hairs on its leaf collars. Some collars may be glabrous—check several. In SW OR see similar *C. koelerioides*, which does not have hairy collars. Hairy collars occasionally occur on *C. canadensis, C. purpurascens,* and *C. stricta* ssp. *inexpansa*, but these species grow in different habitats. Pinegrass is often a community dominant with *Carex geyeri* in dry, montane conifer forests E of the Cascades. Established shaded populations seldom flower, but low-intensity fires can stimulate vigorous flowering.

Calamagrostis stricta (Timm) Koeler
Slimstem Reedgrass

Cespitose perennial with rhizomes < 5 cm; **culms** (10)35–90(120) cm. **Leaf blades** (1)1.5–5(6) mm wide, flat or involute; **ligules** (0.5)1–5.5(10) mm. **Panicles** dense, spike-like, sometimes interrupted; branches usually spikelet-bearing to base. **Glumes** 2–4(5) mm, longer than florets. **Callus hairs** (1)1.5–3(4.5) mm, (50)70–90(130)% as long as lemmas, abundant. **Lemmas** 2–4(5) mm; awns 1.5–2.5 mm, usually attached at or below middle, rarely above middle, not exserted or exserted slightly beyond glumes, straight or bent, stout or slender, usually distinguishable from callus hairs. **Anthers** 0.9–2.4 mm. **Comments**: *Calamagrostis stricta* is a cespitose grass with dense spike-like panicles, abundant callus hairs, and short awns that are often hidden by the glumes. *Calamagrostis canadensis* is similar but has looser, wider panicles and more delicate awns. There are 2 subspecies of *C. stricta,* both growing in the PNW.

Calamagrostis stricta ssp. *inexpansa* (A. Gray) C.W. Greene
Northern Reedgrass

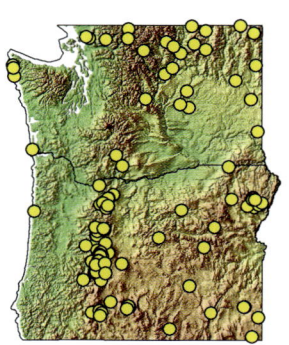

Plants more robust; culms usually scabrous, sometimes smooth. **Leaf blades** (1.5)2–5(6) mm wide, **ligule** of upper culm leaf mostly 4–10 mm, often lacerate. **Panicle branches** 1.5–9.5 cm. **Glumes** 3–4(5) mm. **Rachillas** prolonged 1–1.5 mm. **Callus hairs** (2)2.5–3(4.5) mm. **Habitat**: Montane and coastal moist meadows, sphagnum bogs, riparian grasslands. **Comments**: Northern reedgrass is larger in many of its parts than slimstem reedgrass. It is thought to set seed asexually and its anthers are usually poorly formed, sterile, and indehiscent. It occurs both inland and at the coast.

Calamagrostis stricta ssp. *stricta*
Slimstem Reedgrass

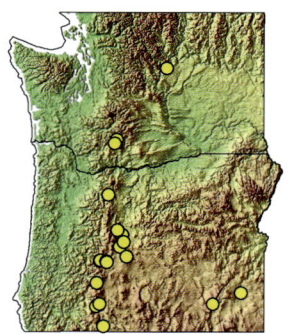

Plants more slender; **culms** smooth, sometimes slightly scabrous. **Leaf blades** (1)1.5–3 mm wide; **ligules** of upper culm leaves 2–4 mm, usually not lacerate. **Panicle branches** 1.4–4 cm. **Glumes** 2–2.5(3) mm. **Rachillas** prolonged 0.5–1 mm. **Callus hairs** (1)1.5–2(3) mm. **Habitat**: Moist meadows and fens, less often in marshes and bogs; not coastal. **Comments**: Slimstem reedgrass is uncommon in the PNW, occurring mainly in and E of the Cascades. It is thought to reproduce sexually, and its anthers are well-formed, fertile, and dehiscent.

SPECIES ACCOUNTS 177

Calamagrostis tacomensis K. Marr & Hebda
Rainier Reedgrass

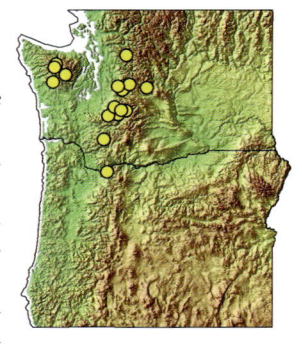

Densely cespitose perennial; **culms** (20)30–55(95) cm; foliage green. **Leaf blades** (1)2–2.5(4) mm, flat; **ligules** (3)3.5–5.5(6) mm. **Panicles** narrow, loosely contracted, branches 2–6 cm, spikelet-bearing on the outer two-thirds, sometimes to base. **Glumes** (4)6–6.5(7) mm, often green with purple patch at base, longer than florets. **Callus hairs** 30–60% as long as lemmas, abundant. **Lemmas** (3.5)4–5(5.5) mm, awns (5.5)7–8.5(10) mm, attached on lower 10–33% of lemma, exserted from glumes > 2 mm, strongly bent, easily distinguished from callus hairs. **Anthers** (1)2–3(3.5) mm. **Habitat**: Montane to alpine, dry to wet meadows, seeps, cliffs. **Comments**: This rare reedgrass is limited to the Olympic Mts. and Cascade Range in WA and N OR. Plants called *C. tacomensis* from Steens Moutain in SE OR have recently been described as a separate species, *C. utsutsuensis*, which has shorter glumes, lemmas, and awns. *Calamagrostis purpurascens* has densely short-hairy upper leaf surfaces and shorter awns.

Calamagrostis tweedyi (Scribn.) Scribn.
Cascade Reedgrass or Tweedy's Reedgrass

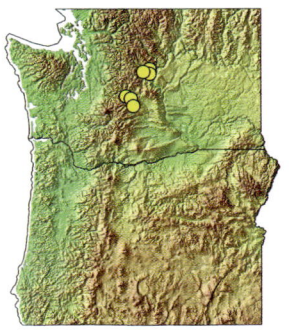

Short-rhizomatous perennial, the rhizomes 1–10 cm; **culms** (47)60–120(150) cm. **Leaf blades** (2)3–13 cm wide, flat; **ligules** (1)3.5–6(8) mm. **Panicles** dense, narrow, spike-like, pale to dark purple; branches spikelet-bearing to the base. **Glumes** (4.5)5.5–8(9) mm, longer than florets. **Callus hairs** 0.8–1 mm, 20–33% as long as lemmas, sparse. **Lemmas** 4–6.5(7.5) mm, awns 6–11 mm, attached on the lower 20–30% of lemma, exserted more than 2 mm beyond glumes, stout, bent, easily distinguished from callus hairs. **Anthers** 2–3.5 mm. **Habitat**: Montane to subalpine moist meadows and conifer forests. **Comments**: Broad leaves and long awns characterize *C. tweedyi*. Similar *C. howellii* has longer awns and is endemic to the Columbia Gorge. Tweedy's reedgrass is a rare plant with scattered populations in mountains of central WA E of the Cascade Crest, and along the ID/MT border.

Calamagrostis utsutsuensis Otting and Wilson
Steens Mountain Reedgrass

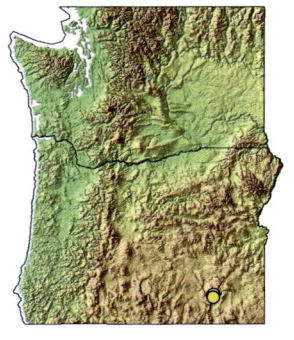

Densely cespitose perennial; **culms** (20)30–55(95) cm, foliage glaucous, obviously blue-green. **Leaf blades** flat, 1–4 mm wide; **ligules** 3–6 mm. **Panicles** loosely contracted; branches 1.5–6 cm, spikelet-bearing to base. **Glumes** 3.3–5.8 mm, green, purple, or brown, longer than florets. **Callus hairs** about 50% as long as lemmas, abundant. **Lemmas** 2.9–4 mm, awns 3.4–5.5 mm, attached on lower (0)10–33% of lemma, exserted 0.5–2.3 mm beyond glumes, strongly bent, easily distinguished from callus hairs. **Anthers** (1)2–3.5 mm. **Habitat**: Rocky stream banks on Steens Mt. from near ridgetops to mouths of gorges. **Comments**: *Calamagrostis utsutsuensis* is a glaucous, cespitose grass known only from Steens Mt. in SE OR. Its glaucous foliage distinguishes it from *C. koelerioides*, which occurs in SW OR. See similar *C. tacomensis*. The species epithet honors the people of the Burns Paiute Tribe, whose name for Steens Mountain is Utsutsu.

Calamovilfa longifolia (Hook.) Scribn. var. *longifolia*
Prairie Sandreed [E]

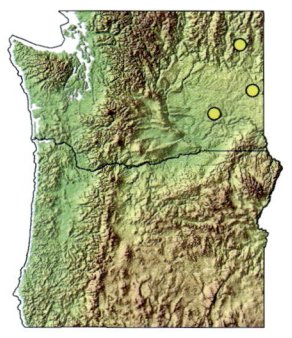

Strongly rhizomatous perennial; rhizomes covered with hard shiny scales; **culms** to 2.4 m. **Leaf sheaths** open with 2–3 mm hairs at the throat; **blades** about 12 mm wide, often rolled lengthwise; **ligules** 0.7–2.5 mm, a fringe of hairs. **Panicles** ± open, branches ascending to erect. **Spikelets** with 1 floret; disarticulation above glumes; **lower glumes** 3.5–6.5 mm, **upper glumes** 5–8.2 mm; **calluses** hairy, the hairs about half as long as lemmas, abundant; **lemmas** 4.5–7 mm, smooth, shiny, awnless. **Anthers** 3.5–4.5 mm. **Habitat**: Dry prairies, roadsides, disturbed areas, usually in sandy soils. Native to the Great Plains. **Comments**: This warm-season (C4) grass persists in NE WA where it was planted for erosion control. The abundant callus hairs of *Calamovilfa* are reminiscent of *Calamagrostis*, which has membranous ligules, equal glumes, and awned lemmas and grows in moister habitats. This genus is sometimes merged into *Sporobolus*.

Catabrosa aquatica (L.) P. Beauv.
Brookgrass, Water Whorlgrass

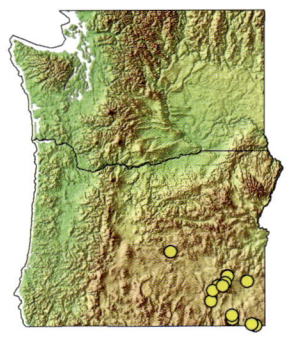

Perennial, often stoloniferous; **culms** 10–60 cm. **Leaf sheaths** closed about 50%; **blades** 2–13 mm wide, flat; **ligules** 1–8 mm, membranous. **Panicles** open, with whorls of 3+ unequal branches at each node. **Spikelets** usually with 2 florets, lower floret sessile, upper floret stalked, falling before lower floret; disarticulation above glumes. **Glumes** < lemmas, lower glumes 0.7–1.3 mm, 1-veined; upper glumes 1.2–2.2 mm. **Lemmas** 2–3 mm, with 3 conspicuous parallel veins, rounded to truncate, tips erose, awnless; **palea** subequal to lemma. **Anthers** 2–3 mm. **Habitat**: Wet meadows, stream and pond shores, often in shallow flowing water. **Comments**: Similar-looking *Glyceria* and *Torreyochloa* have glumes with 3 veins and usually more florets per spikelet. Brookgrass is widespread in the Intermountain West, but rare in the PNW with a handful of records of SE OR. It colonizes newly deposited sediments in streams and is used as an indicator for showing stabilization of degraded riparian zones.

Cenchrus longispinus (Hack.) Fernald
Longspine Sandbur, Mat Sandbur [E]

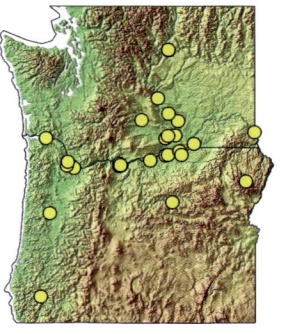

Sprawling, many-branched annual; **culms** 20–90 cm, often decumbent. **Leaf sheaths** open, compressed, keeled; **blades** 1.5–5(7.5) mm wide; **ligules** 0.6–1.8 mm, ciliate, the hairs 2–4× as long as membranous base. **Panicles** narrow, spike-like, composed of 5–10 spiny burs, each 8.3–12 mm × 3.5–6 mm, short-hairy, with 45–75 sharp spines, containing 2–3(4) spikelets. **Spikelets** with 2 florets; lower floret staminate with **anthers** 1.5–2 mm; upper floret bisexual with poorly developed **anthers** 0.7–1 mm. **Habitat**: Sandy soil, disturbed areas, roadsides; mostly Columbia and Snake River basins in PNW; native from central to E North America. **Comments**: Sandbur is a grass you won't forget once you meet it. Spiny burs stick in clothing or feet, and when you try to remove them, they stick painfully in your fingers. The spines are modified, basally fused, sterile panicle branches. Reports of *C. spinifex* in the PNW are misidentified *C. longispinus*. Sandbur flowers late in the growing season. Some previous treatments have called it native in the PNW, but a lack of early collections suggests otherwise. It is on noxious weed lists in WA and CA; ironically, it is listed as threatened in ME and NH.

Chloris verticillata Nutt.
Tumble Windmill Grass [E]

Cespitose perennial; **culms** 14–40 cm. **Leaf sheaths** open, strongly keeled, long-hairy at top; **blades** 2–3 mm wide, hairy at base; **ligules** 0.7–1.3 mm, membranous, ciliate. **Panicles** with 10–16 spike-like branches in several whorls, with a single vertical branch at tip; panicle breaks off intact at the uppermost culm node. **Spikelets** laterally compressed with 2 florets, lower one sessile and fertile, upper one sterile, on short stalk. **Lower glumes** 2–3 mm; **upper glumes** 2.8–3.5 mm. **Lower lemmas** 2–3.5 mm, with awns 4.8–9 mm; **upper lemmas** 1.1–2.3 mm, with awns 3.2–7 mm. **Habitat**: Dry, disturbed areas, sandy or gravelly soils. Native to central Great Plains. **Comments**: *Chloris verticillata* has a distinctive inflorescence of spike-like branches arranged in whorls. It breaks off at the upper culm node and blows away as a tumbleweed. Introduced and spreading near Vale, OR, and adjacent ID. See also *Cynodon dactylon* and *Digitaria*, grasses with a similar-looking inflorescence structure.

Cinna latifolia (Trevir. ex Göpp) Griseb.
Slender Woodreed

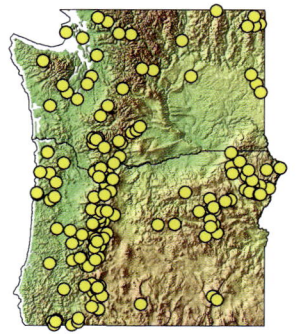

Loosely cespitose perennial, sometimes rhizomatous; **culms** 20–190 cm. **Leaf sheaths** open, glabrous; **blades** (1)7–20 mm wide, flat, lax, scabrous; **ligules** 2–8 mm, membranous, pubescent. **Panicles** nodding, branches spreading to ascending; disarticulation below glumes. **Spikelets** with 1(2) floret, **rachilla** prolonged 0.1–1.3 mm beyond the base of the floret. **Glumes** (1.8)2.5–4(5) mm long, ≥ lemmas; **lemmas** 1.8–3.8 mm, minutely bifid with delicate awns 0.1–2.5 mm, attached at the notch of the lemma tip. **Anthers** 0.4–1 mm. **Habitat**: Moist forests, shaded stream banks and springs. **Comments**: A rather nondescript large grass, *C. latifolia* can leave you wondering, "Why can't I key that huge woodland *Agrostis*?" until you note that its spikelets disarticulate below the glumes, leaving stubby naked pedicels behind on the inflorescence branches. Slender woodreed often grows on lush, mossy boulders in mountain streams or along moist creek shores, its large panicle arching gracefully over the water.

Coleanthus subtilis (Tratt.) Seidl
Mud Grass, Moss Grass [N?]

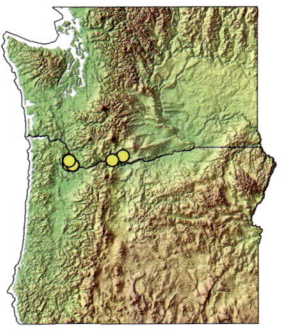

Tufted winter annual; **culms** 2–7(10) cm. **Leaf sheaths** closed, strongly inflated, enclosing base of panicles; **blades** 1–2 cm × 0.5–1.5 mm; **ligules** 1–1.5 mm, membranous. **Panicles** 1–5 cm, with 3–6 umbel-like clusters of branches. **Spikelets** with 1 floret; glumes absent; **lemmas** 0.75–1.5 mm, 1-veined, keels ciliate, with awn-like tip approximately = basal portion; seed 1 mm, longer than lemma. **Anthers** 0.3 mm. **Habitat**: Disturbed, wet, muddy or sandy sites, usually on receding shores of rivers and lakes. **Comments**: Mud grass is a bizarre little grass that flowers in fall and spring. Likely to be overlooked, it would be easy to confuse it with an annual *Juncus*. All PNW records date from the early 1900s; more recent collections have been made in BC. Its range includes disjunct populations in Europe, Russia, and China. It may be native in the PNW or not.

Cortaderia jubata (Lemoine ex Carrière) Stapf
Purple Pampas Grass, Jubatagrass [E]

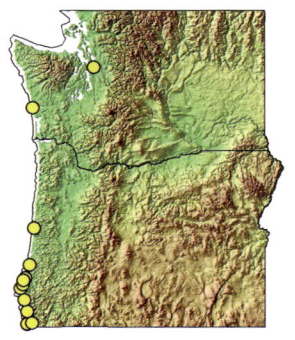

Very large, densely cespitose perennial. Plants only female (apomictic); **culms** 1.5–7 m, 4–7× as long as panicles. **Leaf sheaths** hairy; **blades** 2–10 mm wide. **Panicles** violet when young, becoming pinkish to dingy tan, base of panicle usually 1–2 m above main clump of leaves. **Spikelets** female, with 3–5 florets. **Glume** midribs white or purple. **Lemmas** 10–13 mm including awn, with hairs 6–10 mm, lemma body acute with short, ± abrupt taper to the awn, awns ≤ 1 mm, tapered lemma tips exceeding lemma hairs by 0–5 mm. **Anthers** absent.
Habitat: Grasslands, dunes, clearcuts, roadsides. Native to S America. **Comments**: *Cortaderia jubata* holds its panicles well above the leaves. The abrupt taper of the lemma body to the awn is a tiny but important character separating it from *C. selloana*. All plants are female, so if your plant has anthers, it isn't jubatagrass. Invasive near the coast.

Cortaderia selloana (Schult. & Schult. f.) Asch. & Graebn.
Pampas Grass [E]

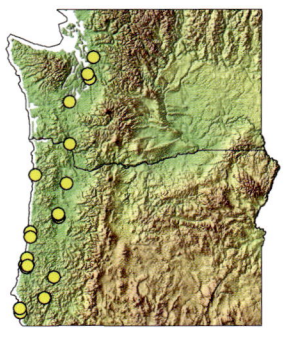

Very large, densely cespitose perennial. Both male and female plants present in North America; **culms** 2–4 m, 2–4× as long as panicles. **Leaf sheaths** mostly glabrous except near collar; **blades** 3–8 mm wide. **Panicles** white to cream in female plants, white to pinkish-tan in male plants, base of panicle in or slightly above main clump of leaves. **Female spikelets** with 5–7 florets; **male spikelets** with 3–5 florets. **Glume** midribs white. **Female lemmas** 15–20 mm including awn, with hairs 1–10 mm, tapering gradually to the awn, awns 2–5 mm, easily broken, tapered lemma tips exceeding lemma hairs by 5–12 mm; **male lemmas** 10–13 mm. **Anthers** sometimes present, 1.5–6 mm. **Habitat**: Grasslands, dunes, clearcuts, roadsides. Native to S America. **Comments**: The panicle bases of *C. selloana* often do not rise above the top of the main clump of leaves. Escaping and invasive, especially at the coast, but increasingly in interior valleys W of the Cascades. Removal can involve chain saws and backhoes.

Corynephorus canescens (L.) P. Beauv.
Gray Hairgrass [E]

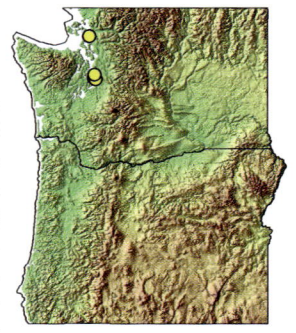

Cespitose glaucous perennial; **culms** 10–60 cm, often geniculate; nodes dark brown to black. **Leaf sheaths** open; **blades** to 1 mm wide, stiff, involute; **ligules** 2–4 mm, membranous, acute. **Panicles** narrow, dense, purple, sometimes with pale green; branches appressed, spreading at anthesis. **Spikelets** with 2 florets that are hidden by 2.9–4.3 mm glumes. **Callus hairs** 0.2–0.4 mm, dense. **Lemmas** 1.4–2 mm, awned from near base; awns 2.3–3.7 mm, bent near middle, with a ring of short projections at the bend, lower segment yellowish to brown, uniform in width, upper segment green to whitish, club-shaped. **Anthers** 1.4–1.6 mm. **Habitat**: Disturbed sandy soils, dunes. Native to Europe. **Comments**: *Corynephorus canescens* has peculiar bent awns with a club-shaped upper segment and a ring of conical protuberances at the bend. It superficially resembles a fine-leaved fescue. It is sold as an ornamental and has escaped at several locations in the Puget Sound area.

Crypsis alopecuroides (Piller & Mitterp.) Schrad.
Foxtail Pricklegrass [E]

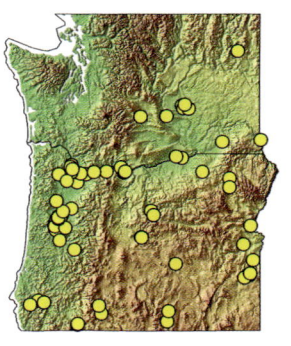

Small annual; **culms** (3)5–75 cm, rarely branching above base. **Leaf sheaths** often hairy along the margins; **collars** sometimes sparsely hairy; **blades** 1.2–2.5 mm wide, not deciduous; **ligules** 0.2–1 mm, primarily a fringe of hairs. **Panicles** narrow, spike-like, 7–8× longer than wide, green to dark purplish, often completely exserted from sheaths at maturity. **Spikelets** 1.5–2.8 mm; **lower glumes** 1.2–2 mm, **upper glumes** 1.4–2.4 mm. **Lemmas** 1.7–2.5 mm. **Anthers** 0.5–0.6 mm. **Habitat**: Receding margins of lakes and stream shores, common around reservoirs. **Comments**: This low-growing annual bears a striking resemblance to small *Alopecurus* species. Flowering time depends on the flood regime: in snowmelt vernal pools it can flower in early summer; around reservoirs, it flowers after drawdown in late fall, but in mild climates fall germinants may flower in early spring before the reservoirs fill. The dark pigment in the inflorescences absorbs heat, speeding maturation of flowers and seeds, and protects against UV light. Some sources merge *Crypsis* into *Sporobolus*.

Crypsis schoenoides (L.) Lam.
Swamp Pricklegrass [E] (yellow dots)

Small annual; **culms** 2–75 cm, often prostrate or sprawling. **Leaf sheaths** glabrous or ciliate at the throat, often inflated; **blades** 2–6 mm wide, not deciduous; **ligules** 0.5–1 mm, mostly a fringe of hairs. **Panicles** congested, spike-like, 1–5× longer than wide, base usually enclosed in upper sheath. **Spikelets** 2.5–3.5 mm; **glumes** unequal, lower glumes 1.8–2.3 mm, glabrous; upper glumes 2.2–2.7 mm. **Lemmas** 2.4–3.0 mm. **Anthers** 0.7–1.1 mm. **Habitat**: Drying reservoir margins, mudflats, vernal pools, playas. **Comments**: The panicles of *C. schoenoides* are chubbier than those of *C. alopecuroides*.

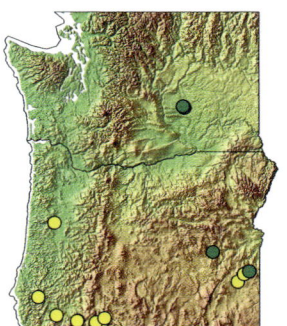

Crypsis vaginiflora (Forssk.) Opiz
Modest Pricklegrass [E] (green dots)

Differs from *C. schoenoides* with hairy **leaf sheath** and **collar** margins; **blades** deciduous at maturity; **panicles** enclosed in sheaths of upper 2 leaves; glumes subequal; hairy **lower glume** margins; **anthers** 0.5–0.9 mm. **Habitat**: Similar to *C. schoenoides*. Leaf blades often detach from the sheaths.

Cynodon dactylon (L.) Pers.
Bermudagrass [E]

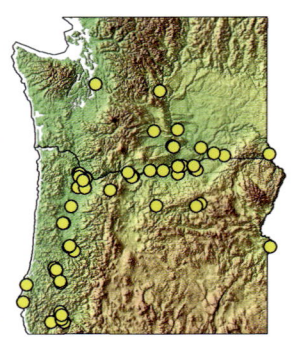

Perennial, stoloniferous, usually rhizomatous, turf forming; **culms** 5–40(50) cm. **Leaf sheaths** open; **blades** (1)2–5(6) mm wide; **ligule** of hairs. **Panicles** with whorl of (3)4–7(9) spike-like branches 1.5–6 cm long, attached at about the same height; branches 1-sided with 2 rows of spikelets. **Spikelets** 2–3.2 mm, laterally flattened, with 1 floret; disarticulation above glumes. **Glumes** unequal, < floret; **lemmas** 1.9–3.1 mm, hairy on keel, awnless. **Anthers** 1 mm. **Habitat**: Lawns, pavement cracks, roadsides, sandy river shores, disturbed sites. Native to Eurasia. **Comments**: Bermudagrass is a warm-season (C4) grass that spreads by stolons and scaly rhizomes. It has a small, whorled, umbrella-like panicle that could almost be called delicate. *Digitaria* species have similar inflorescences but are annuals with longer panicle branches, wider leaves, dorsiventrally compressed spikelets, and membranous ligules. *Eleusine indica*, another warm-season grass found only once in OR, is an annual with 4–10(17) panicle branches attached at different points, 5–7 florets per spikelet, and membranous ligules.

Cynosurus cristatus L.
Crested Dogtail [E]

Cespitose perennial; **culms** 15–75 cm. **Leaf blades** 0.5–2 mm wide, flat; **ligules** 0.5–2.5 mm, truncate. **Panicle** narrow, linear. **Fertile spikelets** with 2–5 florets, **fertile lemmas** awnless or with awns to 3 mm. **Sterile spikelets** strongly laterally compressed, with 6–18 florets. **Anthers** 1.8–3 mm. **Habitat**: Disturbed grasslands and roadsides. Native to Europe. **Comments**: Crested dogtail has a narrow spike-like panicle that vaguely resembles a wheatgrass but is 1-sided. The fan-like sterile spikelets close protectively around the fertile spikelets.

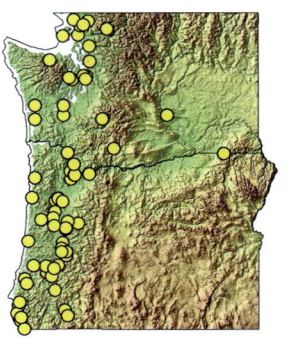

Cynosurus echinatus L.
Bristly Dogtail, Hedgehog Dogtail [E]

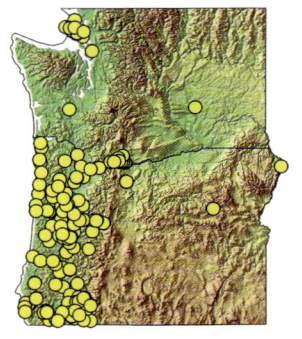

Tufted annual; **culms** 9–70 cm. **Leaf blades** 2.5–14 mm wide, flat, often twisted so bottom side is up; **ligules** 2.5–5 mm, membranous and decurrent. **Panicle** head-like, dense, bristly, 1-sided. **Fertile spikelets** with 1–5 florets, **fertile lemmas** with awns 5–25 mm. **Sterile spikelets** strongly laterally compressed, with 6–18 florets. **Anthers** 1–4 mm. **Habitat**: Disturbed dry grasslands, savannas, woodlands, prairies, mostly W of the Cascades. Native to S Europe. **Comments**: *Cynosurus echinatus* has dense, 1-sided, head-like inflorescences that might remind you of *Polypogon monspeliensis*. For identification of vegetative plants, look for the twisted leaves and the ligule; the back of the ligule is attached higher up on the leaf blade than the sides of the ligule. Bristly dogtail is an aggressive invader of dry grasslands, displacing native upland prairie species.

Dactylis glomerata L.
Orchard Grass [E]

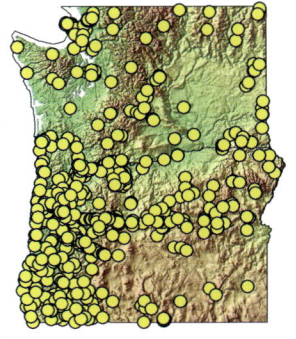

Cespitose perennial; **culms** to 120(210) cm. **Leaves** mostly basal, folded in the bud; **sheaths** closed at least 50% of their length, strongly flattened; **blades** 4–8(10) mm wide, with conspicuous midrib and very narrow, white, ± scabrous margins; **ligules** 3–11 mm, membranous, erose. **Panicles** often with spreading lower branches. **Spikelets** borne in dense, 1-sided clusters, laterally compressed, with 2–6 florets; **glumes** shorter than lemmas, keels ciliate; **lemmas** 4–8 mm, keels often ciliate, with 1 mm awn arising from the tip. **Anthers** 2–3.5 mm. **Habitat**: Meadows, pastures, open forest, roadsides, disturbed areas. **Comments**: Usually, a lower panicle branch sticks out to the side like a hitchhiker's thumb or a hula dancer's arm. For vegetative identification, the soft, folded leaves with the sharp crease down the back are a distinctive character. Native to Eurasia and Africa, widely introduced for erosion control, forage, and hay.

Danthonia californica Bolander
California Oatgrass

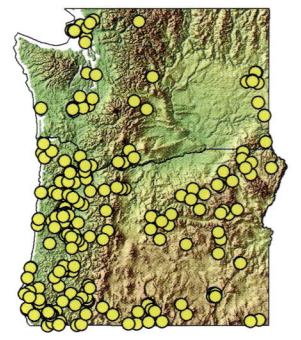

Cespitose perennial; **culms** (10)30–130 cm, disarticulating when mature. **Leaf sheaths** glabrous or hairy; **collar** hairs 1–2 mm; **blades** (1)2–5(6) mm wide; **ligules** a fringe of hairs, 1–2 mm. **Inflorescences** with (1)2–4(6+) spikelets; branches widely spreading to reflexed; pedicels on lowest branch ≥ spikelet length. **Spikelets** (10)14–26(30) mm with 3–6(12) florets. **Glumes** 14–18 mm; **lemma bodies** 5–10 mm, glabrous or sparsely pilose over the back, margins pubescent, teeth (2)4–6(7) mm, awns (7)8–12 mm. **Anthers** 2.5–4 mm.
Habitat: Grasslands, oak savannas. **Comments**: In *D. californica* the culms lean to the side, the leaves spread widely or angle down, and the large spikelets are held out at the tips of stiffly spreading panicle branches. This widespread grass was once a community dominant in mesic prairies. It tolerates grazing better than other native prairie species, but it has grudgingly given ground under the onslaught of disturbance and invasive species that have accompanied livestock and other human activities. In some grasslands W of the Cascades, it is truly the last native plant standing. Compare to *D. unispicata*.

Danthonia decumbens (L.) DC.
Common Heathgrass [E]

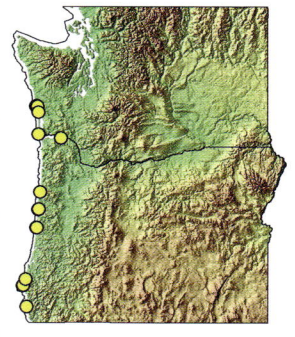

Cespitose perennial; **culms** 8–60 cm, not disarticulating. **Leaf blades** 0.5–4 mm wide; **ligules** a fringe of hairs 0.3–0.7 mm. **Inflorescences** with up to 15 spikelets; branches erect; pedicels < spikelet length except for terminal spikelet on each branch. **Spikelets** 6–15 mm with 3–5 florets. **Glumes** 7–10 mm; **lemma bodies** 5–6 mm, glabrous or margins hairy, teeth short, acute, awns absent or sometimes a short, inconspicuous point between the teeth. **Anthers** 0.2–0.4 mm (rarely about 2 mm in cross-pollinating plants). **Habitat**: Coastal in grasslands, dunes, roadsides. Native to Europe. **Comments**: The smaller, more rounded, awnless lemmas separate *D. decumbens* from our native *Danthonia* species. It might be mistaken for *D. spicata*, but that species is awned and generally is not found in coastal habitats.

Danthonia intermedia Vasey
Timber Oatgrass

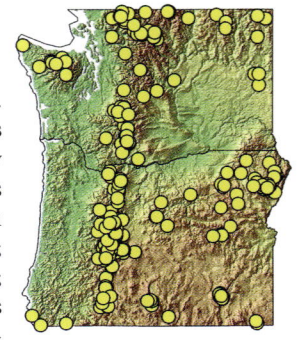

Cespitose perennial; **culms** 20–50 cm, not disarticulating. **Leaf blades** 1–3.5 mm wide, not curling at maturity; **ligules** a fringe of hairs, 0.5–0.7 mm. **Inflorescences** dense, nearly head-like, with (4)5–13 spikelets; branches stiff, erect; pedicels on lowest branch ≤ spikelet length. **Spikelets** 11–15(19) mm with 3–12 florets. **Glumes** (11)13–17 mm, > the florets; **calluses** of middle florets longer than wide, concave dorsally; **lemma bodies** 3–6 mm, nearly glabrous on back, margins densely hairy, teeth acute to awn-like, 1.5–2.5 mm, awns 6.5–8 mm. **Anthers** usually tiny (to 4 mm in rare outcrossing florets). **Caryopses** 2–3 mm. **Habitat**: Montane to alpine, meadows, open woods, rocky outcrops. **Comments**: *Danthonia intermedia* forms a dense head of large spikelets. Similar *D. spicata* usually has a narrower, more spike-like inflorescence, shorter lemmas with shorter teeth, and sparse, long white hairs over the lemma back.

Danthonia spicata (L.) P. Beauv. ex Roem. & Schult.
Poverty Oatgrass

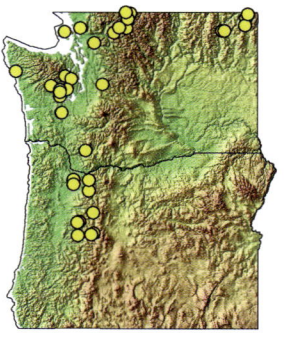

Cespitose perennial; **culms** 10–70 cm, disarticulating at the nodes, nodes with long, stiff white hairs. **Leaf blades** 0.8–3(4) mm wide, usually curled or twisted, except in shade-grown plants; **ligules** a fringe of hairs, 0.2–1 mm. **Inflorescences** narrow, spike-like, with 5–10(18) spikelets; branches stiff, ascending; pedicels on lowest branch ≤ spikelet length. **Spikelets** 7–15 mm with 3–12 florets, glumes > florets. **Glumes** (8)9–12 mm; **calluses** of middle florets about as long as wide, convex dorsally; **lemma bodies** 2.5–5 mm, usually with sparse, long hairs on back, occasionally glabrous, margins hairy on lower half, teeth acute to awn-like, 0.5–2 mm, awns 5–8 mm. **Anthers** to 2.5 mm. **Caryopses** 1.5–2 mm. **Habitat**: Sunny, dry, rocky, or sandy meadows and slopes. **Comments**: Poverty oatgrass has narrow, almost spike-like inflorescences, and lemmas that are usually evenly hairy over the back. It usually grows at lower elevations than similar *D. intermedia*.

Danthonia unispicata (Thurb.) Munro ex Vasey
One-spike Oatgrass

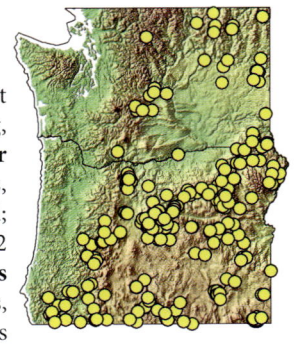

Cespitose perennial; **culms** 15–30 cm, disarticulating at the nodes. **Leaf sheaths** usually densely hairy with long, spreading, often papillose-based hairs, rarely glabrous; **collar hairs** 3–4 mm; **blades** 1–3 mm wide; **ligules** a fringe of hairs, 0.5–1 mm. **Inflorescences** with 1(4) spikelets, unbranched; pedicels ≤ spikelet length. **Spikelets** (8)12–26 mm with 3–12 florets, glumes > florets. **Glumes** 15–23 mm; **lemma bodies** 5.5–11 mm, glabrous on back, rarely with a few scattered hairs, margins usually hairy, teeth acute to awn-like, 1.5–7 mm, awns 5.5–13 mm. **Anthers** to 3.5 mm. **Habitat**: Dry grasslands, shrub steppe, open forest and adjacent scablands, in the mountains to timberline. **Comments**: Easily identified by its densely long-hairy sheaths and blades, and usually single spikelet per culm (sometimes 2 or 3 spikelets). Occasional plants with glabrous leaves or 2–4 spikelets per culm can resemble *D. californica*, which has spreading to reflexed inflorescence branches, pedicels that are ≥ the spikelets on lowest branches, and 1–2 mm collar hairs. Hold the sheaths up to the sun for backlighting to view the long, soft hairs.

Deschampsia bolanderi (Thurb.) Saarela
Scribner's Grass

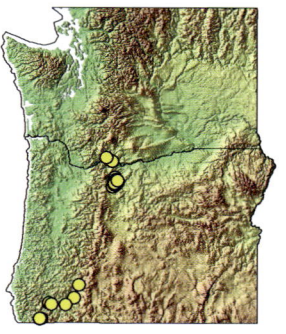

Small tufted annual; **culms** 3–15(45) cm tall, ascending to erect, nodes purple. **Leaf blades** 0.8–1.6 mm wide; **ligule** membranous, slightly hairy 2–4 mm. **Inflorescence** a spike with 1 spikelet at all or most nodes, the spikelets oriented with the flat side against the inflorescence axis and sunken into the axis. **Glumes** stiff, purplish, hiding the floret. **Lemmas** with a hairy callus and awned from between 2 short apical lobes, the awn inconspicuous, 2–4 mm long. **Habitat**: Vernal pools and dry, sandy, and rocky soils. Rare. **Comments**: S WA, N and SW OR. Easily overlooked due to its small stature. Formerly in the genus *Scribneria*, recent work has placed Scribner's grass in *Deschampsia*.

Deschampsia cespitosa (L.) P. Beauv.
Tufted Hairgrass

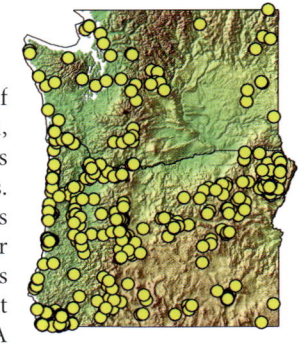

Densely cespitose perennial; culms (7)35–150+ cm. Leaf blades 1–4 mm wide, flat or rolled; ligules 2–13 mm, membranous, acute. Panicles usually open, diffuse, sometimes contracted, tan to silvery pink, or blackish at high elevations. Lower glumes 2.7–7 mm, upper glumes 2–7.5 mm; lemmas 2–5(7) mm, lemma awns 0.5–8 mm, attached from near base to about midlength, straight or slightly bent. Anthers 1.5–3 mm. **Habitat**: Wet prairies, wet meadows, coastal salt marshes, serpentine fens; sea level to alpine. **Comments**: A tan to pinkish cloud of inflorescences shimmering in the sun above densely tufted leaves typifies the look of a stand of tufted hairgrass. This highly variable grass can be a community dominant in a wide array of wet or seasonally wet habitats. Prior to Euro-American settlement it covered vast areas of wet prairie in the Willamette Valley. Tufted hairgrass is a critical species for restoration in all its native habitats, from previously drained coastal salt marshes to degraded wet meadows high in the mountains. When held up to the sun, the leaf blades have a colorless, cellophane-like tissue between the green veins.

Deschampsia danthonioides (Trin.) Munro
Annual Hairgrass

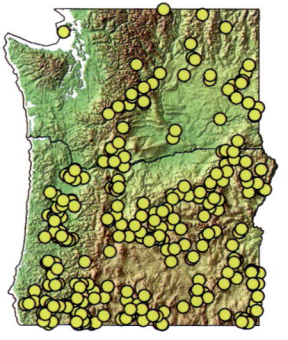

Slender annual; **culms** 10–40(70) cm. **Leaves** mostly cauline, narrow; **blades** 0.3–1.5 mm wide, involute or flat; **ligules** (0.5)1.5–3(4.7) mm, membranous, acute to acuminate. **Panicles** contracted when young, open at maturity, with spikelets borne at tips of branches. **Glumes** 3.5–9 mm, 3-veined; **lemmas** 1.5–3 mm, awns 4–9 mm, attached from near lemma base to about midlength, strongly bent. **Anthers** 0.3–0.5 mm. **Habitat**: Vernal pools, scablands, stream banks, vernally wet disturbed areas. **Comments**: *Deschampsia danthonioides* grows in a variety of vernally wet habitats, including roadside ditches. The weedy, introduced *Ventenata dubia* also invades these habitats but differs in having 2–3 florets per spikelet, glumes with 5–9 veins, longer lemmas, lower lemmas with straight awns that arise near the lemma tip, and longer anthers.

Deschampsia elongata (Hook.) Munro
Slender Hairgrass

Slender cespitose perennial; **culms** (10)30–120 cm. **Leaf blades** 0.2–2 mm wide, usually involute; **ligules** 2.5–8(9) mm, acute to acuminate. **Panicles** very narrow, pale green to silvery, nodding, with erect to ascending branches. **Glumes** 3–5.5(6.5) mm; **lemmas** 1.7–3.3 mm, awns (1.5)3–6 mm, attached near midlength of lemma, straight or slightly bent. **Anthers** 0.3–0.7 mm. **Habitat**: Seasonally moist roadsides, lake and stream shores, disturbed areas where competition is low, often in shade. **Comments**: *Deschampsia elongata* is a slender grass with dense tufts of short basal leaves and narrow panicles that lean to one side. Immature *D. danthonioides* can look similar but has few basal leaves and generally longer glumes. With its small tuft of basal leaves and shallow root system slender hairgrass could easily be mistaken for an annual.

Dichanthelium acuminatum (Sw.) Gould & C.A. Clark ssp. *fasciculatum* (Torr.) Freckmann
Hairy or Western Panicgrass, Tapered Rosette Grass

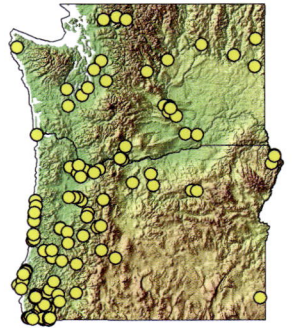

Cespitose perennial; **culms** 15–75 cm. **Leaf sheaths** with ascending to spreading papillose-based hairs; **blades** 6–12 mm wide, glabrous or sparsely hairy, the bases with papillose-based cilia; **ligules** a fringe of hairs, 1–5 mm. **Primary panicles** 3–12 × 0.75–9 cm, open, exserted, branches ascending. **Spikelets** 1.5–2.0 mm; **lower glumes** 25–50% as long as spikelets, acute; **upper glumes** 1.5–1.8 mm, 7-veined; **lower lemmas** similar to upper glumes; **upper lemmas** 1.2–1.5 mm, tough or hard. **Anthers** 0.5–0.8 mm. **Habitat**: Disturbed grasslands, open woodlands, shores of streams, rivers, or lakes, often below highwater line. **Comments**: Similar *D. oligosanthes* has larger spikelets.

Dichanthelium oligosanthes (Schult.) Gould ssp. *scribnerianum* (Nash) Gould
Scribner's Rosette Grass, Scribner's Panicgrass

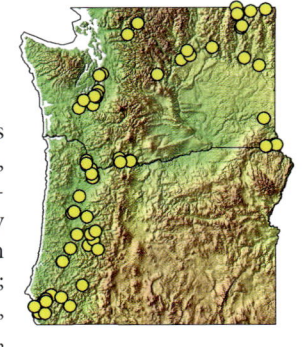

Cespitose perennial; **culms** 20–50 cm. **Leaf sheaths** glabrous or with sparse, papillose-based hairs; **blades** 6–15 mm wide, glabrous, or sparsely hairy below, the bases with papillose-based cilia; **ligules** a fringe of hairs, 1–1.5 mm. **Primary panicles** 5–9 × 3–6 cm, open, sometimes partially hidden in leaf sheaths, branches ascending. **Spikelets** 2.7–3.5 mm; **lower glumes** 1.1–1.6 mm, acute; **upper glumes** 2.7–3.5 mm, 9-veined; **lower lemmas** ± hairy; **upper lemmas** 2.5–3 mm, tough or hard. **Anthers** about 1 mm. **Habitat**: Meadows, prairies, shores of lakes and rivers, disturbed areas. **Comments**: *Dichanthelium oligosanthes* has short, broad leaves and open panicles with small, round spikelets. Formerly in the genus *Panicum*, *Dichanthelium* species are cool-season (C3) grasses. They usually have two seasons of active growth, spring and fall. Our *Panicum* species are warm-season (C4) annuals that lack the winter rosettes of *Dichanthelium*.

Digitaria ischaemum (Schreb.) Schreb. ex Muhl.
Smooth Crabgrass [E]

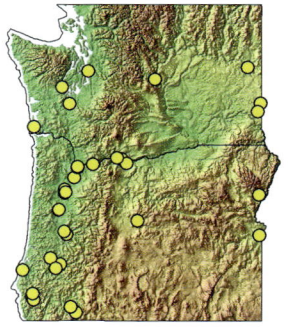

Annual or perennial, tufted, mat-forming, or rhizomatous; **culms** 20–55(70) cm. **Leaf sheaths** nearly glabrous except some hairs on upper margins; **blades** 3–5 mm wide, glabrous or with a few papillose-based hairs on margins near base; **ligules** membranous, 1–2 mm, with even margins, sparse hairs at the base of the leaf sheath. **Panicle branches** 2–7. **Lower glumes** absent or reduced to a membranous rim; **upper glumes** 1.3–2.3 mm, 3-veined, 75–100% as long as spikelet. **Lower lemmas** 1.7–2.3 mm, 7-veined, with 3 widely spaced middle veins, and outer veins tightly spaced on outer margins, hairy; upper lemmas dark brown at maturity. **Anthers** 0.4–0.6 mm. **Habitat**: Weedy, disturbed sites, lawns, roadsides, shorelines. Uncommon; native to Eurasia. **Comments**: Similar *D. sanguinalis* has hairier leaf sheaths and blades.

Digitaria sanguinalis (L.) Scop.
Hairy Crabgrass [E]

Annual, tufted or sometimes mat-forming; **culms** 20–70(112) cm. **Leaf sheaths**, collars, and blades covered with long, papillose-based hairs; **blades** 3–8(12) mm wide; **ligules** membranous, 1–2 mm, with uneven teeth or margins. **Panicle branches** 4–13. **Lower glumes** 0.2–0.4 mm; **upper glumes** 0.9–2 mm, 3-veined, 33–60% as long as spikelet. **Lemmas** 1.7–3 mm; **lower lemmas** 7-veined, with 3 widely spaced middle veins, and outer veins tightly spaced on outer margins, glabrous; **upper lemmas** brown at maturity. **Anthers** 0.5–0.9 mm. **Habitat**: Weedy, disturbed sites, lawns, roadsides, shorelines. Common; native to Eurasia. **Comments**: The spike-like panicle branches of *D. sanguinalis* spread like the spokes of an umbrella and take on a purplish hue. See also *Cynodon dactylon*. *Digitaria* species in the PNW are warm-season (C4) grasses.

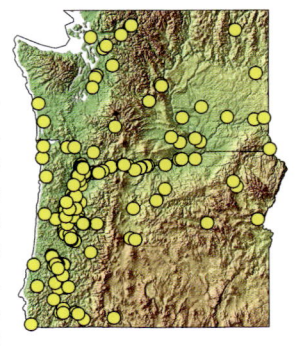

Diplachne fusca (L.) P. Beauv. ex Roem. & Schult. ssp. *fascicularis* (Lam.) P.M. Peterson & N. Snow
Bearded Sprangletop (yellow dots)

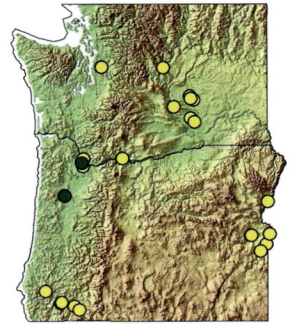

Cespitose annual or perennial; **culms** 5–110 cm, prostrate to erect. **Leaf sheaths** open; **blades** 2–7 mm wide, flat, uppermost blades > panicles; **ligules** 2–8 mm, membranous. **Panicles** with 3–35 spike-like branches with spikelets in 2 rows, lower branches occasionally with secondary branches. **Panicles** usually partially enclosed in uppermost leaf sheaths. **Spikelets** with 6–20 florets, disarticulating above glumes. **Glumes** < florets; **lemmas** 3.5–5 mm, 3(5)-veined, smoky white with dark smudge in basal half, tips acute, awnless or awned, awns to 3.5 mm. **Anthers** 0.2–0.5 mm. **Habitat**: Roadsides, seasonal wetlands, river shores, ditches; transported by migratory waterfowl. **Comments**: The genus name has flip-flopped from *Diplachne* to *Leptochloa* and back again. At the moment, we're calling it *Diplachne*.

Diplachne fusca ssp. *uninervia* (J. Presl) P.M. Peterson & N. Snow
Mexican Sprangletop [E] (green dots)

Similar to *D. fusca* ssp. *fascicularis* but **uppermost leaf blades** < panicles; **panicles** completely exserted from upper sheaths; **lemmas** 2–3.6 mm, light brown, dark green, or lead-colored, usually without basal dark spot, tips usually truncate or obtuse, often notched, awnless, sometimes with short point. **Anthers** 0.2–0.6(1) mm. **Habitat**: Disturbed areas, roadsides. Native from S US to Argentina.

Distichlis spicata (L.) Greene
Saltgrass

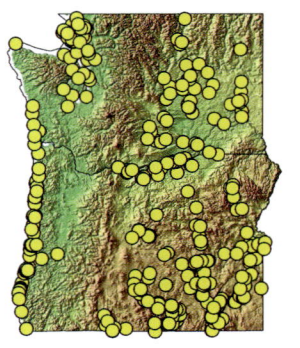

Short, strongly rhizomatous perennial, usually dioecious; **culms** 10–60 cm. **Leaves** stiffly 2-ranked; **sheaths** open, with tuft of hairs at top; **blades** 2–4 mm wide, involute, stiffly spreading; **ligules** about 0.5 mm, membranous, short-serrate, sometimes with hairs to 3 mm at sides. **Panicles** dense, usually unisexual. **Spikelets** laterally compressed, with 5–20 florets, disarticulation above glumes; **lower glumes** 2–3 mm, 3(5)-veined; **upper glumes** 3–4 mm, 5–7(9)-veined. **Lemmas** 3.5–6 mm, awnless; pistillate lemmas firm; staminate lemmas thinner in texture. **Anthers** 3–4 mm. **Habitat**: Coastal salt marshes, interior alkali flats, seeps and lakeshores with alkaline or saline soils. On inland sites, it is often associated with greasewood (*Sarcobatus vermiculatus*). **Comments**: Saltgrass is characterized by stiffly spreading, 2-ranked leaves, strongly rhizomatous growth form, and separate male and female plants. This low-growing grass is considered poor forage, but a valuable soil stabilizer.

Echinochloa crus-galli (L.) P. Beauv.
Barnyard Grass [E]

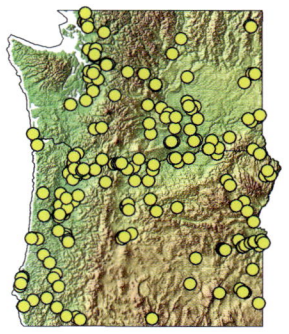

Annual; **culms** 30–200 cm, decumbent to erect. **Leaf blades** 5–30 mm wide; **ligules** absent. **Panicle branches** often with secondary branches. **Spikelets** 2.5–4 mm. **Lower lemmas** awned or awnless, awns to 50 mm. **Upper lemmas** broadly ovate to elliptical, the tough body rounded distally with an abrupt transition to an early-withering, acuminate, membranous tip; tip set off from leathery portion by a line of minute hairs barely visible at 25×. **Anthers** 0.5–1 mm. **Habitat**: Disturbed moist sites, fields, ditches, roadsides, gardens. Native to Eurasia. **Comments**: This common, weedy grass is highly variable in size, color, inflorescence branching, and awn presence. The lemma narrows abruptly to a withering tip. The line of hairs between the lemma body and tip are virtually invisible as a field character. See comments under *E. muricata*.

Echinochloa muricata (P. Beauv.) Fernald var. *microstachya* Wiegand
American Barnyard Grass

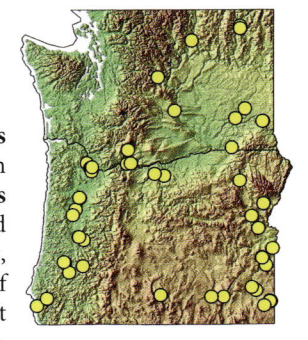

Annual; **culms** (10)80–160 cm, erect to spreading. **Leaf blades** 5–30 mm wide; **ligules** absent. **Panicle branches** often with secondary branches. **Spikelets** 2.5–5 mm. **Lower lemmas** awnless or with awns to 6(10) mm. **Upper lemmas** obovoid to orbicular, tapering gradually to a stiff, non-withering, acuminate tip; tip not set off from lemma body by a line of minute hairs. **Anthers** 0.4–1.1 mm. **Habitat**: Shorelines, wet ditches, disturbed moist sites, reservoir drawdown zones.
Comments: This grass is a true native weed. Similar *E. crus-galli* differs in having upper lemmas that abruptly narrow to a withering tip. Because the growth habit of the two species is so similar, field photos are not adequate for identification. Collect a sample for examination with a microscope.

Ehrharta erecta Lam.
Panic Veldtgrass [E]

Weakly cespitose perennial; **culms** 30–100 cm, erect or ascending from a decumbent base. **Leaf sheaths** open, **blades** lax, 2–15 mm wide, margins often wavy; **ligules** to 3 mm, membranous, auricles ciliate with long slender hairs. **Panicles** erect or nodding, open to contracted, bearing spikelets with 3 florets, the lower 2 sterile, the uppermost bisexual. **Glumes** unequal. **Sterile lemmas** with a basal appendage, the upper transversely rugose in distal half and longer than the bisexual floret. **Bisexual lemmas** 2.5–3.5 mm, firm, glabrous, obscurely 5–7-veined, awnless. **Anthers** 0.7–1.2 mm. **Habitat**: Moist, often shady sites. Recently found near Reedsport, Douglas County, OR. **Comments**: Introduced to CA from South Africa, also weedy in Australia. Given its reputation of weediness, this species may be more widespread than reported.

Elymus canadensis L. var. *canadensis*
Canada Wildrye

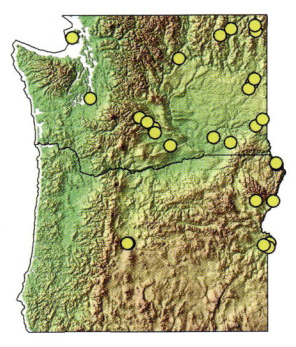

Loosely cespitose perennial, rarely with rhizomes to 4 cm; **culms** (40)60–150(180) cm, plants often glaucous. **Leaf blades** (3)4–15(20) mm wide, often ascending and somewhat involute; **ligules** to 1(2) mm, membranous, truncate, ciliolate; **auricles** 1.5–4 mm, brown or purplish-black. **Spikes** nodding to strongly curved, often strongly glaucous, (1)3(5) spikelets per node; disarticulation above (rarely below) glumes, not in inflorescence axis. **Spikelets** 12–20 mm, ± divergent, with (2)3–5(7) florets. **Glume body** 6–13 × 0.5–1.6 mm, linear-lanceolate, broadest above base, 3–5-veined, gradually tapering into straight or outcurving awns, the awns (5)10–25(27) mm. **Lemmas** 8–15 mm, margins glabrous or with uniform hairs, awns 15–40 mm, moderately to strongly outcurved. **Anthers** 2–3.5 mm. **Habitat**: Prairies, stream banks, roadsides, disturbed areas. **Comments**: This robust, broad-leaved wildrye has a thick, nodding to drooping spike and conspicuous, outcurving awns. Native in E WA and in the Snake River drainage in OR; seeded elsewhere in our area.

Elymus curvatus Piper
Awnless Wildrye

Cespitose perennial; **culms** 60–110 cm, plants often glaucous. **Leaf sheaths** glabrous; **blades** 5–15 mm wide, flat to somewhat involute, glabrous; **ligules** < 1 mm; **auricles** to 1 mm or absent. **Spikes** erect, 2 spikelets per node; disarticulation below glumes, but not in inflorescence axis. **Spikelets** 9–15 mm, with (2)3–4(5) florets. **Glumes** 7–15 mm, the basal 2–3 mm cylindrical, hardened, without evident veins, strongly bowed out to expose the enclosed floret bases; body linear-lanceolate, broadest above base, 3–5-veined, awns 0–3(5) mm. **Lemmas** 6–10 mm, glabrous, rarely hairy, awns (0.5)1–3(4) mm, rarely 5–10 mm on lemmas of distal spikelets, straight. **Anthers** 1.5–3 mm. **Habitat**: Moist sites in open forest, thickets, grasslands, disturbed areas. **Comments**: This species is probably extirpated in WA due to dams on the Pend Oreille River, but there is a recent record in SE BC. The hardened, bowed-out glume bases are distinctive.

Elymus elymoides (Raf.) Swezey ssp. *elymoides*
Common Squirreltail, Bottlebrush Squirreltail

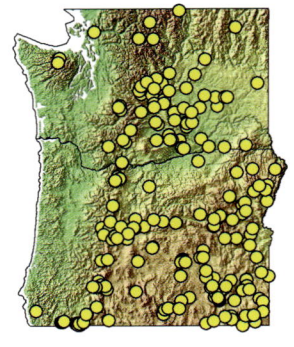

Cespitose perennial; **culms** 8–45 cm. **Leaf sheaths** and blades glabrous to densely white-hairy; **blades** (1)2–4(6) mm wide; **ligules** < 1 mm; **auricles** usually present to 1 mm. **Spikes** erect to slightly nodding, spikelets 2 per node; internodes 3–10(15) mm; disarticulation in the axis. **Spikelets** with 2–4(5) florets, the lowest 1–2 florets sterile and glume-like. **Glumes** linear, often awn-like throughout; glume awns 15–85 mm, often split into 2 or 3 unequal divisions, widely divergent. **Fertile lemmas** 6–12 mm, with 2 lateral veins extending into bristles to 10 mm; **lemma awns** 25–75 mm × 0.4 mm at base, widely divergent; **paleas** often with veins extended as 1–2 mm bristles. **Anthers** 0.9–2.2 mm. **Habitat**: Dry, rocky soils, shrub steppe, grasslands, open forest, disturbed sites, to alpine. **Comments**: The great challenge with all *E. elymoides* subspecies is interpreting the visual complexity exploding from all the long, spreading glume and lemma awns. Subspecies *elymoides* often appears to have 5 glumes at each node. One of them arises between 2 real glumes and is actually an awn-like sterile lemma. This is often the last native bunchgrass standing on a disturbed or overgrazed site.

Elymus elymoides ssp. *brevifolius* (J.G. Sm.) Barkworth
Longleaf Squirreltail

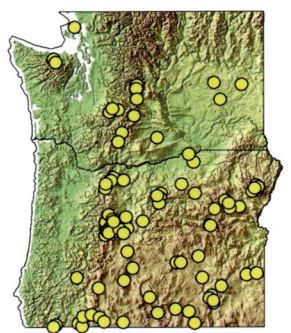

Like *E. e.* ssp. *elymoides* (see above) but **culms** 25–65(77) cm; **lowermost floret** in each spikelet fertile, not glume-like; **glume awns** 50–125 mm, entire; **lemma awns** 50–120 mm, **paleas** usually not with veins extended as bristles. **Habitat**: Low to alpine elevations, often at higher elevations than the other subspecies.

Elymus elymoides ssp. *hordeoides* (Suksd.) Barkworth
Barley Squirreltail

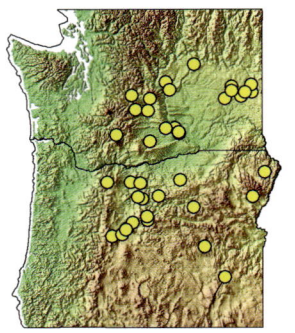

Like the other *E. elymoides* subspecies but **culms** 10–20 cm; **spikelets** 3 per node; central spikelet with 2 fertile florets, lateral spikelet with rudimentary or awn-like florets; **glume awns** 15–50 mm, usually entire; **fertile lemma awns** 15–30 mm; paleas with or without distinct bristles. **Habitat**: Not extending to alpine elevations. **Comments**: Dwarf alpine *E. elymoides* usually are ssp. *brevifolius*. Subspecies *hordeoides* is sometimes confused with *Elymus-Hordeum* hybrids. See also *E. multisetus*.

Elymus glaucus Buckley ssp. *glaucus*
Blue Wildrye (yellow dots)

Loosely cespitose to short-rhizomatous perennial; usually glaucous; **culms** 30–140(200) cm. **Leaf sheaths** and blades glabrous or hairy; **blades** 4–17 mm wide, usually flat; **ligules** to 1 mm; **auricles** to 2.5 mm. **Spikes** erect to slightly curved; (1)2(3) spikelets per node; disarticulation above glumes. **Spikelets** 8–25 mm, with (1)2–4(6) florets. **Glumes** flat, thin, linear-lanceolate, 3–5-veined to the base, awns (0.5)1–5(9) mm. **Lemmas** (8)9–14(16) mm, glabrous to evenly short-hairy, awns (5)10–25(35) mm, straight to slightly curving. **Anthers** 1.5–3.5 mm. **Habitat**: Lowland to montane meadows, savannas, and open woodlands. Frequently planted in habitat restoration projects.

Elymus glaucus ssp. *virescens* (Piper) Gould
Coastal Blue Wildrye (green dots)

Like *E. g.* ssp. *glaucus* but **leaf blades** 2–10 mm wide, **glume awns** 0–2 mm, and **lemma awns** (0)1–5(7) mm. **Habitat**: Coastal cliffs, thickets, and forest edges. **Comments**: *E. g.* ssp. *virescens* occurs near the coast. It differs from ssp. *glaucus* with its awnless or short-awned glumes and lemmas. Awnless or short-awned inland plants are abnormal subspecies *glaucus*.

Elymus hirsutus J. Presl
Northwestern Wildrye

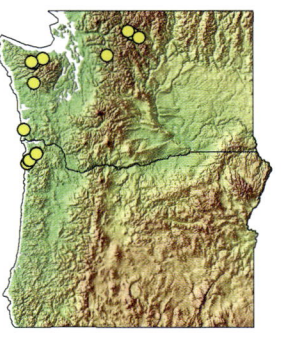

Loosely cespitose to short-rhizomatous perennial; **culms** 40–140 cm. **Leaf sheaths** usually glabrous; **blades** 4–12 mm wide, flat, lax; **ligules** to 1 mm, **auricles** lacking or to 1.5 mm. **Spikes** nodding to pendent; 2(3) spikelets per node; disarticulation above glumes. **Spikelets** 12–20 mm with 2–4(7) florets. **Glumes** flat, linear-lanceolate, 3–5-veined, awns 1–10 mm, straight. **Lemmas** 7–14 mm, smooth to scabrous; lateral veins hairy; margins hairy beyond midlength, the hairs > than those elsewhere on lemma, awns (2)8–30 mm. **Anthers** 2–3.5 mm.
Habitat: Moist forest edges, thickets, and grasslands; coastal mountains, NW OR to AK; and Cascades of N WA. **Comments**: Differs from *E. glaucus* in its nodding spike and lemmas with long marginal hairs and outcurving awns.

Elymus lanceolatus (Scribn. & J.G. Sm.) Gould ssp. *lanceolatus*
Thickspike Wheatgrass

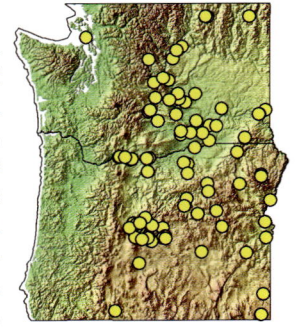

Strongly rhizomatous perennial; **culms** 60–130 cm, often clustered on long rhizomes; plants sometimes glaucous. **Leaf sheaths** glabrous or pubescent; **blades** 1.5–6 mm wide, often mostly basal or more evenly distributed, loosely involute, outer surface glabrous, inner surface with straight hairs, veins ± equal; **ligules** 0.1–0.5 mm; **auricles** 0.5–1.5 mm. **Spikes** erect to slightly nodding; 1 spikelet per node; disarticulation above glumes. **Spikelets** 10–28 mm, with 3–11 florets; 1.5–3× internodes. **Glumes** lanceolate, acute, tapering from midlength or above, 50–75% the length of adjacent lemmas, flat or with off-center keel, 3–5-veined, midvein straight, sometimes short-awned. **Lemmas** 7–12 mm, with stiff hairs < 1 mm, awnless or with awns < 2 mm. **Anthers** 2.5–6 mm. **Habitat**: Dry grasslands, shrub steppe, dry forest openings, on clay, sand, loam, or rocky soils. **Comments**: This common, widespread wheatgrass is an important soil binder. Similar *Pascopyrum smithii* has acuminate glumes that taper from below midlength, and a midvein that "leans" to one side. *Elymus trachycaulus* is generally cespitose.

Elymus lanceolatus ssp. *psammophilus* (J. M. Gillett & H. Senn) Á. Löve
Sand-dune Wheatgrass

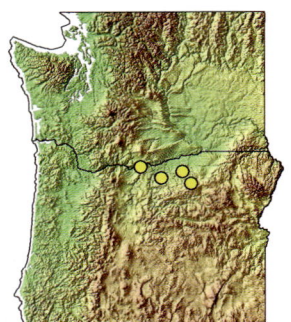

Like *E. lanceolatus* ssp. *lanceolatus* (see above) but with **culms** 20–95 cm and **lemmas** with softer, denser, longer hairs ≥ 1 mm. **Habitat**: Sandy soils in the Columbia Plateau. **Comments**: Also see *Leymus flavescens*.

Elymus lanceolatus ssp. *riparius* (Scribn. & J.G. Sm.)
Streambank Wheatgrass

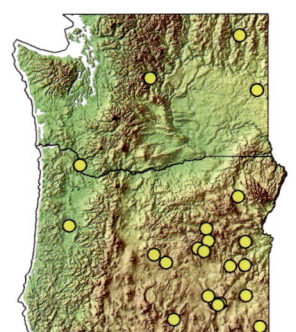

Like *E. lanceolatus* ssp. *lanceolatus* but with **culms** 22–60 cm and **lemmas** that are smooth or scabrous distally, mostly glabrous, but the margins sometimes hairy proximally. **Habitat**: Clay soils in mesic to moist alkaline meadows, riparian terraces, sagebrush steppe. **Comments**: Widespread, often in moister habitats than the other subspecies.

Elymus multisetus M.E. Jones
Big Squirreltail

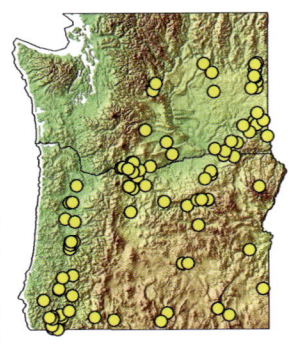

Cespitose perennial; **culms** 15–65 cm. **Leaf sheaths** and blades glabrous to white-hairy; **blades** (1)2–4(6) mm wide; **ligules** to 1 mm; **auricles** 0.5–1.5 mm, or absent. **Spikes** erect, spikelets 2 per node; internodes 3–5.5(8) mm; disarticulation in the axis. **Spikelets** with 2–4 florets, the lowest 1–2 florets sterile and glume-like. **Glumes** ± awn-like; **glume awns** (8)25–90 mm, split above the base into 3–9 unequal divisions, outcurving from near base at maturity. **Fertile lemmas** 8–10 mm, with 2 lateral veins extending into bristles to 10 mm; **lemma awns** (10)20–110 mm × 0.2 mm at base, divergent to arching; **paleas** often with veins extended as 1–2 mm bristles. **Anthers** 1–2 mm. **Habitat**: Dry, rocky grasslands, savanna, juniper/sagebrush steppe. **Comments**: With its glumes split into up to 9 divisions there are a whole lot of awny things sticking out of *E. multisetus* spikelets. The awns are narrower at the base than those of *E. elymoides*. With its 6 awn-like glumes *E. e.* ssp. *hordeoides* can be confused with *E. multisetus*, which has 2 rather than 3 spikelets per node. Widespread, but often mistaken for *E. elymoides*. All Willamette Valley records are from before 1940.

Elymus repens (L.) Gould
Quackgrass [E]

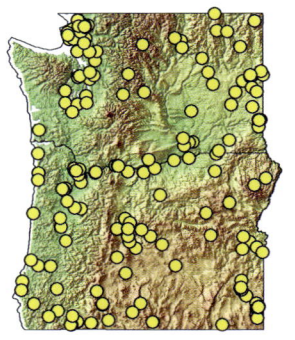

Strongly rhizomatous perennial; **culms** 50–100 cm. **Leaf sheaths** glabrous or lower ones with spreading hairs; **blades** 6–10 mm wide, flat, green, or glaucous, glabrous or sparsely hairy, veins unequal, with primary veins separated by narrower secondary veins; **ligules** 0.25–1.5 mm; **auricles** 0.3–1 mm. **Spikes** erect; 1(2) spikelets per node; disarticulation above glumes. **Spikelets** 10–27 mm, with 4–7 florets, > 1.25× internodes. **Glumes** oblong, glabrous, midveins scabrous and keeled near tip, smooth and ± flat near base, awnless or with awns to 3 mm. **Lemmas** 8–12 mm, glabrous, awnless or with straight awns 0.2–4(10) mm. **Anthers** 4–7 mm. **Habitat**: Disturbed grasslands, ditches, streamsides, gardens, roadsides. Highly invasive; native to Eurasia. **Comments**: In drier regions E of the Cascades, quackgrass is generally restricted to moister habitats. When drought-stressed, quackgrass leaves roll lengthwise, making it difficult to recognize. Native *E. lanceolatus* usually has narrower, basally concentrated leaves. People have eaten quackgrass rhizomes in times of famine. Quackgrass is the nemesis of many a gardener. The sharp-tipped rhizomes can pierce and grow through a potato.

Elymus scribneri (Vasey) M.E. Jones
Scribner's Wheatgrass

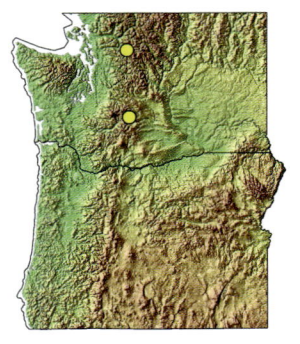

Cespitose perennial; **culms** 15–35(55) cm, prostrate to strongly decumbent. **Leaf sheaths** glabrous or short-hairy; **blades** 1.5–4 mm wide, usually involute, upper surfaces prominently ribbed; **ligules** short, 0.2–0.4(0.7) mm; **auricles** usually present, 0.5–1 mm. **Spikes** with 1 spikelet per node; disarticulation in the inflorescence axis. **Spikelets** 9–15 mm, with 3–6 florets. **Glumes** narrowly lanceolate, mostly glabrous, tapering to a divergent, 12–30 mm awn. **Lemmas** 7–10 mm, usually glabrous, awns 15–30 mm, divergent. **Anthers** 1–1.6 mm. **Habitat**: Open, rocky, windswept, subalpine and alpine sites. **Comments**: This inconspicuous bunchgrass usually has only a few decumbent culms. With its long-spreading awns the spikes resemble those of *E. elymoides*, which has 2–3 spikelets per node and narrower glumes. *Elymus scribneri* spikes disarticulate in the inflorescence axis, but more slowly than in *E. elymoides*. Scribner's wheatgrass is rare in our area, known from two sites in the WA Cascades. It is more common in the Rocky Mountains.

Elymus trachycaulus (Link) Gould ssp. *trachycaulus*
Slender Wheatgrass

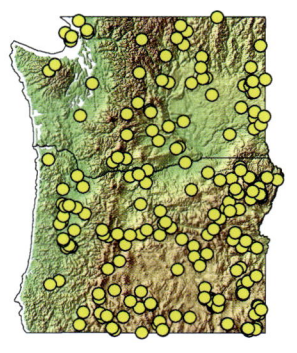

Cespitose perennial, sometimes weakly rhizomatous; **culms** 30–150 cm. **Leaves** mostly basal; **sheaths** usually glabrous, sometimes retrorsely hairy; **blades** 2–5(8) mm wide, flat or involute, glabrous or hairy; **ligules** 0.2–0.8 mm; **auricles** lacking or to 1 mm. **Spikes** erect; 1 spikelet per node; disarticulation above glumes. **Spikelets** 9–17(20) mm, with 3–9 florets, ≥ 2× internodes. **Glumes** lanceolate to narrowly ovate, green (purplish at higher elevations), ≥ 75% the length of adjacent lemmas, 3–7-veined; hyaline margins 0.2–0.5 mm wide, ± equal in width; glumes widest at or slightly beyond midlength, tips acute, awnless or with awns to 2 mm. **Lemmas** 6–13 mm, glabrous, awnless or with straight awns to 5 mm. **Anthers** (0.8)1.2–2.5(3) mm. **Habitat**: Grasslands, sagebrush steppe, open forest, from near sea level to alpine. Widespread, common. **Comments**: *Elymus lanceolatus* is strongly rhizomatous, has shorter anthers, and shorter glumes relative to adjacent florets. A number of cultivars have been released in the US and Canada since the 1940s.

Elymus violaceus (Hornem.) Feilberg
Arctic Wheatgrass

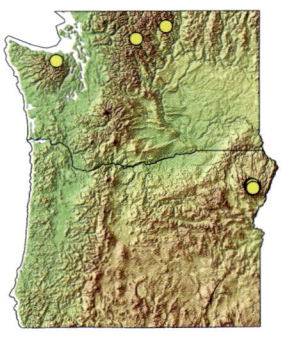

Cespitose perennial; **culms** 18–75 cm, often decumbent or geniculate. **Leaf sheaths** glabrous; **blades** 3–4 mm wide, flat, glabrous or hairy; **ligules** 0.5–1 mm; **auricles** ± 0.5 mm. **Spikes** erect, usually purplish; 1 spikelet per node; disarticulation above glumes. **Spikelets** (9)10–15(17) mm, with (3)4–5 florets, ≥ 2× internodes. **Glumes** narrowly ovate to obovate, often purplish, ≥ 75% the length of adjacent lemmas, 3(5)-veined; hyaline margins unequal in width, the wider one 0.3–1 mm wide, glumes usually widest in distal 33%; tips acute to rounded, awnless or with awns to 2 mm. **Lemmas** (6.4)8–9.5 mm, glabrous or pubescent, usually with straight awns 0.5–3 mm. **Anthers** 0.7–1.3 mm. **Habitat**: Dry to moist alpine meadows, talus, lake and creek shores. **Comments**: Arctic wheatgrass differs from *E. trachycaulus* by the wider, unequal glume margins and glumes that are widest in the distal 33%. Intermediate forms occur that are hard to assign to one species or the other. Arctic wheatgrass is rare in OR and WA.

Elymus wawawaiensis J.R. Carlson & Barkworth
Snake River Wheatgrass

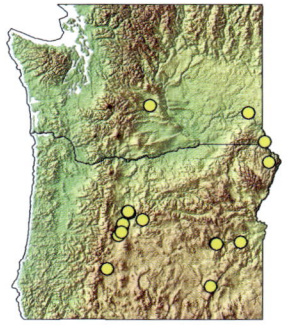

Cespitose perennial, sometimes weakly rhizomatous; **culms** (15)50–130 cm. **Leaf blades** 1.7–5 mm wide, involute when dry, inner surface usually densely pubescent; **ligules** 0.1–1.1 mm, **auricles** to 1.2 mm. **Spikes** erect to slightly nodding; 1 spikelet per node. **Spikelets** 10–22 mm, with 4–10 florets; about 2× internodes. **Glumes** narrowly lanceolate, 0.5–1.3 mm wide, widest at or below middle, 1–3-veined; **awns** 0–6 mm. **Lemmas** 6–12 mm, awns 9–28 mm, strongly divergent. **Anthers** 3.5–6 mm. **Habitat**: Dry, rocky canyon slopes, disturbed sagebrush steppe, roadsides. Endemic to Snake, Yakima, and Salmon drainages of OR, WA, and ID. Planted elsewhere E of the Cascades. **Comments**: Resembles a robust *Pseudoroegneria spicata*, but the spikelets are more overlapping, the glumes are stiffer and narrower, with 1–3 veins, and the leaves are more evenly distributed. "Secar" is an *E. wawawaiensis* cultivar.

Eragrostis cilianensis (All.) Vignolo ex Janch.
Stinkgrass [E]

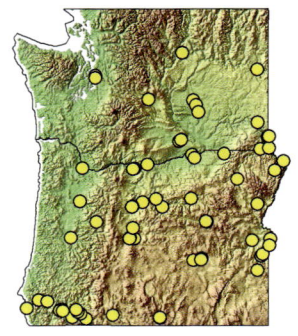

Unpleasant-smelling annual; **culms** 15–45(65) cm, tufted, erect to decumbent, often with saucer-like glands below nodes. **Leaf sheaths** occasionally glandular, top with hairs to 5 mm; **blades** (1)3–5(10) mm wide, flat to involute, margins glandular, glabrous or hairy above, sometimes glandular below; **ligules** 0.4–0.8 mm, membranous, ciliate. **Panicles** condensed to open. **Spikelets** 6–20 × 2–4 mm, with 10–40 florets. **Glumes** usually glandular; lower glumes 1.2–2 mm, 1-veined; upper glumes 1.2–2.6 mm, 3-veined; **lemmas** 2–2.8 mm, keels with 1–3 saucer-like glands, obtuse to acute; **paleas** often ciliate on keels, falling after the lemmas. **Anthers** 0.2–0.5 mm. **Seeds** globose-ellipsoid, not grooved.
Habitat: Disturbed sites, pavement cracks, gardens, roadsides. Native to Europe.
Comments: *Eragrostis cilianensis* inflorescences often seem too large for the short plants. Stinkgrass differs from our other annual *Eragrostis* in having globose seeds, yellow anthers, and wider spikelets.

Eragrostis curvula (Schrad.) Nees
Weeping Lovegrass [E]

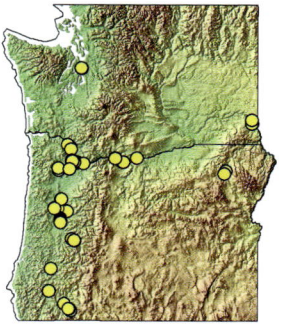

Cespitose perennial without glands; **culms** 60–150 cm. **Leaves** mostly basal; **sheaths** with scattered hairs to 9 mm; **blades** 1–3 mm wide, involute or flat, mostly glabrous but upper surfaces with scattered hairs to 7 mm near base; **ligules** 0.6–1.3 mm, ciliate. **Panicles** 16–35(40) × (4)8–24 cm, open, nodding, usually exceeding the basal leaves, branches ascending to divergent. **Spikelets** 4–8.2(10) mm, gray to yellowish, with 3–10 florets. **Lower glumes** 1.2–2.6 mm; **upper glumes** 2–3 mm. **Lemmas** 1.8–3 mm, veins conspicuous, acute. **Paleas** falling after or with the lemmas; **rachilla** not persistent. **Anthers** 0.6–1.2 mm. **Seeds** 1–1.7 mm, ellipsoid to obovoid, dorsally compressed, with or without a broad, shallow groove. **Habitat**: Roadsides, weedy meadows, ditches, shorelines, and disturbed areas. Native to southern Africa. **Comments**: Resembling a large, fine-leaved fescue, *E. curvula* is unique among our lovegrasses. Initially planted for soil stabilization, it has spread and is showing potential to be invasive in disturbed grasslands.

Eragrostis hypnoides (Lam.) Britton, Sterns & Poggenb.
Teal Lovegrass

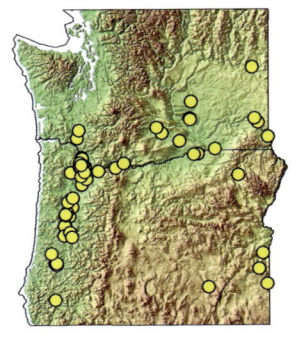

Stoloniferous mat-forming annual without glands; **culms** (2)5–12(20) cm, decumbent, rooting at nodes, often branched. **Leaf sheaths** short-hairy on margins and collars; **blades** 1–2 mm wide, flat to involute, glabrous below, short-hairy above; **ligules** 0.3–0.6 mm, ciliate. **Panicles** open to somewhat congested. **Spikelets** 4–13 × 1–1.5 mm, with 12–35 florets, greenish-yellow to purplish. **Lower glumes** 0.4–0.7 mm; **upper glumes** 0.8–1.2 mm; **lemmas** 1.4–2 mm, strongly 3-veined, tips acuminate; **paleas** persistent on the rachilla after lemmas fall. **Anthers** 0.2–0.3 mm. **Seeds** 0.3–0.5 mm, ellipsoid. **Habitat**: Receding shorelines of lakes, rivers, and ponds; native to both N and S America. **Comments**: This low, creeping shoreline grass is usually found late in the growing season as stream and lake levels recede.

Eragrostis lutescens Scribn.
Sixweeks Lovegrass

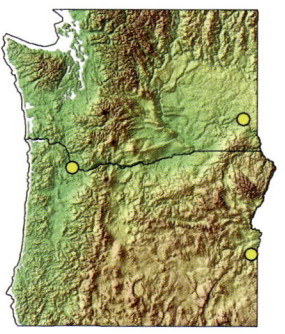

Tufted annual; **culms** (2)6–25 cm, with yellowish glandular pits below nodes. **Leaf sheaths** and blade bases with elliptical, glandular pits; **sheaths** with sparse hairs to 2 mm at the throat; **blades** 1–3 mm wide, flat to involute; **ligules** 0.2–0.5 mm, ciliate. **Panicles** 4–10(15) × 0.5–2 cm, contracted, branches appressed to diverging 30°; dense, glandular pits on inflorescence axis and branches. **Spikelets** 3.6–7.5 × 1.2–3 mm, with 6–11(14) florets, light yellowish, sometimes mottled with reddish-purple, glandless. **Lower glumes** (0.7)0.9–1.4 mm; **upper glumes** 1.2–1.8 mm; **lemmas** 1.5–2.2 mm, veins conspicuous, acute; **paleas** persistent, remaining on rachilla after lemmas fall. **Anthers** 0.2–0.3 mm. **Seeds** 0.5–0.8 mm, pear-shaped, slightly flattened on 1 side. **Habitat**: Sandy river banks and alkaline flats. Rare. **Comments**: *Eragrostis lutescens* is a rare native lovegrass of the W United States. The handful of OR and WA records are from along the Columbia and Snake Rivers and their tributaries; most are more than 80 years old.

Eragrostis mexicana (Hornem.) Link ssp. *virescens* (J. Presl) S.D. Koch & Sánchez Vega
Mexican Lovegrass [E]

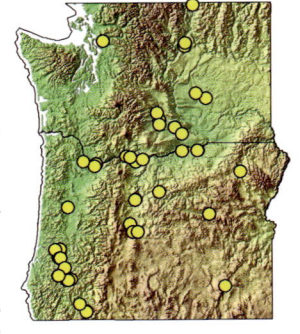

Tufted annual; **culms** 10–130 cm. Glands lacking. **Leaf sheaths** with papillose-based hairs to 4 mm at mouth; **blades** 2–7(9) mm wide, flat; **ligules** 0.2–0.5 mm, ciliate. **Panicles** (5)10–40 × (2)4–18 cm, open, primary branches solitary, paired, or whorled, appressed or diverging up to 80°. **Spikelets** 5–10 × 0.7–1.5 mm, with 5–11(15) florets, gray-green to purplish. **Glumes** 0.7–1.7 mm, subequal; **lemmas** 1.2–2.4 mm, acute; **rachilla** persistent, lemmas falling before paleas. **Anthers** 0.2–0.5 mm. **Seeds** 0.5–1 mm, with a shallow ventral groove on the side opposite the embryo. **Habitat**: Disturbed areas, roadsides. Native to CA and S America. **Comments**: Seeds of the more common *E. pectinacea* are not grooved opposite the embryo. Seeds of both *E. mexicana* and *E. pectinacea* may have a shallow depression over the embryo. This is not the groove!

Eragrostis minor Host
Little Lovegrass [E]

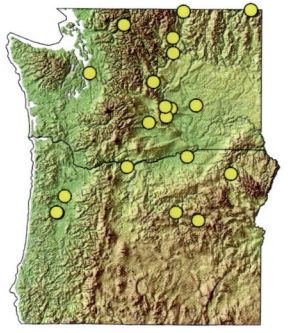

Tufted annual; **culms** 10–45 cm, with or without a ring of glandular tissue below nodes. **Leaf sheaths** sometimes glandular on midveins, tops with hairs to 4 mm; **blades** 1–3(4) mm wide, margins with saucer-shaped glands; **ligules** 0.2–0.5 mm, ciliate. **Panicles** 4–20 × 2.2–8(10) cm, open to contracted, glandular below nodes of axis, branches ascending to divergent to 100°; pedicels usually with a distal ring of saucer-shaped glands. **Spikelets** 4–7(11) × 1.1–2.2 mm, with 7–12(20) florets, reddish-purple to greenish or gray. **Lower glumes** 0.9–1.4 mm; **upper glumes** 1.2–1.6 mm; **lemmas** 1.4–1.8 mm, occasionally with 1–2 saucer-like glands, acute to obtuse; **paleas** persistent on rachillas after lemmas fall. **Anthers** 0.2–0.3 mm. **Seeds** ellipsoid, not grooved. **Habitat**: Disturbed roadsides, pavement cracks. Native to Europe. **Comments**: Similar *E. pectinacea* lacks glands; *E. cilianensis* has denser inflorescences and wider spikelets.

Eragrostis pectinacea (Michx.) Nees var. *pectinacea*
Tufted Lovegrass (yellow dots)

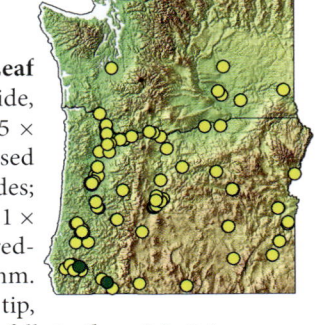

Tufted annual, without glands; **culms** 10–80(90) cm. **Leaf sheaths** with hairs to 4 mm at top; **blades** 1–4.5 mm wide, flat to involute; **ligules** 0.2–0.5 mm, ciliate. **Panicles** 5–25 × 3–12(15) cm, usually open, with primary branches appressed or diverging to 80°, solitary or paired at lowest 2 nodes; pedicels appressed, rarely spreading to 20°. **Spikelets** 3.5–11 × 1.2–3.5 mm, with 6–22 florets; gray, yellow-brown, or dark red-purple. **Lower glumes** 0.5–1.5 mm; **upper glumes** 1–1.7 mm. **Lemmas** 1–2.2 mm, gray-green at base, reddish-purple at tip, acute; **paleas** persistent, remaining on rachilla after lemmas fall. **Anthers** 0.2–0.4 mm. **Seeds** 0.5–1.1 mm, pear-shaped, not grooved opposite embryo. **Habitat**: Pavement cracks, river shorelines, disturbed areas. **Comments**: Our most common lovegrass. Variety *miserrima* (E. Fourn.) Reeder (not pictured; green dots), native from the S US to S America, was collected twice in SW OR, is expected to move northward. Its spikelets are borne on widely divergent (not appressed) pedicels.

Eragrostis pilosa (L.) P. Beauv. var. *pilosa*
India Lovegrass [E]

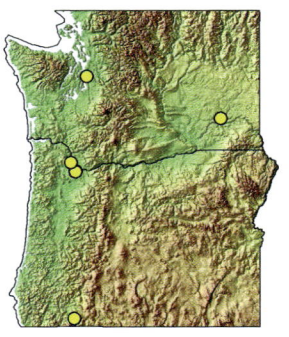

Tufted annual with few or no glands; **culms** 8–45(70) cm. **Leaf sheaths** with hairs to 3 mm at the top; **blades** 1–2.5(4) mm wide, flat; **ligules** 0.1–0.3 mm, ciliate. **Panicles** open, 4–20(28) × 2–15(18) cm, branches diverging 10–80(110°), whorled on lowest 2 nodes. **Spikelets** (2)3.5–6(10) × 0.6–1.3 mm, with (3)5–17 florets, gray. **Lower glumes** 0.3–0.6(0.8) mm; **upper glumes** 0.7–1.2(1.4) mm; **lemmas** 1.2–1.8 mm, lateral veins inconspicuous, acute; **paleas** tardily deciduous, leaving zigzag rachillas. **Anthers** 0.2–0.3 mm. **Seeds** 0.5–0.9 mm, obovoid to prism-shaped. **Habitat**: Disturbed sites, roadsides, lake shores. Uncommon; native to Eurasia. **Comments**: *Eragrostis pilosa* has whorled lower panicle branches and deciduous paleas that leave bare, zigzag rachillas. Similar *E. pectinacea* has persistent paleas and only 1 or 2 panicle branches at each lower node.

Eremopyrum triticeum (Gaertn.) Nevski
Annual Wheatgrass [E]

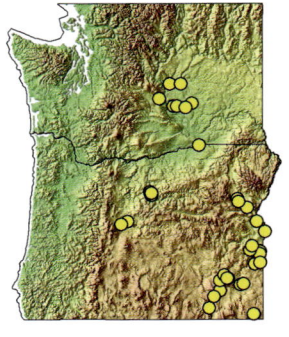

Tufted annual; **culms** 10–30 cm, decumbent to geniculate. **Leaf sheaths** mostly open, upper sheaths inflated; **blades** 1–3(6) mm wide; **ligules** 0.4–2 mm, membranous, truncate. **Spikes** 1.3–2.4 cm × 0.8–2 cm, strongly 2-sided; 1 spikelet per node and crowded in 2 rows on opposite sides of axis. **Spikelets** 6–12 mm with 2–3 florets. **Glumes** usually 2-keeled, with curved, inflated bases. **Lemmas** 5–7.5 mm, leathery, glabrous, except lowest lemma pubescent on lower half; tip acute to short-awned. **Anthers** 0.4–1.3 mm. **Habitat**: Disturbed, often alkaline sites. Native to Central and W Asia. **Comments**: This small wheatgrass has short, strongly flattened spikes that are a little longer than wide.

Eriocoma hendersonii (Vasey) Romasch.
Henderson's Ricegrass

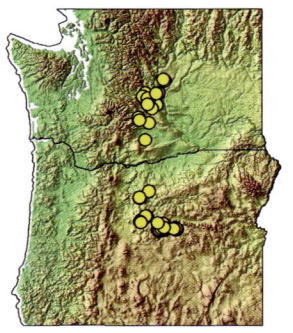

Cespitose perennial; **culms** 10–35 cm. **Leaf blades** tightly rolled or folded, 0.5–1 mm wide, hairy on the inner surface; **ligules** 0.4–1 mm. **Panicles** narrow, with erect to ascending branches. **Glumes** 3–5 mm, tips rounded to obtuse. **Florets** 3.5–4.5 mm, 1–1.5 mm wide, elliptic, laterally compressed; mature **lemmas** dark brown to black, leathery, shiny, glabrous, **awns** 6–10 mm, deciduous, scabrous, not geniculate. **Anthers** about 2.5 mm, tips hairy. **Habitat**: Shallow, rocky soils in scablands, sagebrush, or ponderosa pine forest. **Comments**: Rare, endemic to E OR and WA; Forest Service sensitive species and BLM special status species. Similar *E. wallowaensis* has spreading, drooping panicle branches.

Eriocoma hymenoides (Roem. & Schult.) Rydb.
Indian Ricegrass

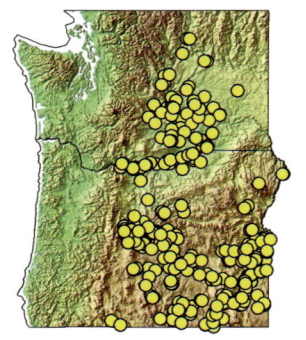

Cespitose perennial; **culms** 25–70 cm. **Leaf blades** tightly rolled, 0.1–1 mm wide, **ligules** 1.5–4 mm, glabrous. **Panicles** diffusely open with strongly divergent branches. **Glumes** 5–9 mm, bulging below, tapering above midlength to acuminate tips. **Florets** 3–4.5 mm, 1–2 mm wide, elliptic; **mature lemmas** hard, dark brown, with long, dense, white hairs, awns 3–6 mm, deciduous, not geniculate. **Anthers** 1.5–2 mm, tips hairy. **Habitat**: Dry, well-drained, usually sandy soils in sagebrush or bitterbrush steppe, juniper woodland, and open ponderosa pine forest. **Comments**: Slimy secretions form a sheath around roots and protects them from abrasion and hosts nitrogen-fixing organisms. Indian ricegrass is used for habitat restoration projects, particularly on sandy soils, and is planted as an ornamental in xeriscaped settings E of the Cascade Range. Seeds of Indian ricegrass were used as food by Native Americans, and more recently to make gluten-free flour.

Eriocoma lemmonii (Vasey) Romasch. ssp. *lemmonii*
Lemmon's Needlegrass

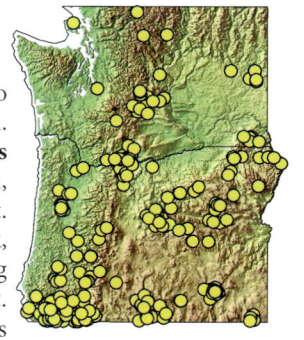

Cespitose perennial; **culms** 15–90 cm. **Leaf blades** folded to convolute, 0.5–2.5 mm wide, **ligules** membranous, to 2.5 mm. **Panicles** contracted with ascending to erect branches. **Glumes** 7–11.5 mm. **Florets** 5.5–7 mm, 0.8–1.3 mm wide, fusiform, somewhat laterally compressed. **Calluses** 0.4–1.2 mm, blunt. **Lemmas** leathery, evenly hairy, hairs 0.4–1 mm, with a thick, stiff apical lobe 0.1 mm, awns 16–30 mm, persistent, appearing off-center on the lemma tip, bent twice, scabrous throughout. **Paleas** 75–100% as long as lemma. **Anthers** 3.2–3.5 mm, tips glabrous. **Habitat**: Grassy balds, oak savanna, upland prairie, sagebrush steppe, ponderosa pine forest. **Comments**: Mature *Eriocoma lemmonii* florets are laterally compressed (roll a floret between your fingers) and have a distinctive, though subtle plumpness that contrasts with the slender, almost cylindric florets of other PNW needlegrasses that have persistent awns. Vegetative *E. lemmonii* is similar to cespitose fine-leaved fescues but has longer ligules and harsher, stiffer leaves.

Eriocoma nelsonii (Scribn.) Romasch.

Cespitose perennial; **culms** 40 to 175 cm. **Leaf blades** flat (becoming rolled or folded when dry), 1.2–5 mm wide, **ligules** 0.2–1.5 mm, truncate to acute. **Panicles** contracted with erect branches. **Glumes** 6–12.5 mm. **Florets** 4.5–7 mm, 0.6–0.9 wide, fusiform, round in cross section. **Calluses** 0.2–1 mm, blunt to sharp. **Lemmas** hairy, hairs 0.5–2 mm, lemma apex with flexible, membranous lobes 0.1–0.4 mm, awns 19–45 mm, persistent, bent twice, first 2 segments scabrous or short-hairy, with hairs < 0.5 mm, hairs of basal segment < hairs of the lemma apex, terminal awn segment glabrous. **Palea** 33–67% as long as lemma. **Anthers** 2–3.5 mm, tips glabrous. **Comments**: See also similar species *E. lemmonii*, *E. nevadensis*, and *E. occidentalis*. Intermediate forms occur between the 2 subspecies of *E. nelsonii*.

1a Calluses blunt, the glabrous area on the dorsal side about as long as wide, with a straight to rounded boundary between the glabrous tip and the hairy portion, glabrous tip 0.02–0.06 mm; awns 19–31 mm; E of the Cascades, NW WA and SW OR..**ssp. *dorei***

1b Calluses sharp, the glabrous area on the dorsal side longer than wide, with an acute boundary between the glabrous tip and the hairy portion, glabrous tip 0.05–0.15 mm; awns 19–45 mm; in and E of the Cascades................ **ssp. *nelsonii***

Eriocoma nelsonii ssp. *dorei* (Barkworth & J. Maze) Romasch.
Dore's Needlegrass

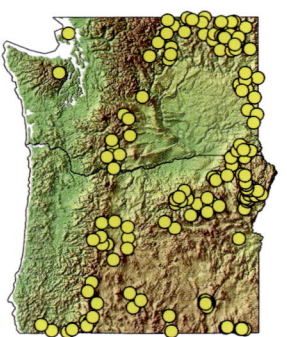

Calluses blunt, the glabrous dorsal part 0.02–0.06 mm, about as long as wide, with a straight to rounded boundary with the hairy part. **Habitat**: Montane to subalpine meadows and forest. **Comments**: Common. Generally at higher elevations than Columbia needlegrass.

Eriocoma nelsonii ssp. *nelsonii*
Columbia Needlegrass

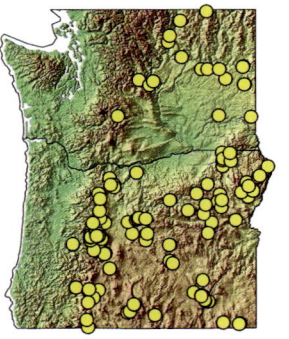

Calluses sharp, the glabrous dorsal part 0.05–0.15 mm, longer than wide, with an acute boundary with the hairy part. **Habitat**: Mesic to moist areas in meadows and forest openings, sagebrush steppe, juniper woodland. **Comments**: Widespread but less common than Dore's needlegrass. Similar *E. occidentalis* ssp. *californica* usually has spreading hairs on the lower awn segments, but occasional plants with scabrous or short-hairy basal awn segments can be difficult to distinguish from *E. nelsonii* ssp. *nelsonii*. *Eriocoma occidentalis* ssp. *californica* has sharper calluses, with the glabrous part extending farther into the hairy part of the callus, and narrower leaf blades than *E. nelsonii* ssp. *nelsonii*.

Eriocoma nevadensis (B.L. Johnson) Romasch.
Nevada Needlegrass

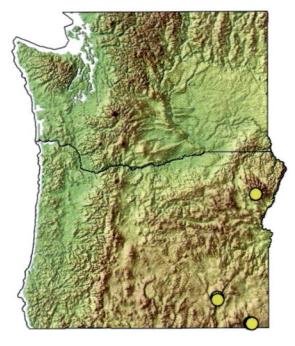

Cespitose perennial; **culms** 20–85 cm, retrorsely hairy below lower nodes. **Leaf blades** usually involute, 1–3 mm wide, **ligules** 0.2–1 mm. **Panicles** contracted, with appressed branches. **Glumes** 7–14 mm. **Florets** 5–6.5 mm, 0.6–1 mm thick, fusiform, round in cross section. **Calluses** 0.5–0.7 mm, sharp, with rounded to acute boundary dorsally between glabrous tip and hairy part. **Lemmas** hairy, hairs 0.5–2 mm, lemma tips with membranous lobes 0.4 mm; **awns** 20–30 mm, persistent, bent twice, the first 2 segments with spreading hairs of mixed lengths, 0.5–1.5 mm, hairs of basal segment shorter than hairs of the lemma apex, terminal awn segment scabrous or smooth. **Palea** 50–75% as long as lemma. **Anthers** about 2.5 mm, tips glabrous. **Habitat**: Sagebrush steppe and open woodland. **Comments**: Similar to *E. occidentalis* ssp. *californica*, differing in the length of hairs at the top of the lemma and on the lower part of the awn. Rare in OR. This is an intermountain species, more common from the E slope of the Sierra Nevada across Nevada into S Idaho.

Eriocoma occidentalis (Thurb. ex S. Watson) Romasch.

Cespitose perennial bunchgrass, **culms** 14–120 cm. **Leaf blades** flat or convolute, 0.8–2 mm wide, **ligules** 0.2–1.5 mm. **Panicles** contracted, with branches appressed. **Glumes** 9–15 mm. **Florets** 5.5–7.5 mm, 0.5–0.9 mm thick, fusiform, round in cross section. **Calluses** 0.8–1.2 mm, sharp, with narrowly acute boundary dorsally between glabrous tip and hairy part. **Lemmas** hairy, hairs 0.5–1.5 mm, lemma tips with membranous lobes 0.3–0.5 mm, awns 15–55 mm, persistent, bent twice, the first 2 segments usually hairy, sometimes scabrous, terminal segment glabrous, scabrous, or hairy. **Palea** 40–60% as long as lemma. **Anthers** 2.5–3.5 mm, tips glabrous.

1a First 2 awn segments scabrous or with hairs of mixed lengths; hairs at the tip of the lemma longer than the hairs on the lower part of the awn ..... **ssp. *californica***
1b First 2 awn segments with hairs that gradually and evenly become shorter toward the first bend; hairs at the tip of the lemma about as long as the hairs on the lower part of the awn .... **ssp. *pubescens***

Eriocoma occidentalis ssp. *californica* (Merr. & Burtt Davy) Romasch.
California Needlegrass

First 2 awn segments with hairs of mixed lengths or scabrous; hairs at lemma tip longer than hairs on basal awn segment. **Habitat**: Sagebrush steppe, coniferous forest. **Comments**: *Eriocoma occidentalis* ssp. *californica* may have originated as a hybrid between *E. nelsonii* and *E. occidentalis*. Similar *E. nelsonii* ssp. *nelsonii* has wider leaves and calluses that are less sharp. See also *E. nevadensis*.

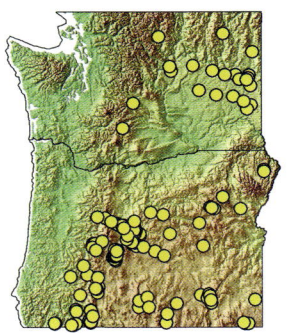

E. occidentalis ssp. *pubescens* (Vasey) Romasch.
Western Needlegrass

First 2 awn segments with hairs that gradually and evenly become shorter toward the first bend; hairs at lemma tip about equal to hairs of the basal awn segment. **Habitat**: Dry grasslands, rock outcrops, sagebrush steppe, coniferous forest. **Comments**: Intergrades with *E. occidentalis* ssp. *californica* and with *E. nelsonii*, which has scabrous to short-hairy basal awn segments.

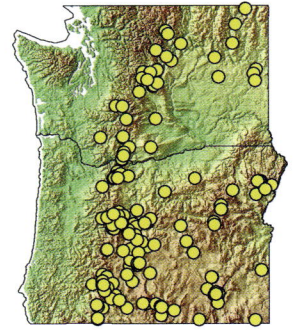

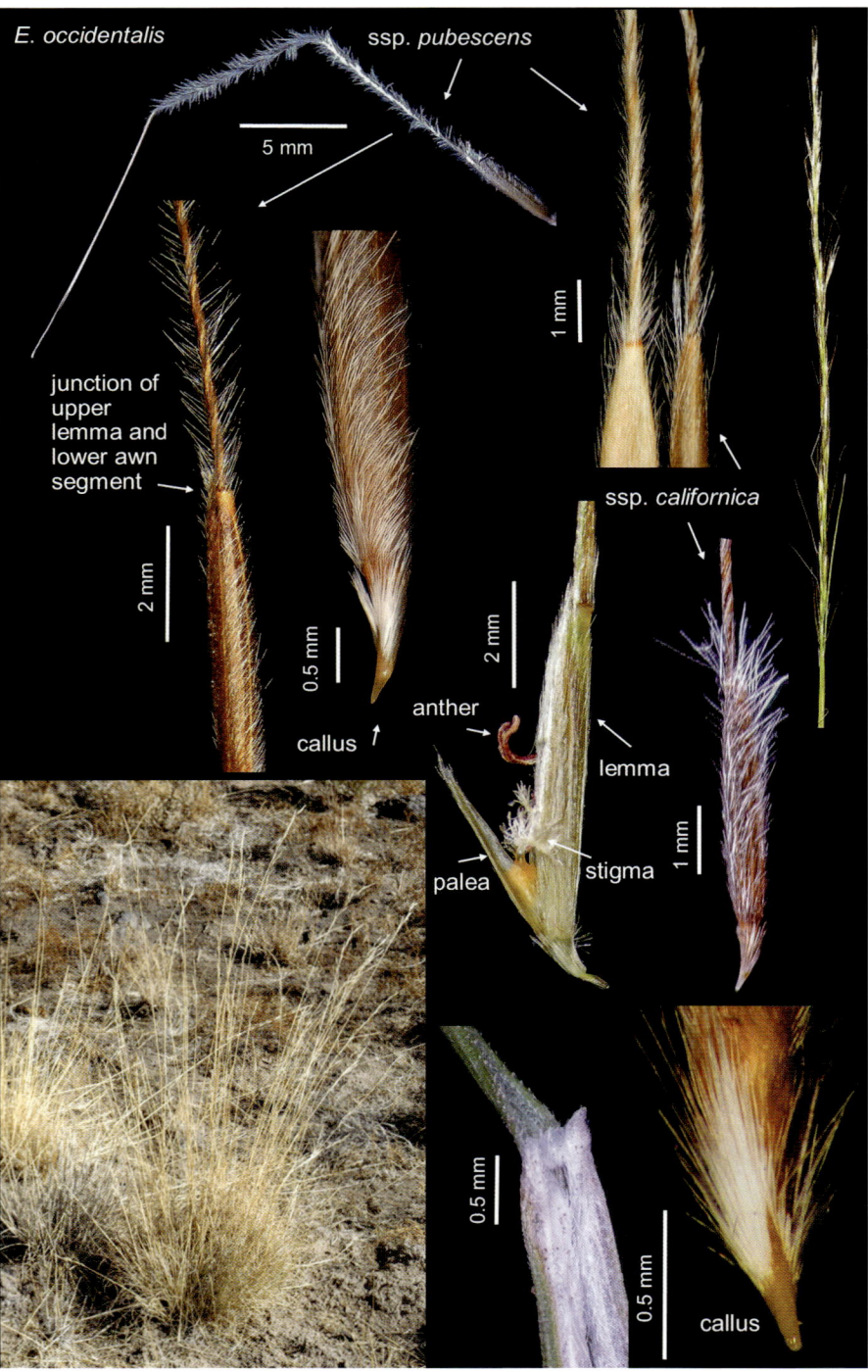

Eriocoma pinetorum (M.E. Jones) Romasch.
Pine Needlegrass

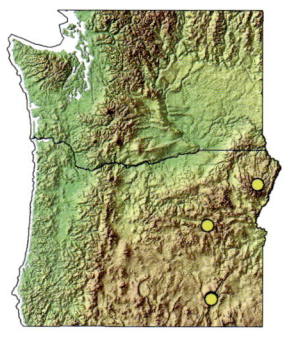

Cespitose perennial; **culms** 14–50(80) cm. **Leaf blades** very narrow, usually rolled, 0.2–0.4 mm in diameter; **upper ligules** to 2 mm, rounded. **Panicles** contracted; with branches appressed. **Glumes** 7–11 mm. **Florets** 3.5–5.5 mm, 0.6–0.8 mm thick, fusiform, round in cross section. **Calluses** 0.4–0.6 mm, sharp. **Lemmas** densely long-hairy, the hairs 1.5–5 mm; tips with thin lobes to 0.3 mm, awns 13–25 mm, persistent, bent twice, the first 2 segments scabrous. **Palea** 67–100% as long as lemma. **Anthers** 1.8–2.6 mm, tips glabrous. **Habitat**: Dry, rocky subalpine to alpine ridges. **Comments**: Dense, long lemma hairs are obvious at arm's length. Rare. In the PNW, known only from Steens Mt. and the Strawberry and Wallowa Mts. in OR.

Eriocoma richardsonii (Link) Romasch.
Richardson's Needlegrass

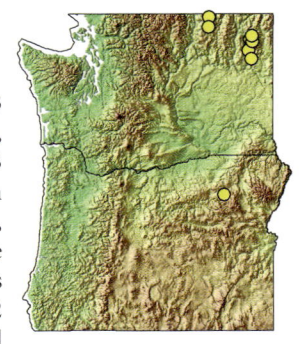

Cespitose perennial; **culms** 30–100 cm. **Leaf blades** 0.8–3 mm wide, **ligules** 0.1–0.5 mm. **Panicles** open with drooping, flexuous panicle branches. **Lower glumes** 7.5–11 mm, 2–3 mm longer than upper glumes. **Florets** 5–6 mm, 0.6–0.9 mm wide, fusiform, round in cross section. **Calluses** 0.4–0.7 mm, blunt. **Lemmas** hairy, becoming glabrous toward the tip, the hairs 0.2–0.5 mm; lemma tips without lobes or with tiny lobes to 0.1 mm, awns 15–25 mm, persistent, bent twice, the first 2 awn segments with very short hairs to 0.1 mm, the terminal awn segment straight. **Palea** 50–60% as long as the lemma. **Anthers** 2.5–3 mm, tips hairy. **Habitat**: Gravelly or sandy soils in open woodlands and montane grasslands. Uncommon in NE WA; rare in NE OR. **Comments**: Easily damaged by livestock grazing, this grass is found only in undisturbed areas. It is listed as a sensitive species in WA and OR.

Eriocoma thurberiana (Piper) Romasch.
Thurber's Needlegrass

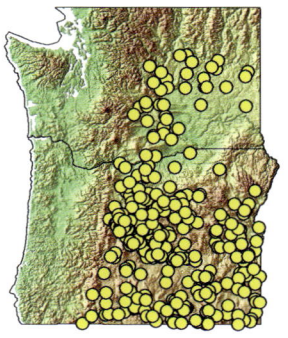

Cespitose perennials with pale roots, **culms** to 30–75 cm. **Leaf blades** convolute, becoming distinctively curly with age, 0.5–2 mm wide; **ligules** of flag leaves membranous, acute, to 8 mm. **Panicle** contracted, often partially enclosed in upper leaf sheath at flowering, branches appressed or strongly ascending. **Lower glumes** 10–15 mm, to 2 mm longer than upper glumes. **Florets** fusiform, round in cross section. **Calluses** 0.9–1.5, sharp. **Lemmas** 6–9 mm, leathery, hairy, sometimes glabrate on upper part of back, hairs 0.5–0.8 mm, lemma tips with a thick 0.1 mm lobe on 1 margin, awns 30–60 mm, persistent, bent twice, the first 2 awn segments with obvious spreading white hairs to 2 mm. **Palea** 75–90% as long as the lemma. **Anthers** 2.5–3.5 mm, tips glabrous. **Habitat**: Dry, open pine or juniper woodlands, shrub steppe. **Comments**: The distinctive long ligules make *E. thurberiana* easy to identify in the field. It is often codominant with Idaho fescue, which has black roots and older leaves that do not curl. Thurber's needlegrass is excellent for restoration and as a garden subject and is available at PNW native plant nurseries.

Eriocoma wallowaensis (J. Maze & K. Robson) Romasch.
Wallowa Ricegrass

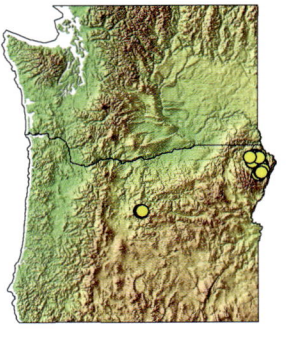

Cespitose perennial; **culms** (10)15–40(45) cm. **Leaf blades** tightly folded or rolled, hairy on the inner surface, 0.5–1 mm wide; **ligules** 0.8–1.6 mm, hairy on the inner surface. **Panicles** open, branches spreading to divergent, often drooping. **Glumes** 3–7 mm, tips obtuse to acute. **Florets** 3–5.5 mm, 1–1.5 mm wide, elliptic, ± round in cross section. **Lemmas** dark brown to black, leathery, shiny, glabrous, awns 5–11 mm, deciduous, scabrous, not or weakly geniculate. **Anthers** 1.6–1.8 mm, tips glabrous. **Habitat**: Dry, shallow, rocky (scabland) soils on nonforested sites with other bunchgrasses and forbs and generally a low density of other vegetation. **Comments**: Wallowa ricegrass is a rare endemic of the Wallowa and Ochoco Mts. in E OR; similar Henderson's ricegrass has appressed panicle branches. Wallowa ricegrass is listed as a sensitive species by both the Forest Service and the Bureau of Land Management; nearly all documented populations occur on land managed by these agencies.

Eriocoma webberi Thurb.
Webber's Needlegrass

Cespitose perennial; **culms** 12–35 cm. **Leaf blades** usually rolled, about 0.5 mm wide, 0.5–1.5 mm wide when flattened, densely hairy on the inner surface; **upper ligules** 1–2 mm, acute. **Panicles** contracted, branches appressed, the base often enclosed in the upper sheath. **Glumes** 6–10 mm. **Florets** 4.5–6 mm, 0.7–1 mm wide, fusiform, round in cross section. **Calluses** 0.3–0.8 mm, blunt. **Lemmas** densely long-hairy, the hairs 2.5–5 mm, tips with membranous lobes 0.6–1.9 mm, awns 4–11 mm, deciduous, straight to bent once, scabrous.
Habitat: Dry open flats, rocky slopes, often with sagebrush. Uncommon. **Comments**: Distinctive long lemma hairs are visible at arm's length; similar *E. pinetorum* has longer, persistent awns. Pine needlegrass grows at high elevations and Webber's needlegrass is considered its desert counterpart.

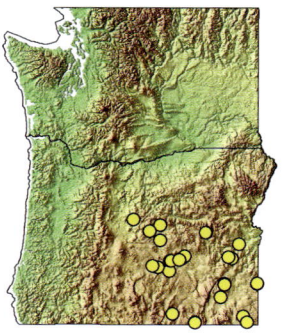

Festuca ammobia Pavlick
Sand Fescue

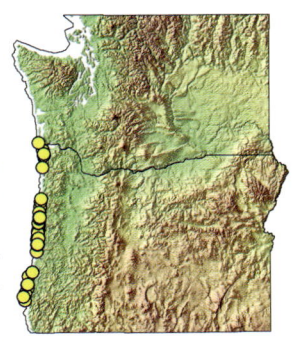

Perennial with short, cespitose shoot clusters arising from delicate rhizomes; **culms** often < 20 cm. Foliage green to strongly glaucous, often in the same population; **sheaths** closed about 75% their length, usually red-brown with contrasting white veins; **blades** 0.1–1.5 mm in diameter, ± diamond-shaped in cross section, **fiber bundles** < 2× as long as wide, strongly developed, sometimes fused with neighboring bundles (see illustration on p. 63); **ligules** 0.1–0.5 mm. **Panicles** very dense, nearly white or marked with turquoise, green, or purple. **Spikelets** with 3–10 florets; **lemmas** smooth, appearing veinless, mucronate or very short-awned. **Anthers** 2.4–3.1 mm. **Habitat**: Sandy soils on coastal bluffs and meadows. **Comments**: Sand fescue is often treated as a subspecies of *F. rubra*. Taxonomy of this group is unsettled, and a variety of names have been applied to this species including *F. r.* var. *densiuscula*, *F. r.* var. *juncea*, and *F. r.* var. *pruinosa*. Expect additional name changes in the future.

Festuca brachyphylla Schult. & Schult. f. ssp. *brachyphylla*
Alpine Fescue (yellow dots)

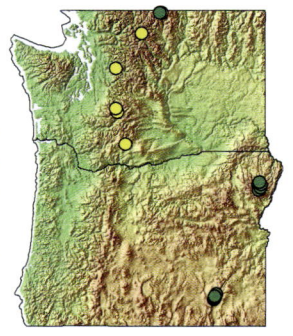

Densely cespitose perennial; **culms** 2.5–15(35) cm, usually > 2× as long as the vegetative shoot leaves. **Leaf sheaths** closed about half their length; **blades** 0.3–0.5(1) mm in diameter, yellow-green or glaucous; ± round in cross section (see illustration on p. 63), inner surface glabrous or nearly so; **ligules** 0.1–0.5 mm. **Inflorescence** contracted; **spikelets** with 2–4(6) florets. **Lemmas** (3)3.5–4.5(6) mm, **awns** 0.7–2 mm. **Anthers** 0.4–0.8(0.9) mm dry, 0.9–1.2 mm fresh or rehydrated, retained long past anthesis. **Habitat**: Alpine rocky slopes and ledges. Mountains of WA. **Comments**: Alpine fescue can be confused with *F. saximontana*, which has longer anthers and thicker leaf fiber bundles.

Festuca brachyphylla ssp. *coloradensis* Fred.
Colorado Fescue (green dots)

Differs from ssp. *brachyphylla* in having **culms** about twice as long as the vegetative shoot leaves; **lemmas** 3–4(4.5) mm. Grows in similar habitats in mountains of N WA, and Wallowa Mts. and Steens Mt., OR.

Festuca bromoides L.
Brome Fescue, Rattail Fescue [E]

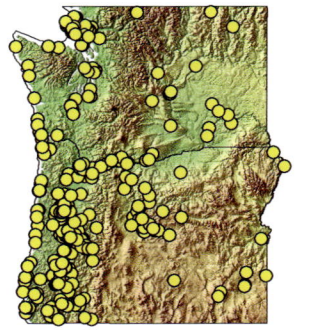

Annual; **culms** 5–75 cm. **Leaf blades** 0.5–2.5 mm wide, flat or rolled; **ligules** to 0.5(1) mm. **Panicles** 1.5–15 × 0.5–3 cm, very narrow, with branches strongly ascending, **pulvini** absent. **Spikelets** 5–10 mm, with 4–8 florets, ascending, not closely overlapping. **Lower glumes** 3.5–5 mm, 50–80% as long as upper glumes; **upper glumes** 4.5–9.5 mm; **lemmas** 4–8 mm, scabrous distally, awns 2–13 mm. **Anthers** 0.4–0.6(1.5) mm. **Habitat**: Moist to dry disturbed areas, roadsides, cultivated grass fields. Native to Europe. **Comments**: *Festuca bromoides* is very similar to *F. myuros*, which has shorter lower glumes that are < half as long as the upper glumes. Strip the florets out of some spikelets so you can see the glumes. Immature *F. microstachys* can be confused with *F. bromoides*, but *F. microstachys* has pulvini in the panicle branch axils.

Festuca californica Vasey
California Fescue

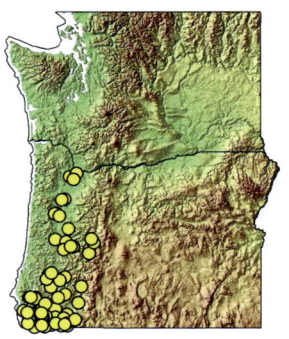

Densely cespitose perennial; **culms** 60–150(200) cm. **Leaf sheaths** closed < one-third their length; **collars** usually hairy at least on margins (check several); **blades** 1.8–5 mm wide, folded or flat, stiff, strongly scabrous; **ligules** 0.5–1.5 mm, ciliate. **Panicles** large, open, with spikelets at ends of spreading to drooping branches. **Spikelets** with 4–6 florets; **lemmas** 7.5–11 mm, hairy or scabrous, awns 1.5–3 mm. **Anthers** 4.5–6(8.5) mm. **Ovary and seed apex** minutely hairy. **Habitat**: Dry grasslands, savannas, open woodlands, especially with Pacific madrone and CA black oak. **Comments**: California fescue is a very large fescue common in savannas and dry forest openings in SW OR, extending to the N end of the Willamette Valley. This grass might remind you of Sitka or California brome, but those species have mostly cauline leaves, fully closed sheaths, and longer lemma awns.

Festuca campestris Rydb.
Mountain Rough Fescue

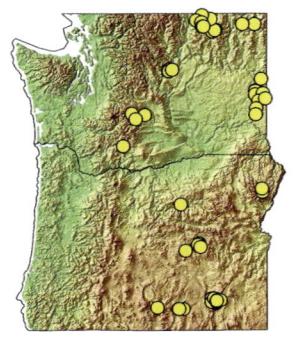

Densely cespitose perennial; **culms** (20)40–90(140) cm. **Leaf sheaths** closed < one-third their length, pale, persisting more than a year, forming a dense tuft of split tubes at plant base; **collars** glabrous; **blades** 0.5–1.1 mm wide when folded, 1.6–2.5(3) mm wide when flat, strongly scabrous; **ligules** 0.1–0.5 mm. Panicle branches erect to stiffly spreading. **Spikelets** with (3)4–5(7) florets. **Lemmas** (6.2)7–8.5(10) mm, scabrous, rarely glabrous, mucronate or with awns 0.5–1.5 mm. **Anthers** (3.3)4.5–6 mm. **Ovary and seed apex** minutely hairy. **Habitat**: Dry grasslands, sagebrush steppe, often on deep soils on N aspects. **Comments**: This uncommon grass has shorter lemma awns and stiffer leaves than *F. idahoensis*. It differs from *F. californica* with its glabrous leaf collars and range E of the Cascades.

Festuca elmeri Scribn. & Merr.
Elmer's Fescue

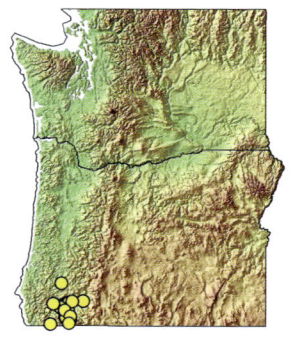

Loosely cespitose perennial; **culms** 40–135 cm. **Leaves** mainly cauline; **sheaths** open; **blades** 1.8–7 mm wide, flat or loosely rolled, upper surface scabrous or pubescent on the veins; **ligules** 0.1–0.5 mm. **Panicles** open, branches spreading; **spikelets** with 2–6 florets. **Calluses** glabrous, wider than long. **Lemmas** 4.8–7 mm, (3)5-veined, scabrous or short-hairy throughout, awns (1)2–5(8) mm, the longest < or > lemma body, straight to slightly bent or wavy. **Anthers** 2–4 mm. **Ovary and seed apex** minutely hairy. **Habitat**: Mesic sites often at the ecotone between savanna and woodland, often on N aspects or on deeper soils under white oak. **Comments**: Elmer's fescue occurs from SW OR to S CA; rare in OR, persisting only in areas protected from livestock grazing. *Festuca subuliflora* has hairy calluses that are longer than wide and longer lemma awns; *F. subulata* lemmas are usually veinless and usually have longer awns.

Festuca filiformis Pourr.
Hair Fescue [E]

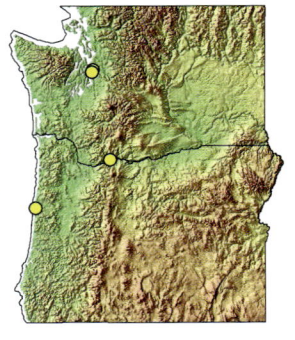

Cespitose perennial; **culms** 18–40 cm. **Leaves** mainly basal, green or glaucous; **sheaths** open; **blades** 0.2–0.6 mm in diameter, folded, elliptic in cross section (see illustration on p. 63), inner surface with fine short hairs; **ligules** 0.1–0.4 mm. **Panicles** narrow with erect branches. **Spikelets** 3–6(6.5) mm, with 2–6 florets; **lemmas** 2.3–4.4 mm, scabrous distally, awnless or with awns to 0.6 mm, awns < 10% as long as lemma body. **Anthers** 1.5–2.2 mm. **Ovary and seed apex** glabrous. **Habitat**: Lawns, roadsides, beaches, disturbed areas. Native to Europe. **Comments**: Hair fescue is a diminutive bunchgrass with small, virtually awnless lemmas and very narrow leaves. Because it rarely flowers it is easily overlooked, and it may be more widespread than we know. It is used in lawn seed mixes.

Festuca idahoensis Elmer
Idaho Fescue

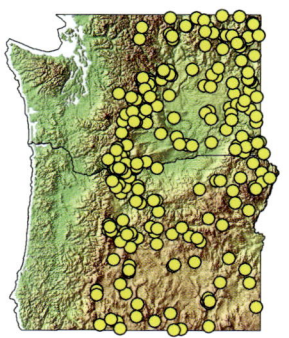

Densely cespitose perennial; roots black; **culms** 30–100 cm. Foliage usually glaucous; **sheaths** open, basal sheaths light brown, lacking contrasting veins, usually crumbling or splitting but not shredding; **blades** 0.3–1 mm in diameter, scabrous, ± round in cross section, spinning readily between the fingers; **ligules** 0.3–0.6 mm. **Panicles** with branches spreading at anthesis, often appressed after flowering. **Spikelets** with 3–7(9) florets; **lemmas** 6–8(10) mm, hairy distally, awns 2–6(7) mm, ≤ lemma body. **Anthers** (2.5)3.2–4.5 mm. **Ovary and seed apex** glabrous. **Habitat**: Grasslands, open juniper, pine or Douglas fir woodlands, shrub steppe; high desert, montane, subalpine. **Comments**: Idaho fescue is an iconic grass of the E Cascades and the shrub steppes commonly called the "high desert." It is often a community dominant in juniper savanna and open ponderosa pine forest, as well as in sagebrush and bitterbrush steppe. Introduced *F. valesiaca* and *F. trachyphylla* have shorter lemmas and denser panicles. Vegetative Idaho fescue can be confused with *Eriocoma thurberiana* (roots pale) or *Carex filifolia* (shoot bases reddish). *Festuca roemeri* is its counterpart W of the Cascade Range.

Festuca lemanii Bastard
Confused Fescue [E]

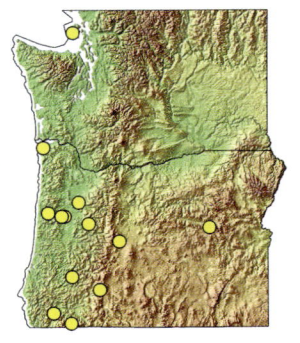

Densely cespitose perennial; **culms** 28–66 cm, erect. Foliage green, usually not glaucous; **basal sheaths** closed < 50% their length, pale and loose, the blade often falling off; **blades** 0.6–1.1 mm in diameter, smooth to finely scabridulous, inner surfaces with 2(4) grooves; **ligules** short, usually minutely ciliate; **sclerenchyma** forming a continuous or nearly continuous band, uniform in thickness. **Panicles** erect, branches ascending to spreading. **Spikelets** 6.5–7.5 mm, with (2)3–6(8) florets; **lemmas** 4–4.9 mm, scabrid or minutely hairy distally, awns 0.3–1.8 mm, **anthers** 1.8–2.5 mm. **Habitat**: Roadsides, disturbed areas. **Comments**: *Festuca lemanii* is a member of the fine-leaved sheep fescue (*F. ovina*) group, which also includes *F. trachyphylla* and *F. valesiaca*. Any of these 3 species might be sold as *F. ovina* or *F. glauca*. Botanists only recently realized that *F. lemanii* occurs in the PNW, although herbarium specimens show that it has been here for decades. Examine a leaf cross section (see illustration on p. 63) to find the continuous, uniformly thick fiber bundles of *F. lemanii*.

Festuca microstachys Nutt.
Small Fescue, Desert Fescue

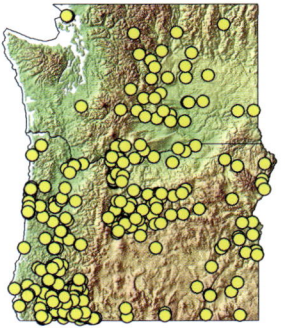

Annual; **culms** 15–75 cm. **Leaf blades** 0.5–1 mm wide, usually rolled; **ligules** 0.5–1 mm. **Inflorescences** open panicles or spike-like racemes, 2–24 × 0.8–8 cm; branches appressed to erect when immature, spreading to reflexed at maturity; **pulvini** present in branch axils. **Spikelets** 4–10 mm, with 1–6 florets. **Lower glumes** 1.7–6.5 mm, 50–75% as long as upper glumes; **upper glumes** 3.5–7.5 mm; **lemmas** 3.5–9.5 mm, smooth, scabrous or evenly hairy, awns (3)6–20 mm. **Anthers** 0.7–3 mm, usually retained inside lemma and plants self-pollinated, occasionally large and exserted, allowing cross-pollination. **Habitat**: Grasslands, scablands, sagebrush steppe, disturbed areas. **Comments**: *Festuca microstachys* varies strikingly in the hairiness of glumes and lemmas. At maturity the lowest panicle branch spreads 90° or more. Young plants with appressed or ascending branches are easily confused with *F. bromoides*. This is our only annual fescue with pulvini that expand to spread the panicle branches at flowering time to facilitate the exchange of pollen.

Festuca myuros L.
Rattail Fescue, Foxtail Fescue [E]

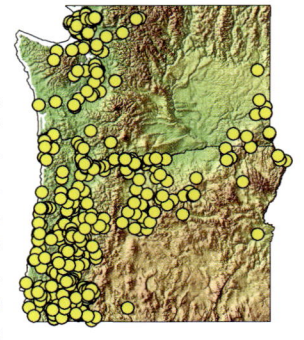

Annual; **culms** 10–75(90) cm. **Leaf blades** 0.4–3 mm wide, usually rolled; **ligules** 0.3–0.5 mm. **Inflorescences** very narrow, dense panicles or spike-like racemes, 3–25 × 0.5–1.5(2) cm; branches strongly ascending, **pulvini** absent. **Spikelets** 5–12 mm, with 3–7 florets. **Lower glumes** 0.5–2 mm, 20–50% as long as upper glumes; **upper glumes** 2.5–5.5 mm; **lemmas** 4.5–7 mm, glabrous, usually scabrous distally, margins sometimes ciliate, awns 5–15(22) mm. **Anthers** 0.5–1(2) mm. **Habitat**: Degraded grasslands, disturbed areas with well-drained soils. Native to Europe, N Africa. **Comments**: *Festuca myuros*, one of our 2 rattail fescues, is similar to *F. bromoides,* which has proportionately longer lower glumes. *Festuca myuros* is a weedy grass of well-drained sites and is a common component of degraded grassland communities that are dominated by weedy annuals.

Festuca nigrescens Lam.
Chewings Fescue [E]

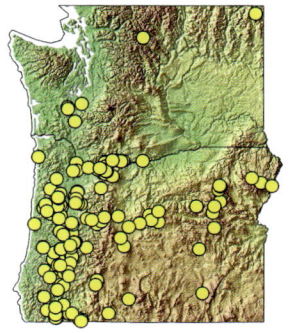

Cespitose perennial; 25–90 cm. Foliage green; **sheaths** closed nearly 100% their length, **basal sheaths** brown with contrasting pale veins, sometimes shredding; **blades** 0.3–0.7(1) mm wide, folded, smooth, usually diamond-shaped to triangular in cross section; **ligules** 0.1–0.5 mm; **fiber bundles** < 2× as wide as thick. **Panicles** open to loosely contracted. **Spikelets** with 3–9 florets; **lower glumes** 2.5–4 mm, 1-veined; **upper glumes** 3.5–5 mm, 3-veined; **lemmas** 4.5–6 mm, usually glabrous, awned, awns 1–3.3 mm. **Anthers** 1.8–2.2(3) mm. **Ovary and seed apex** glabrous. **Habitat**: Roadsides, lawns, meadows. Sometimes treated as a variety of *F. rubra* which differs in its rhizomatous habit. Similar *F. roemeri* can grow mixed with *F. nigrescens* and wild-collected *F. roemeri* seed is often contaminated with *F. nigrescens* seed. *Festuca roemeri* has more scabrous and often glaucous leaves, pale basal sheaths, and lemmas usually > 6 mm long.

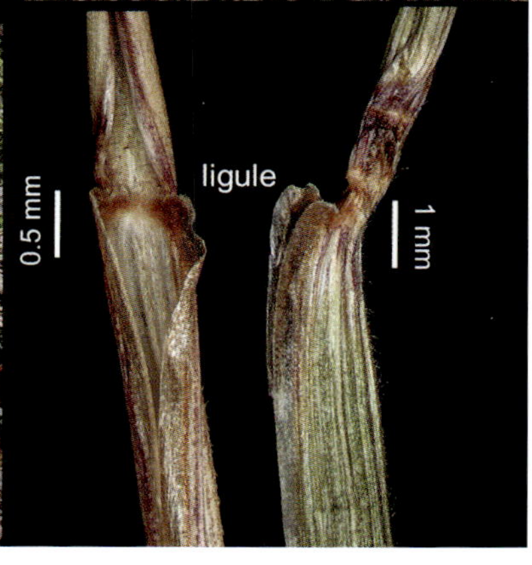

Festuca occidentalis Hooker
Western Fescue

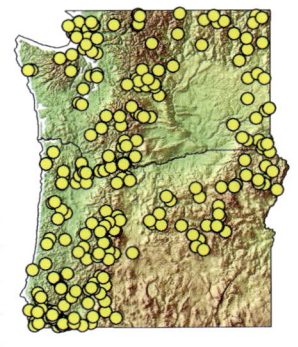

Cespitose perennial with a weak tuft of narrow basal leaves; **culms** (25)45–110 cm, slender. Foliage green; **sheaths** open; **blades** 0.3–0.75 mm in diameter, folded, diamond-shaped in cross section (see illustration on p. 63); **ligules** 0.1–0.4 mm. **Panicles** sparse, open, with reflexed lower branches. **Spikelets** with 2–5(7) florets; **lemmas** 4–6.5(8) mm, usually hairy near tip only (rarely hairy throughout), awns 3–12 mm; some awns > lemma bodies. **Anthers** (1)1.5–2(3) mm. **Ovary and seed apex** with dense minute hairs. **Habitat**: Mesic to dry forest, in full or partial shade. **Comments**: This widespread fine-leaved fescue is easily identified by lemma awns that are longer than the lemma bodies, and the hairy ovary and seed apex. It is common in mesic to dry forests throughout the PNW and is consistently found growing in the shade of a forest canopy. Compare with *F. idahoensis*, *F. roemeri*, and *F. rubra*.

Festuca octoflora Walter
Sixweeks Fescue

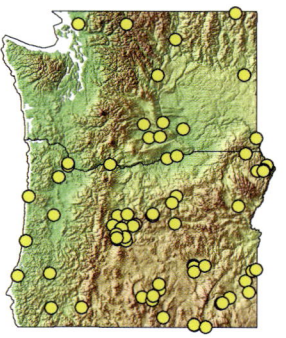

Annual; **culms** 5–60 cm. **Leaf blades** 0.5–1 mm wide, flat or rolled; ligules 0.3–1 mm. **Panicles** 1–7(20) × 0.5–1.5 cm; branches appressed to somewhat spreading, **pulvini** absent. **Spikelets** (4)5–11(13) mm, with (5)7–12 florets. **Lower glumes** 1.7–4.5 mm, 50–67% as long as upper glumes; **upper glumes** 2.5–7.2 mm; **lemmas** 2.7–6.5 mm, smooth, scabrous, or hairy, **awns** 0.2–9 mm. **Anthers** 0.3–1.5 mm. **Habitat**: Disturbed sandy soils in grasslands, sagebrush steppe, open forests. **Comments**: This native annual is being displaced by introduced competitors like *Bromus tectorum* and *Taeniatherum caput-medusae*. Small *F. octoflora* is easily confused with *F. microstachys*, especially when young. Even collections of mature *F. octoflora* often contain both species. *Festuca octoflora* has more-condensed panicles, more florets per spikelet, and shorter awns than *F. microstachys*. Three weakly defined varieties have been described, but we find them to be of questionable taxonomic significance and have not included them.

Festuca roemeri (Pavlick) Alexeev
Roemer's Fescue, Oregon Fescue

Densely cespitose perennial; **culms** 35–100 cm. Foliage glaucous or green; **sheaths** open, basal ones usually pale, not conspicuously splitting between veins; **blades** 0.5–1.2(2.5) mm wide, folded or occasionally flat, elliptic or L-shaped in cross section, not spinning easily between the fingers; **ligules** 0.1–0.5 mm. Panicle branches spreading at anthesis, often appressed after flowering (lower ones sometimes reflexed). **Spikelets** with 3–6 florets; **lemmas** (5.8)6.5–8 mm, scabrous distally, awns 2–4(5) mm, ≤ lemma body. **Anthers** 3–5 mm. **Ovary and seed** apex glabrous. **Habitat**: Prairies, savannas, and oak woodlands W of the Cascade Crest. Often on serpentine substrates in SW OR. **Comments**: Formerly included in *F. idahoensis*, which has narrower leaves that spin readily between the fingers; the 2 species blend in the Columbia Gorge and near the Jackson-Klamath county line in SW OR. *Festuca roemeri* is often confused with cespitose *F. nigrescens*, which has smooth leaves, darker, shredding sheaths, and shorter lemmas. Historically, Roemer's fescue was a community dominant in the upland prairies and savannas of W OR and WA. There are 2 varieties.

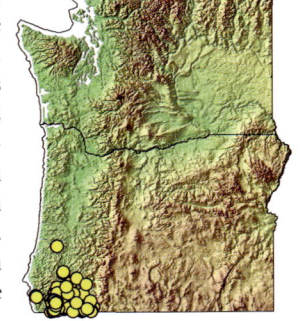

Festuca roemeri var. *klamathensis* B.L. Wilson
Inland Roemer's Fescue

Inner surfaces of leaf blades have dense, longer hairs that can be seen by unfolding the leaf, bending it around a finger, and using a hand lens to view the inner leaf surface; SW OR to CA, not coastal. See illustration of leaf cross sections, p. 63.

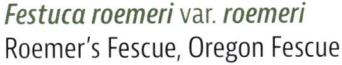

Festuca roemeri var. *roemeri*
Roemer's Fescue, Oregon Fescue

Inner surfaces of leaf blades have sparse, shorter, harder-to-see hairs. Widespread W of the Cascades.

Festuca rubra L.
Red Fescue [N and E]

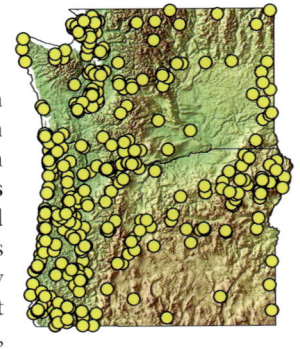

Rhizomatous perennial; **culms** 10–120(130) cm. Foliage green or glaucous; **sheaths** closed about 75% their length (assess on second from top leaf of vegetative shoot), basal sheaths brown with contrasting white veins, shredding lengthwise; **blades** 0.3–2.5 mm wide, folded (less often 1.5–7 mm wide and flat), smooth, usually diamond-shaped to triangular in cross section (see p. 63); **ligules** 0.1–0.5 mm; **fiber bundles** usually 5+, <2× as wide as thick (outer bundles sometimes confluent in serpentine habitats). Pani**cles** open to loosely contracted, sometimes dense. **Spikelets** with 3–10 florets; **lemmas** 4–8(9.5) mm, usually glabrous, sometimes hairy, usually awned, awns 0.4–4.5 mm, < lemma body. **Anthers** 1.8–4.5 mm. **Ovary and seed apex** glabrous. **Habitat**: Moist to mesic meadows, prairies, coastal dunes and headlands, salt marshes, lawns, roadsides, sometimes on serpentine substrates. **Comments**: *Festuca rubra* is a variable and diverse complex of introduced and native forms with unresolved taxonomy. We recognize two species split from this confusion: *F. ammobia* and *F. nigrescens*. If you want to delve into the subspecies, consult the *Flora of North America* or the *Flora of Oregon*.

Festuca saximontana Rydb.
Rocky Mountain Fescue

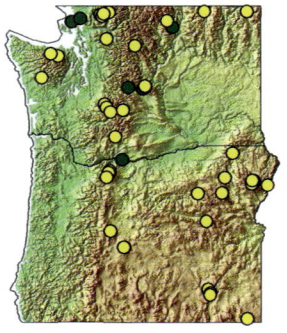

Densely cespitose perennial; culms 5–50(60) cm. **Foliage** usually glaucous; **sheaths** closed about half their length, usually persistent; **blades** 0.5–1.2 mm in diameter, ± round in cross section (see illustration on p. 63), hairy on inner surface; **ligules** 0.1–0.5 mm. **Panicles** contracted. **Spikelets** with (2)3–5(7) florets; **lemmas** (3)3.4–4(5.6) mm with awns (0.4)1–2(2.5) mm. **Anthers** (0.8)1.5–1.7(2) mm. **Ovary and seed apex** glabrous. **Comments**: *F. idahoensis* has longer lemmas and anthers; see also *F. brachyphylla*. Two varieties.

Festuca saximontana var. *purpusiana* (St.-Yves) Fred. & Pavlick (yellow dots)

Culms 5–25 cm, usually 2–3× taller than leaves; **leaves** smooth or scabrous; **fiber bundles** in 5–7 narrow strands; **lemmas** usually scabrous near tips. **Habitat**: Subalpine to alpine, rocky slopes and ledges.

Festuca saximontana var. *saximontana* (green dots)

Culms 25–50(60) cm, usually 3–5× taller than leaves; **leaves** usually scabrous; **fiber bundles** in 3–5 strands or continuous; **lemmas** smooth or scabrous. **Habitat**: Lowland to montane, rocky areas.

Festuca subulata Trin.
Bearded Fescue

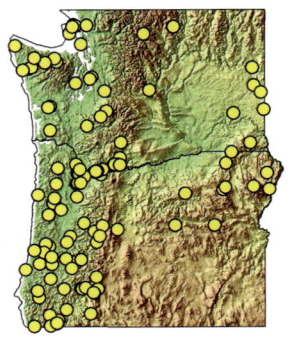

Loosely cespitose perennial; **culms** 35–120 cm. Foliage green, leaves mostly cauline; **sheaths** glabrous to scabrous, partially closed; **collars** glabrous; **blades** 3–8(10) mm wide, flat, lax, upper surface glabrous to sparsely hairy, lower surface glabrous or soft-hairy; **ligules** 0.2–0.7(1) mm, ciliate, cilia much < membranous part. **Panicle branches** spreading or drooping. **Spikelets** with 3–4(5) florets; **lemmas** (5)6–7.5(9) mm, 0(1–3)-veined, usually glabrous, sometimes scabrous or short-hairy, with awns (2.5)5–10(17) mm, awns sometimes curving, longest awns > lemma body. **Calluses** 0.1–0.3 mm, wider than long, not hairy. **Anthers** 1.5–2.5(3) mm. **Ovary and seed** apex minutely hairy. **Habitat**: Moist to mesic forest, lowlands to montane. **Comments**: Bearded fescue is a common forest grass with leafy stems and nodding inflorescences. Similar *F. subuliflora* has velvety-pubescent upper leaf surfaces and hairy calluses longer than wide. *Graphephorum canescens* and *G. cernuum* have lemmas awned from the back. In SW OR compare with *F. elmeri*, which usually has longer awns and veinier lemmas. *Melica subulata* and *M. smithii* have drooping panicles, but their sheaths are closed to the top.

Festuca subuliflora Scribn.
Crinkle-awn Fescue

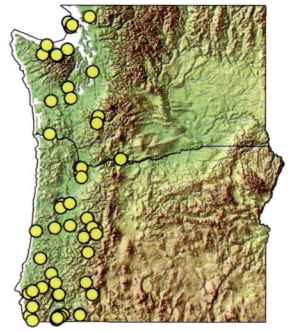

Loosely cespitose perennial; **culms** (40)60–100(125) cm. Foliage green, leaves mainly cauline; **sheaths** glabrous to densely hairy; **blades** 2–6(10) mm wide, upper surface with dense, soft, velvety hairs, lower surface usually glabrous; **collars** with sparse long hairs; **ligules** 0.1–0.5 mm, ciliate, cilia ≥ membranous part. Panicle branches spreading to drooping. **Spikelets** with (2)3–4(5) florets; **lemmas** 6–9 mm, (3)5-veined, hairy; awns 10–15 mm, crinkled, curved, longest awns much > lemma bodies; **calluses** (0.3)0.4–0.6 mm, longer than wide, hairy. **Anthers** 2.5–4 mm. **Ovary and seed apex** hairy. **Habitat**: Mesic to dry forests in and W of the Cascades. **Comments**: *Festuca subuliflora* is a leafy-stemmed forest grass similar to *F. subulata* but with velvety-hairy upper leaf surfaces and hairy, elongated calluses. *Festuca subulata* grows in moister microsites. The oddly elongated, hairy calluses may aid in seed dispersal.

Festuca trachyphylla (Hackel) Krajina
Hard Fescue, Sheep Fescue [E]

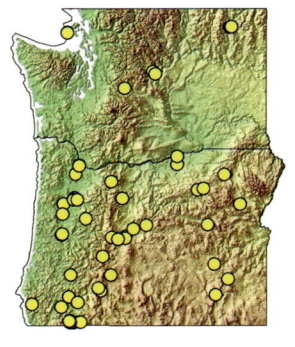

Densely cespitose perennial; **culms** (20)30–50(70) cm. Foliage green or glaucous; **sheaths** closed < 33% their length, not loose; **blades** (0.5)0.6–1.1 mm in diameter, folded, smooth, or scabridulous on distal 33%, elliptic or ovate in cross section (see illustration on p. 63); **fiber bundles** usually > 3, at midrib and margins and in between, sometimes continuous but uneven in thickness; **ligules** 0.1–0.5 mm. **Panicles** dense. **Spikelets** with 4–7(8) florets; **upper glumes** 3.4–4.5 mm; **lemmas** 4.2–5(6.5) mm, glabrous, scabrous or hairy on margins near tip, awns 0.5–2.5 mm, usually < 50% as long as lemma body. **Anthers** (1.6)2–3(3.4) mm. **Ovary and seed apex** glabrous. **Habitat**: Roadsides, disturbed areas, ski slopes. Native to Europe. **Comments**: *Festuca trachyphylla* is a fine-leaved bunchgrass with small lemmas and dense inflorescences. Similar-looking native species *F. idahoensis* and *F. roemeri* have longer lemmas and more-open panicles. *Festuca trachyphylla* is one of 3 species introduced as "sheep fescue" (none were actually *F. ovina*) for erosion control and as a turf grass. *Festuca valesiaca* has finer vegetative shoot leaves; *F. lemanii* has loose lower leaf sheaths and continuous fiber bundles of uniform thickness.

Festuca valesiaca Schleich. ex Gaud.
Covar Sheep Fescue [E]

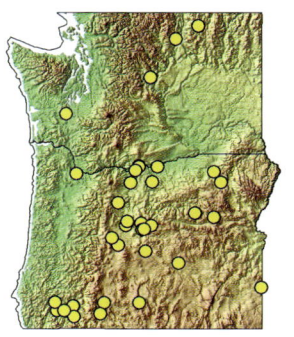

Densely cespitose perennial; **culms** 20–50(75) cm. Foliage green or glaucous; **sheaths** closed < half their length, not loose, **blades** (0.2)0.3–0.5(0.6) mm in diameter, folded, scabrous, elliptic in cross section; **fiber bundles** 3, sometimes with smaller ones between, > 2× as wide as thick (see illustration on p. 63); **ligules** 0.1–0.5 mm. **Panicles** dense. **Spikelets** with 2–5 florets; **upper glumes** 2.5–3.9 mm; **lemmas** 3.4–4.9(5.2) mm, glabrous, scabrous, or ciliate near tip, awns (0.5)1.5–2.2 mm, usually < 50% as long as lemma body. **Anthers** 2.1–2.6 mm. **Ovary and seed** apex glabrous. **Habitat**: Roadsides, dry pastures, scablands, shrub steppe. Native to Eurasia. **Comments**: One of the "sheep" fescues, *F. valesiaca* is a very fine-leaved bunchgrass that is planted for erosion control. In seeded stands on road cuts the clumps tend to have a very evenly spaced appearance. *Festuca valesiaca* is more drought tolerant than *F. idahoensis* or *F. trachyphylla* and thrives on dry, S-facing slopes. As a roadside ground cover it effectively outcompetes most native plants, as well as weeds. See also *F. lemanii* and *F. trachyphylla*.

Festuca viridula Vasey
Green Fescue

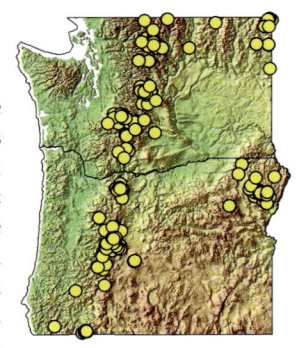

Densely cespitose perennial; **culms** (35)50–100 cm. Foliage green, not glaucous; **sheaths** closed < half their length; **blades** 0.5–1.3 mm in diameter when rolled, to 2.5 mm wide when flat, flexible, inner surface scabrous or pubescent, elliptic in cross section (see illustration on p. 63); lower cauline blades much reduced; **ligules** 0.3–0.8 mm. **Panicles** open to somewhat contracted. **Spikelets** with (2)3–6(7) florets; **lemmas** (5.5)6.5–8 mm, 0(1)-veined, glabrous or nearly so, awnless or with awns to 0.5(1.4) mm. **Anthers** (2)3–4(5) mm. **Ovary and seed apex** minutely hairy. **Habitat**: Montane to alpine, dry meadows, balds, open forest. **Comments**: Green fescue is often the dominant species in dry montane meadows. It has excellent value for restoration projects in high elevation grasslands. *Festuca idahoensis* has thinner leaves and longer awns, and *F. rubra* generally grows in moister habitats. Both of those species have well-developed blades on lower cauline leaf sheaths.

Festuca washingtonica Alexeev
Washington Fescue

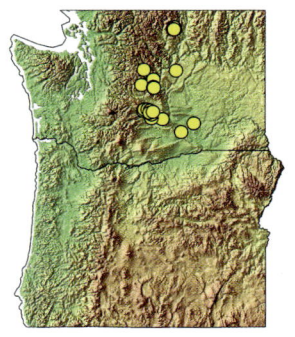

Loosely to densely cespitose perennial; **culms** 60–70 cm. Foliage green, senescing reddish-brown in late summer; **sheaths** closed > half their length, scabrous on veins, smooth and splitting between veins; **blades** 1.5–3 mm wide, flat, or folded after drying, lower surface scabrous on veins; **ligules** 0.2–0.5 mm. **Fiber bundles** broad or continuous in leaf cross sections, sometimes forming pillars (see illustration on p. 63). **Panicles** loosely contracted. **Spikelets** with 4–6(10) florets; **lemmas** 6.5–8(12) mm, scabrous, especially distally, awns (1)2–4.3 mm, < lemma bodies. **Anthers** 3–4(4.7) mm. **Ovary and seed apex** minutely hairy. **Habitat**: Open conifer forest, shrub steppe. **Comments**: *Festuca washingtonica* resembles *F. rubra* but differs in having scabrous leaves and hairy ovary and seed tips. It is a rare endemic that grows in dry habitats in and E of the Cascades in WA and BC. It persists only at sites protected from livestock grazing.

Gastridium phleoides (Nees & Meyen) C.E. Hubb.
Nit Grass [E]

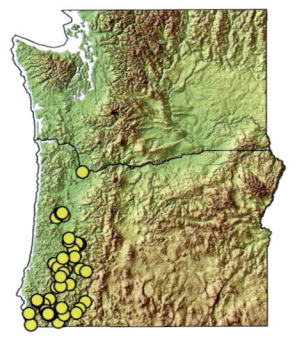

Annual; **culms** 7–70 cm. **Leaf sheaths** open; **blades** flat, 1.5–4 mm wide; **ligules** 1–7 mm, membranous. **Panicles** dense and spike-like. **Spikelets** 3–8 mm, laterally compressed, with 1 floret. **Lower glumes** 3–7 mm; **upper glumes** 2.7–5.5 mm; glumes slightly swollen at the base where distended by the floret. **Lemmas** 1–1.5 mm, oval, 5-veined, short-appressed-hairy, awnless or with twisted awn to 6 mm, awn arising in distal 33% and exserted from glumes. **Anthers** 0.5–0.9 mm. **Habitat**: Disturbed, seasonally moist to dry sites, roadsides. Native to SW Asia, N Africa. **Comments**: Nit grass glumes have a distinctive swollen pouch at the base and enclose a very small, long-awned lemma. Similar *Apera interrupta* has smaller spikelets and a looser panicle.

Glyceria borealis (Nash) Batch.
Small Floating Mannagrass, Boreal Mannagrass

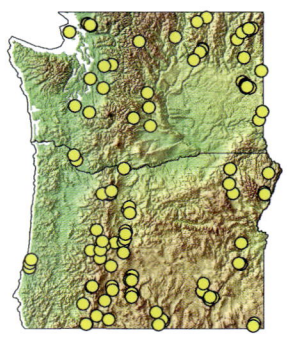

Rhizomatous perennial; **culms** 60–100 cm, often rooting at lower nodes. **Leaf blades** 9–25 cm, 2–7 mm wide, midstem leaves with densely papillose upper surfaces that are nonwettable, helping them float; **ligules** membranous, 4–12 mm. **Panicles** narrow, branches ascending, 5–10(15) cm long with 3–6 spikelets. **Spikelets** cylindrical, 9–22 mm × 0.8–2.5 mm. **Lemmas** 2.7–5.4 mm, midvein ending 0.1–0.2 mm below the tip, veins smooth or minutely scabrous, area between veins smooth (sometimes scabridulous between the veins with the scabers shorter than those on the veins), tips usually acute (to obtuse), entire or nearly so. **Anthers** 3, 0.4–1.5 mm. **Habitat**: Emergent in ponds, streams, sloughs, irrigation canals. **Comments**: Common in mountains in and E of the Cascades, occasional W of the Cascades. See also *G.* × *occidentalis* and *G. leptostachya*.

Glyceria canadensis (Michx.) Trin. var. *canadensis*
Rattlesnake Mannagrass, Canadian Mannagrass [E]

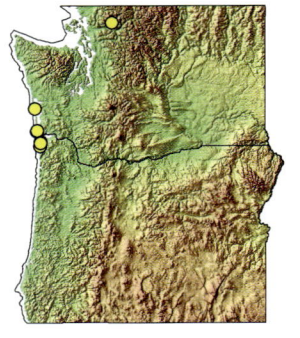

Rhizomatous perennial; **culms** in large clumps, 60–160 cm. **Leaves** 3–8 mm wide; **ligules** 2–6 mm. **Panicles** open with drooping branches. **Spikelets** laterally flattened, 5–8 × (2.5)3–5 mm, oval in outline. **Upper glumes** 1.5–2.5 mm, veins not extending to the tip. **Lemmas** 2.4–4 mm, veins not raised distally, tips prow-shaped. **Paleas** 1.5–1.8× as long as wide. **Anthers** 2, 0.4–0.5 mm. **Habitat**: Coastal cranberry bogs. **Comments**: Native to E North America. Introduced in cranberry bogs, spreading to marshes in W WA and NW OR. Native *G. grandis* has lemma tips that are not prow-shaped, glume veins that extend to the tip, and 3 anthers. *Glyceria maxima*, a rare introduction in NW WA, is taller and has upper glumes 3–4 mm long.

Glyceria declinata Brébiss.
Waxy Mannagrass, Low Glyceria [E]

Rhizomatous perennial; **culms** ascending to erect from decumbent bases, (10)20–90 cm. **Leaf blades** 3–12 cm, 4–8 mm wide; **ligules** membranous, 4–9 mm. **Panicles** narrow, branches ascending, 1.5–9.5 cm with 1–5 spikelets. **Spikelets** cylindrical (slightly laterally flattened at flowering time), 11–24 × 1.3–3 mm. **Lemmas** (3)4–6 mm, scabridulous on and between the veins, acute to blunt, tips 3(5)-toothed. **Anthers** 3, 0.5–1.4 mm. **Habitat**: Marshes, wet ditches, disturbed wet meadows, vernal pools, low elevations W of the Cascades. Native to Europe. **Comments**: The dentate lemma tips, short leaves and panicle branches, and the shorter stature of *G. declinata* distinguish it from similar *G. × occidentalis*.

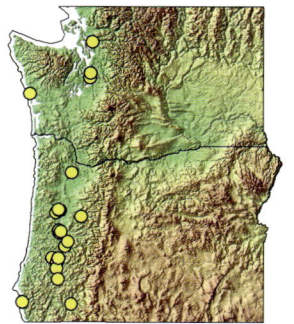

Glyceria elata (Nash ex Rydb.) M.E. Jones
Tall Mannagrass

Rhizomatous perennial; **culms** to 150 cm, 2.5–8 mm thick. **Leaves** 6–15 mm wide; **ligules** membranous, 2.5–4(6) mm. **Panicles** pyramidal, open. **Spikelets** 3–6 × 1.5–2.8 mm, oval in outline. **Upper glumes** 1–1.5 mm, veins not extending to the tip. **Lemmas** 1.7–2.2 mm, veins raised throughout, tips prow-shaped. **Paleas** ≥ lemmas, 2.4–3× as long as wide. **Anthers** 2, 0.5–0.8 mm. **Habitat**: Emergent in ponds, wet meadows, stream banks, moist woods, irrigation canals. **Comments**: Common and widespread. See *G. striata*.

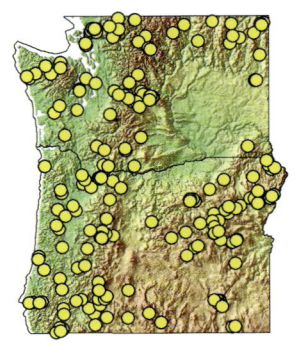

Glyceria grandis S. Watson var. *grandis*
American Mannagrass

Rhizomatous perennial; **culms** 50–150 cm. **Leaves** 4.5–15 mm wide; **ligules** membranous, 1–5 mm. **Panicle** open, widely divergent to drooping. **Spikelets** 3.2–6.4 × 2–3 mm, oval in outline; **upper glumes** 1.5–2.7 mm, mostly hyaline, veins of 1 or both glumes extending to the acute tip. **Lemmas** 1.8–3 mm, tips almost flat, not prow-shaped. **Anthers** 3, 0.5–1.2 mm. See comments under *G. canadensis* and *G. maxima*. **Habitat**: Wetlands, stream banks, lakeshores, irrigation canals. **Comments**: Widespread, but not in W OR except historic collections near Portland and the lower Columbia River.

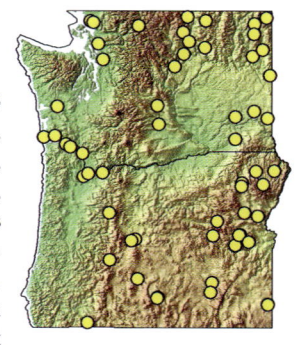

Glyceria fluitans (L.) R. Br.
Water Mannagrass [E]

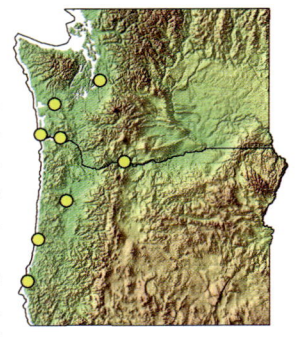

Rhizomatous perennial; **culms** to 150 cm, erect or spreading, sometimes decumbent and rooting from the lower nodes, sometimes floating in shallow water. **Leaf blades** 5–25 cm, 3–10 mm wide; **ligules** 5–15 mm. **Panicles** narrow, branches 3–5 cm, appressed to ascending (spreading at anthesis), with 1–4 spikelets. **Spikelets** cylindrical, (15)18–39 × 1.7–3.3 mm. **Lemmas** 5.2–8 mm, midveins ending within 0.1 mm of the tip, lemmas scabrous on and between the veins, tips acute, usually entire. **Anthers** 3, 1.5–3 mm. **Habitat**: Emergent in streams, lakes, irrigation canals. **Comments**: Uncommon, W of the Cascades, perhaps often overlooked or misidentified. Introduced from Eurasia. Hybridizes with *G. leptostachya*. See *G.* × *occidentalis*.

Glyceria leptostachya Buckley
Narrow Mannagrass

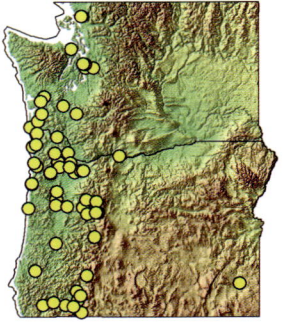

Rhizomatous perennial; **culms** 50–200 cm, erect or decumbent and rooting from the lower nodes. **Leaf blades** 12–30 cm, 3.5–11 mm wide; **ligules** membranous, 4.5–12 mm. **Panicles** ± narrow, branches 4.2–14.7 cm, appressed to ascending with 3–8(10) spikelets. **Spikelets** cylindrical, 9–20 × 0.4–3 mm. **Lemmas** 2.6–4.5 mm, midveins ending within 0.1 mm of the tip, lemmas scabrous or scabridulous on and between the raised veins, the scabers all about the same length; tips truncate to obtuse, crenulate. **Anthers** 3, 0.3–0.9 mm. **Habitat**: Emergent in springs, swamps, lakeshores, stream banks, and irrigation canals. **Comments**: Common W of the Cascades. Hybridizes with *G. fluitans*. See *G.* × *occidentalis*.

Glyceria × *occidentalis* (Piper) J.C. Nelson
Northwestern Mannagrass [N&E]

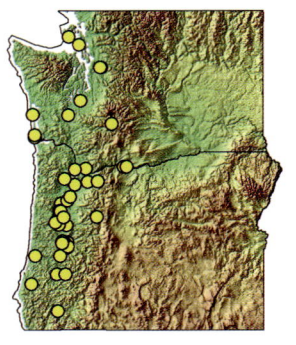

Rhizomatous perennial; **culms** to 160 cm, erect, or decumbent and rooting from the lower nodes. **Leaf blades** 20–30 cm, (2.5)4–12 mm wide; **ligules** membranous, 7–12 mm. **Panicles** narrow, branches 4.5–18 cm with 2–8 spikelets. **Spikelets** cylindrical, 13–23 × 1.5–3.5 mm. **Lemmas** 4.5–5.9 mm, midveins extending to within 0.1 mm of the tip, minutely scabrous on and between the veins, the scabers all about the same length; tips acute, usually irregularly crenate. **Anthers** 2, 0.6–1.6 mm. **Habitat**: Shores of lakes, ponds, streams, vernal pools, marshes. **Comments**: Mainly in lowlands W of the Cascades, with a few records scattered E of the Cascades. Long thought to be a native species, *G.* × *occidentalis* is now considered to be a complex of hybrids and backcrosses between native *G. leptostachya* and introduced *G. fluitans*. *Glyceria leptostachya* has shorter anthers and shorter lemmas with blunt tips. *Glyceria fluitans* has longer anthers and usually entire lemma tips. *Glyceria* × *occidentalis* is also confused with *G. declinata*, which has shorter panicle branches and lemma tips with 3(5) distinct teeth.

Glyceria striata (Lam.) Hitchc.
Fowl Mannagrass

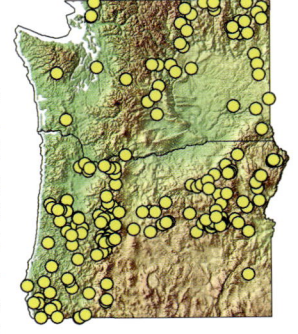

Rhizomatous perennial; **culms** to 80 cm, (1.5) 2–3.5 mm thick. **Leaves** 2–6 mm wide; **ligules** membranous, 1–4 mm. **Panicles** pyramidal, open. **Spikelets** 1.8–4 × 1.2–2.9 mm, oval in outline. **Upper glumes** 0.6–1.6 mm, veins not extending to the tip. Glume tips often splitting with age. **Lemmas** 1.2–2 mm, veins raised throughout, tips prow-shaped; **paleas** ≤ lemmas, 1.5–3× as long as wide. **Anthers** 2, (0.2)0.4–0.6 mm. **Habitat**: Emergent in ponds, wet meadows, stream banks, lake margins, and irrigation canals. **Comments**: Common and widespread. Similar to *G. elata*, but smaller with narrower leaves, thinner culms, and smaller anthers.

Graphephorum canescens (Buckley) Röser & Tkach
Tall Trisetum

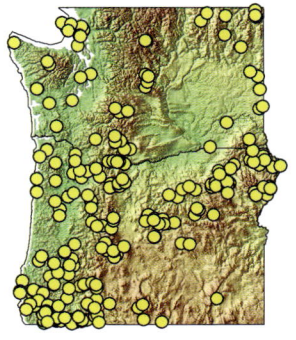

Cespitose perennial; **culms** 40–120 cm. **Leaf sheaths** hairy, scabrous or smooth; **blades** (3)7–10 mm wide, flat, soft; **ligules** (1.5)3–6 mm, membranous, erose. **Panicles** 10–25 cm, erect or sometimes nodding a bit at the tips, with branches ascending or somewhat divergent, usually spikelet-bearing to the base, sometimes the lower ones spikelet-bearing only distally. **Spikelets** 7–9 mm, with 2–4 florets. **Lower glumes** 3–5 mm; **upper glumes** (3.4)5–7(9) mm, widest at or below middle. **Lemmas** 5–7 mm, glabrous, with 2 awn-like teeth to 2.5(3.2) mm, awns 7–14 mm, bent, arising on upper third of lemma. **Callus hairs** about 0.5 mm. **Anthers** 1–3 mm. **Habitat**: Dry to moist forest, stream banks. **Comments**: Tall trisetum has stiffly ascending to spreading panicle branches and spikelets. It is similar to *G. cernuum*, which has nodding inflorescences, longer panicle branches lacking spikelets in the lower half, smaller lower glumes, and slightly moister habitat.

Graphephorum cernuum (Trin.) Röser & Tkach
Nodding Trisetum, Nodding Oatgrass

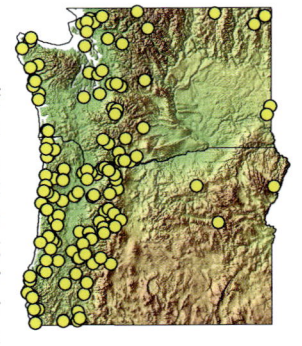

Loosely cespitose perennial; **culms** (30)50–110 cm. **Leaf sheaths** hairy or minutely scabrous; **blades** flat, 7–12 mm wide, lax and soft; **ligules** 1.5–3 mm, erose. **Panicles** open, 10–30 cm, nodding, with curved branches, at least the lowest spreading or drooping, spikelet-bearing mostly toward the ends, except sometimes the upper branches spikelet-bearing to near the base. **Spikelets** 6–12 mm, with 2–4 florets. **Lower glumes** 0.75–2(3) mm; **upper glumes** 3.5–5 mm, widest at or above the middle. **Lemmas** 5–6 mm, glabrous, with 2 teeth to 1.3 mm, awns (7)9–14 mm, arising from just above midlength to just below the teeth, curved. **Callus hairs** to 1 mm. **Habitat**: Moist forests, streamsides, lakeshores. **Comments**: Nodding trisetum is a common grass in moist forest understories, mainly W of the Cascades. See *G. canescens*, *Festuca subulata*, and *F. subuliflora*.

Graphephorum wolfii (Vasey) Vasey ex Coult.
Wolf's False Oat

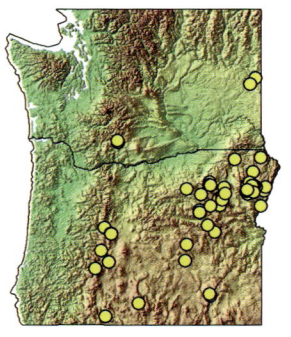

Short-rhizomatous perennial (often appearing cespitose); **culms** 20–80(100) cm. **Leaves** basally concentrated; **sheaths** open; **blades** 2–7 mm wide, flat; **ligules** (1.2)2.5–4(6) mm, membranous, truncate, erose. **Panicles** (10)20–40(50) × 1–1.5 cm, condensed, narrow, green or purplish, shiny, branches appressed or ascending. **Spikelets** 4–7(8) mm, with 2(3) florets; **glumes** 4–7 mm, usually > lowest floret; **calluses** with hairs < 0.5 mm; **lemmas** 4–6.5 mm, obscurely 2-lobed, awnless or with awns to 2 mm arising just below and rarely exceeding lemma tips. **Anthers** (0.6)1(1.5) mm. **Habitat**: Montane to subalpine meadows and riparian areas. **Comments**: This uncommon grass decreases with livestock grazing. *Graphephorum canescens* and *G. cernuum* have longer lemma awns.

Hainardia cylindrica (Willd.) Greuter
Thintail, Barbgrass [E]

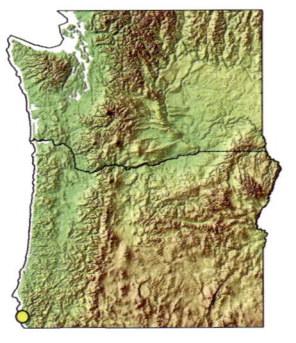

Annual; **culms** 5–45 cm, erect or ascending, internodes solid. **Leaf sheaths** open, glabrous, the uppermost enclosing bottom of spike; **blades** 1.5–2.5 mm wide; **ligules** 0.2–1 mm, membranous. **Spikes** cylindrical, stiff, straight to somewhat curved, with spikelets sunken into thickened inflorescence axis. **Spikelets** 5–8 mm; oriented edgewise to axis; **glumes** 1 per spikelet (terminal spikelet has 2 glumes), rigid; **florets** 1 or 2, the second floret (if present) reduced and sterile, floret(s) hidden by glume and rachis; **fertile lemmas** 4–6 mm, membranous, awnless; disarticulation in the rachis. **Anthers** 2–3.5 mm. **Habitat**: Coastal salt marshes, alkaline soils, vernal pools, disturbed areas; native to Europe; in our area known only from S OR coast. **Comments**: *Hainardia cylindrica* spikelets have a single, stiff, pointed glume that spreads widely, giving the spike a barbed appearance. *Lolium* species also have 1 glume per spikelet, but the 2+ florets are usually visible. *Parapholis* and *Deschampsia bolanderi* have 2 glumes per spikelet and spikelets oriented flatwise to the axis; *Nardus* has a 1-sided inflorescence with awned lemmas.

Hesperostipa comata (Trin. & Rupr.) Barkworth ssp. *comata*
Needle-and-Thread

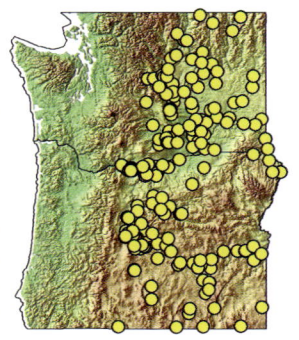

Cespitose perennial; **culms** 12–110 cm. **Leaf sheaths** open; **blades** 0.5–4 mm wide, usually rolled; **ligules** 1–7 mm, membranous. **Panicles** open, drooping. **Spikelets** with 1 floret; **glumes** 16–35 mm; **callus** 2–4 mm, sharp; **lemmas** 7–13 mm, with hairs about 1 mm, sometimes glabrous just above callus, awns 75–225 mm, scabrous to hairy, the terminal segment 40–120 mm, wavy to curled when mature; **paleas** ± = lemmas. **Anthers** 3 mm, tips hairy. **Habitat**: Well-drained, usually sandy soil in shrub steppe, juniper, and open pine forest. **Comments**: Strikingly long awns make this a lovely and unmistakable needlegrass. The awn twists with changes in humidity, pushing the sharp callus at the base of the seed (where the radicle emerges) into the ground for optimum seedling survival. Needle-and-thread is often used in restoration projects on sandy soils. It makes an attractive ornamental, but tends to self-seed, so plan to find homes for the offspring.

Hierochloë occidentalis Buckley
California Sweetgrass

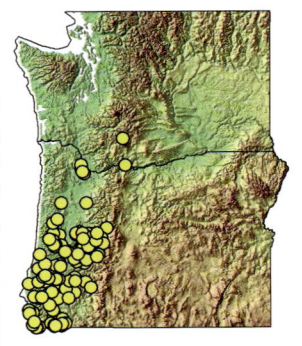

Loosely cespitose perennial with short rhizomes 2–4 mm thick; **culms** (40)60–90 cm. **Leaf blades** (3)5–15 mm wide; twisted at base so the shiny, dark green dorsal surface faces up and the glaucous ventral surface faces down; **ligules** membranous, 1.5–4(6) mm; **flag leaves** (1.5)3–13.5 cm. **Panicles** open. **Spikelets** 4.5–6 mm; **glumes** 3.5–4.5 mm, slightly < or = florets, ovate, green with wide hyaline margins. **Lower lemmas** 4–5 mm, staminate, mostly glabrous, with spreading hairs on margins and near tip, tips rounded and bilobed, often with awn to 1 mm; **upper lemmas** 3.5–4.5 mm, bisexual, shiny, golden-brown, with spreading hairs on margins, especially near tip. **Anthers** 2–3.5 mm. **Habitat**: Moist forests and forest openings W of the Cascade Crest. **Comments**: The longer, wider flag leaves and larger, rounded lemmas separate this species from *H. odoratum*. Sometimes placed in *Anthoxanthum*, where it is called *A. occidentale*. The attractive leaves give off a sweet vanilla fragrance; this grass would make a lovely addition to a moist, shady garden.

Hierochloë odorata (L.) P. Beauv.
Vanilla Grass, Hairy Sweetgrass

Loosely cespitose perennial with short rhizomes 0.7–2 mm thick; **culms** 40–85(110) cm. L**eaf blades** 2.5–5.5 wide; **ligules** membranous, 2.5–5.5 mm. **Flag leaves** reduced, 1–3(6) cm. **Panicles** open. **Spikelets** 5–6 mm; **glumes** 4–6.3 mm, > florets, ovate, translucent. **Lower lemmas** 3–5 mm, staminate, ± evenly hairy with longer hairs on margins, acute, awnless or with awns 0.1–1 mm; **upper lemmas** 2.9–3.5 mm, bisexual, golden-brown, shiny, hard, hairy toward tip, the hairs 0.5–1 mm. **Anthers** 1.5–2.1 mm in staminate florets, 1.2–1.3 mm in bisexual florets. **Habitat**: Wet meadows, marshes, shores of streams and lakes, mostly E of the Cascade Crest. Rare in OR. **Comments**: Vanilla grass has ± round spikelets with almost transparent glumes surrounding hairy brown lemmas. The solitary, nearly leafless stems can be difficult to spot. Sometimes placed in *Anthoxanthum,* where it is called *A. hirtum* or *A. nitens.*

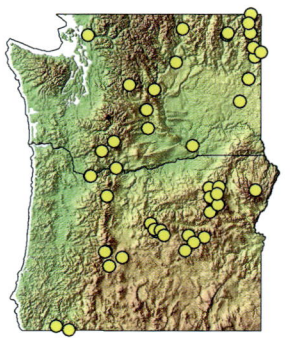

Holcus lanatus L.
Common Velvetgrass, Yorkshire Fog [E]

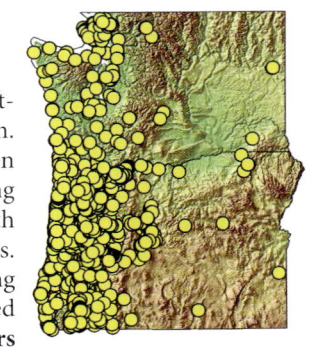

Cespitose perennial; **culms** 20–100 cm; plants densely soft-hairy. **Leaves** 5–10 mm wide, **ligules** membranous, 1–4 mm. Young **panicles** soft, purplish-red and dense, becoming open and pyramidal and pink to white at anthesis, then closing again after anthesis, becoming dense and mostly white with age. **Glumes** ciliate on keel and veins, fully enclosing 2 florets. **Awn of upper lemma** 1–2 mm, straight when young, curving into a distinctive hook when mature, hidden or exserted sideways from the glumes; **lower lemma** awnless. **Anthers** (1.2)2–2.5 mm. **Habitat**: Moist meadows, wetlands, and along streams and irrigation canals. Introduced from Europe. Invasive. **Comments**: *Holcus lanatus* is covered with soft, velvety hairs on the culms, sheaths, and blades. *Holcus mollis* is less hairy, except on culm nodes, and its lemma awn does not curl into a tight hook at maturity.

Holcus mollis L. ssp. *mollis*
Creeping Velvetgrass [E]

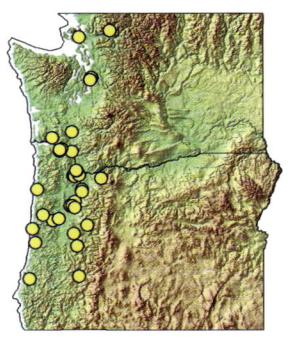

Rhizomatous perennial; **culms** 20–100 cm, often decumbent at the base, glabrous or nearly so except for densely hairy nodes; **blades** pubescent, 3–10 mm wide; **ligules** 1–5 mm. Young **panicles** whitish-green, becoming straw-colored with age, contracted to open. **Glumes** short-ciliate on keel and veins, > florets. **Awn of upper lemma** 3–5 mm, straight when young, bent but not hooked when mature, exserted from glumes; **lower lemma** awnless. **Anthers** about 2 mm. **Habitat**: Moist soils in disturbed sites, roadsides, lawns, meadows, oak savannas, W of the Cascades. Introduced from Europe. Invasive. **Comments**: The densely hairy nodes contrasting with the glabrous internodes distinguish *H. mollis* from *H. lanatus*. Creeping velvetgrass is uncommon but is expanding its range, and it is invasive in mesic to moist habitats.

Hordeum brachyantherum Nevski
Meadow Barley

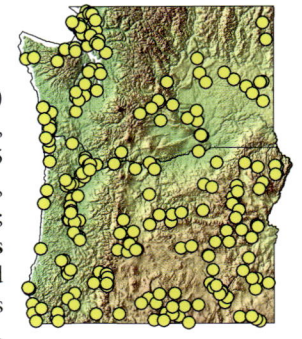

Loosely to densely cespitose perennial; **culms** (20)40–80(100) cm. **Leaf blades** 2–6(9) mm wide; **ligules** membranous, truncate, to 0.5 mm, **auricles** absent. **Spikes** 3–8.5 × 0.7–1.5 cm, erect; 3 spikelets per node. **Central spikelet** bisexual, with 1 floret; **glumes** 9–19 × 0.2 mm, awn-like throughout; **lemmas** 5.5–10 mm, with **awns** 3.5–14 mm. **Lateral spikelets** staminate; **glumes** 7–19 mm, awn-like throughout or flattened near base; **lemmas** reduced or well developed, with awns 0–7.5 mm. **Anthers** 0.8–4 mm. **Habitat**: Moist meadows, wet prairies, salt marshes, shorelines, from sea level to alpine. **Comments**: *Hordeum brachyantherum* is characterized by long, narrow, neat spikes. Like *Deschampsia cespitosa*, *Hordeum brachyantherum* grows on sites ranging from alpine to coastal salt marshes, exhibiting an amazing ecological amplitude that probably indicates the evolution of ecotypes. Meadow barley is often planted for wetland restoration.

Hordeum depressum (Scribn. & J.G. Sm.) Rydb.
Low Barley

Annual; **culms** 10–30(55) cm. **Leaf blades** to 4.5 mm wide; **ligule** 0.3–0.8 mm; **auricles** lacking. **Spikes** erect; **glumes** 5–20 mm, all linear, not or only slightly widened above the base, generally < 0.7 mm wide. **Pedicels of lateral spikelets** ± 1 mm, generally straight. **Lemmas** of lateral spikelets awnless or with awns to 1 mm. **Anthers** 0.5–1.5 mm. **Habitat**: Seasonally moist, often alkaline sites, vernal pools, disturbed areas. **Comments**: Rare in OR and WA; most records from the early 1900s. *Hordeum depressum* is one of several small annual grasses that may not have survived the disturbances of European settlement and the introduction of exotic annual grasses.

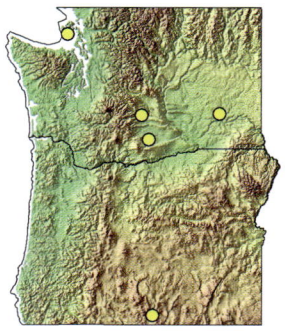

Hordeum jubatum L. ssp. *jubatum*
Foxtail Barley

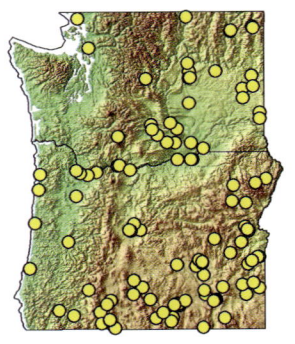

Cespitose perennial, sometimes appearing annual; **culms** 20–80 cm. **Leaf blades** to 5 mm wide; **ligules** to 0.8 mm; **auricles** absent (rarely to 0.5 mm). **Spikes** 3–15 × 0.7–1.2 cm, nodding at anthesis; 3 spikelets per node. **Central spikelet** bisexual, with 1 floret; **glumes** 35–85 mm, awn-like throughout, strongly spreading at maturity; **lemmas** 4–8 mm, with ± straight awns 35–90 mm. **Anthers** 0.6–1.2 mm. **Lateral spikelets** staminate or sterile; **glumes** 17–83 mm, awn-like; **lemmas** 4–6.5 mm, with divergent awns 2–15 mm. **Anthers** 1–1.5 mm. **Habitat**: Seasonally moist meadows, salt marshes, lakeshores, disturbed sites, often on alkaline soils. **Comments**: The bright green to reddish spikes of *H. jubatum* shining in the sun look like rippling water when a breeze blows. The awns spread as the spikes mature, giving them the appearance of scared cats' tails. The long awns are a hazard to grazing animals, significantly reducing palatabilty to livestock as the season progresses. Sometimes confused with *Elymus elymoides*.

Hordeum marinum Huds. ssp. *gussoneanum* (Parl.) Thell.
Mediterranean Barley [E]

Tufted annual; **culms** 10–30(50) cm. **Leaf blades** 1.5–4 mm wide; **ligules** 0.2–0.5 mm; **auricles** lacking. **Spikes** 1.5–7 × 0.5–1(2) cm, erect; 3 spikelets per node. **Glumes** to 26 mm. **Awns** of lateral lemmas 3–8 mm. **Anthers** 0.8–1.3 mm. **Habitat**: Seasonally moist disturbed sites, compacted soils. **Comments**: Similar *H. murinum* has prominent auricles; lateral lemmas of *H. depressum* are awnless or have awns to 1 mm.

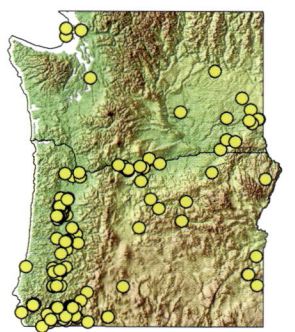

Hordeum vulgare L. ssp. *vulgare*
Barley [E] (not mapped)

Annual; **culms** to 100(150) cm. **Leaf blades** 5–15 mm wide; **ligules** 0.5–1 mm; **auricles** to 6 mm, well developed. **Spikes** erect to nodding; usually 2 or all 3 spikelets fertile. **Lemmas of central spikelets** 3+ mm wide with awns 30–180 mm or absent and lemmas 3-lobed. **Anthers** 6–10 mm. **Habitat**: Cultivated fields, occasionally escaping on roadsides; native to Eurasia. **Comments**: One of the earliest domesticated grains, barley is cultivated for food and brewing malt, and planted for erosion control. It generally does not persist more than a few seasons. Hooded barley lacks awns and has weird 3-lobed lemmas.

Hordeum murinum L.
Mouse, Wall, Smooth, Foxtail, or Hare Barley [E]

Tufted annual; **culms** 15–110 cm. **Leaf blades** ≤ 5(7) mm wide, **ligules** 1–4 mm, **auricles** to 8 mm, well developed. **Spikes** 3–8 × 0.7–1.6 cm; ± erect; 3 spikelets per node. **Glumes** to 25 mm, margins ciliate. **Lemma of central spikelet** ≤ 2 mm wide with awns 20–40 mm. **Anthers** 0.2–3.2 mm. **Habitat**: Disturbed sites, fields, roadsides. Native to Eurasia. **Comments**: *H. murinum* is distinguished from other annual barleys (except *H. vulgare*) by its well-developed auricles. The 3 subspecies (*glaucum*, *leporinum*, and *murinum*) have sometimes been treated as species. Differences among them are subtle (see key on next page). Subspecies *murinum* usually grows in moister habitats than the other subspecies.

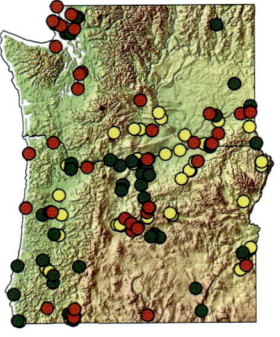

Key to subspecies of *Hordeum murinum*

1a Floret of central spikelet sessile or subsessile, the rachilla joint < 0.6 mm; inner glume of the lateral spikelets narrower than those of the central spikelet; palea of lateral floret ± glabrous (wall barley) ....**ssp. *murinum*** (red dots)

1b Floret of central spikelet with a distinct rachilla joint equal or subequal to the pedicel of the lateral spikelets, or the spikelet itself with a distinct pedicel; inner glume of the lateral spikelets as broad as those of the central spikelet; palea of lateral floret pubescent

    2a Rachises rarely as much as 2 mm; leaf sheath glabrous; floret of central spikelet slightly < florets of lateral spikelets; anthers of central florets to 0.6 mm (smooth barley)...**ssp. *glaucum*** (yellow dots)

    2b Rachises at least 2(3) mm; leaf sheath glabrous or hairy; floret of central spikelet much < florets of lateral spikelets; anthers of central florets 0.9–3.2 mm (hare barley) .... **ssp. *leporinum*** (green dots)

Koeleria macrantha (Ledeb.) Schult.
Junegrass

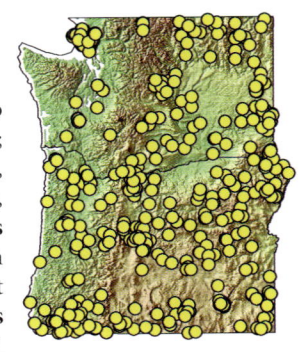

Cespitose perennial; **culms** (10)20–85(130) cm; glabrous to densely hairy. **Leaf sheaths** open, **blades** 0.5–3(4.5) mm wide; **ligules** 0.5–2 mm, membranous. **Panicles** 4–17 × 0.5–2 cm, narrow and dense, interrupted near base, shiny, usually pale, axis and branches usually densely soft-pubescent. **Spikelets** 2.5–6.5 mm, with 2(3) florets. **Glumes** 2.5–5 mm, differing in length, width, and sometimes shape with lower glume widest near base and upper glume widest near midlength. **Lemmas** 2.5–6.5 mm, 5-veined, usually awnless, or with awns to 1 mm. **Anthers** 1–3 mm. **Habitat**: Upland meadows, prairies, oak and pine savannas, sagebrush/juniper steppe; lowlands to alpine. **Comments**: Widespread and variable, *K. macrantha* can be a challenge to identify. Panicle color and openness vary with maturity and between plants. The shiny, silvery, narrow panicles, and panicle branches with short, soft hairs distinguish it from *Festuca* and *Poa*. *Sphenopholis obtusata* has a very similar-looking inflorescence, but its upper glumes are ± hooded and it grows in wetter habitats.

Leersia oryzoides (L.) Sw.
Rice Cutgrass

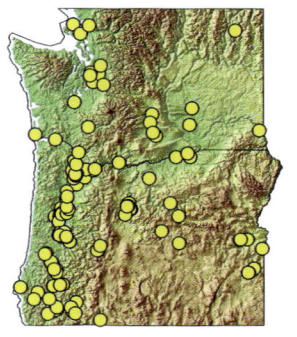

Rhizomatous perennial; **culms** 35–150 cm, decumbent, rooting at nodes. **Leaf sheaths** open; **blades** 5–15 mm wide, margins sharply scabrous, apt to cut skin and stick to clothing; **ligules** 0.5–1 mm, membranous. **Panicles** terminal and open, also sometimes axillary and partially enclosed in upper sheath; spikelets borne on outer 67% of each branch. **Spikelets** with 1 floret; **glumes** lacking; **lemmas** 4.2–6.5 mm; margins and keels ciliate; awnless. **Seeds** 2–3.5 mm, resembling rice grains. **Anthers** 0.4–2(3) mm. **Habitat**: Shores of lakes, streams, irrigation ditches; often emergent. **Comments**: Cutgrass is an excellent name for this grass with leaves that draw blood. It is one of the few grasses in OR and WA that do not have glumes. Rice cutgrass is a valuable food source for waterfowl, shorebirds, and small mammals. Vegetatively it can resemble *Phalaris arundinacea*, which usually has longer ligules and also lacks the sharply scabrous leaf margins.

Leucopoa kingii (S. Watson) W.A. Weber
Spike Fescue

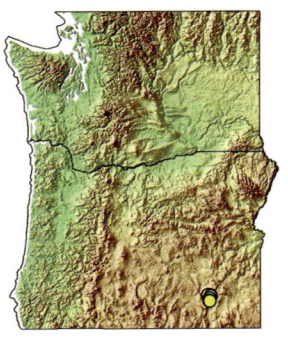

Strongly rhizomatous tufted perennial, the tufts arising from deep rhizomes; plants unisexual; **culms** 30–100(120) cm. **Sheaths** open; **blades** 1.5–7(10) mm wide, flat or loosely rolled; uppermost blade often short, stiff, erect; **ligules** 0.8–2(4) mm, membranous. **Panicles** contracted, often interrupted, branches erect to ascending. **Spikelets** laterally compressed, with (2)3–4(6) florets. **Glumes** nearly transparent; lower glumes 3–5.5(6.5) mm, upper glumes 4–6.5(7.5) mm; **lemmas** 4.5–8(10) mm, glabrous, scabrous, or short-hairy, short-awned or awnless. **Anthers** (2.5)3.5–5(6) mm, vestigial in female plants. **Ovary and seed apex** hairy, ovaries vestigial in male plants. **Habitat**: Alpine grasslands and rocky slopes. **Comments**: Spike fescue clumps grow from deep-seated rhizomes that are seldom collected. The female and male flowers are borne on separate plants. In OR and WA *L. kingii* is known only from Steens Mountain in SE OR. *Leucopoa* is a small, otherwise Asiatic genus of dioecious, fescue-like grasses; this is the only American *Leucopoa*. The population on Steens Mountain is an outlier of its Basin and Range distribution; *L. kingii* is a dominant grass species in the high, flat-topped mountains in central NV, such as the Monitor Range.

Leymus cinereus (Scribn. & Merr.) Á. Löve
Great Basin Wildrye

Densely cespitose perennial; **culms** 100–270 cm, 2–5 mm thick. **Leaf blades** 3–12 mm wide, not glaucous, involute to flat; **ligules** 1.5–8 mm; **auricles** to 1.5 mm. **Spikes** 8–17 mm wide; 2–7 spikelets per node, disarticulation above glumes. **Glumes** awl-shaped, keeled; **lemmas** 6.5–12 mm, glabrous or hairy, tips acute or with awns to 3 mm. **Anthers** 4–7 mm. **Habitat**: Streamside terraces, toe slopes, mesic valleys in sagebrush steppe and juniper woodlands, usually on deeper, sandy or gravelly soils. **Comments**: *Leymus cinereus* is a very large, dense bunchgrass common E of the Cascades. It is frequently confused with *L. condensatus*, which has branched inflorescences and wider, usually glaucous leaves and grows in more disturbed habitats. For a grass this large, *L. cinereus* looks oddly "delicate" when compared with *L. condensatus*. Great Basin wildrye decreases under grazing pressure. It is a desirable species for restoration of deep floodplain soils, as it provides good cover for birds.

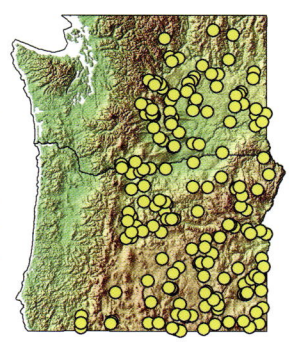

Leymus condensatus (J. Presl) Á. Löve
Giant Wildrye [E]

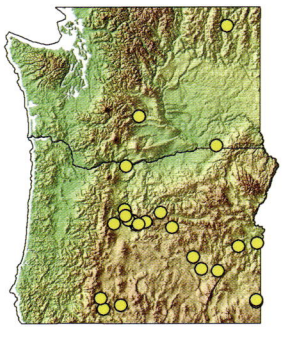

Cespitose to weakly rhizomatous perennial; **culms** 115–350 cm, 6–10 mm thick. **Leaf blades** 10–30 mm wide, often glaucous; **ligules** 0.7–7.5 mm; auricles absent. **Inflorescences** dense, ± spike-like panicles 2–6 cm wide, branched at the lower nodes, with 5–35 spikelets per node, including side branches; disarticulation above glumes. **Glumes** narrowly lanceolate, keeled; **lemmas** 7–14 mm, usually glabrous, tips acute or with awns to 4 mm. **Anthers** 3.5–7 mm. **Habitat**: Disturbed sagebrush steppe, roadsides. Native to coastal mountains of CA. **Comments**: *Leymus condensatus* is a big, broad-leaved, glaucous bunchgrass with branched inflorescences. It is often misidentified as native *L. cinereus* and mistakenly included in revegetation plantings. It usually grows in more disturbed habitats. Giant wildrye hybridizes with *L. triticoides*.

Leymus flavescens (Scribn. & J.G. Sm.) Pilg.
Yellow Wildrye

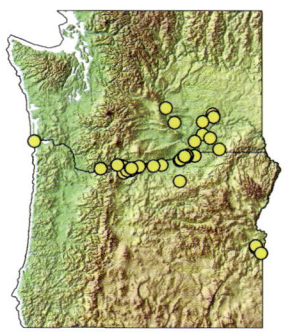

Strongly rhizomatous perennial, with shoots arising in tufts; **culms** 40–120 cm, 2–4 mm thick. **Leaf blades** 3–4 mm wide, involute; **ligules** 0.3–1.5 mm, membranous; **auricles** absent. **Inflorescences** spikes or dense spike-like racemes 1.2–2 cm wide; 2 spikelets per node, 1 spikelet of each pair sessile, the other on a pedicel to 15 mm. **Glumes** awl-like to narrowly lanceolate, keeled distally, hairy; **lemmas** 10.5–15 mm, with dense hairs 2–3 mm, awnless or with awns to 2 mm. **Anthers** 4.5–7 mm. **Habitat**: Sand dunes and disturbed sandy soils along the Snake and Columbia Rivers. **Comments**: *Leymus flavescens* is an uncommon rhizomatous wildrye of dunes and shifting sandy soils in the Columbia Basin. Look for the pair of hairy spikelets at each node, 1 sessile and 1 pedicellate. *Elymus lanceolatus* ssp. *psammophilus* has hairy lemmas but only 1 spikelet per node. *Leymus flavescens* forms sterile hybrids with *Elymus lanceolatus* and *L. triticoides*.

Leymus mollis (Trin.) Pilg. ssp. *mollis*
American Dunegrass

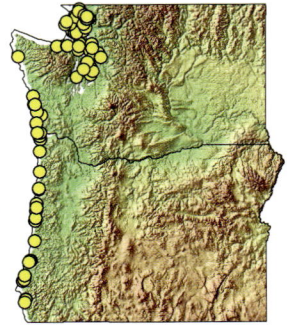

Strongly rhizomatous perennial; **culms** (50)70–170 cm. Foliage glaucous; **leaf blades** 5–15 mm wide, flat; **ligules** 0.2–2.5 mm; **auricles** to 0.7 mm. **Spikes** 1–2 cm wide; 2 spikelets per node, each with 3–6 florets. **Glumes** lanceolate, tapering from midlength or above, flat or rounded on the back, not keeled, hairy; **lemmas** 11–20 mm, with dense, soft hairs 0.5–1 mm, tips acute, awnless. **Anthers** 4–9 mm. **Habitat**: Coastal sands. **Comments**: Prior to European settlement, *L. mollis* was a dominant species in coastal dunes; it is an important species for restoration of those habitats. It has been largely displaced by European beachgrass, *Calamagrostis arenaria*, which was introduced for dune stabilization. European beachgrass differs in having narrower leaves and 1 floret per spikelet.

Leymus racemosus (Lam.) Tzvelev
Mammoth Wildrye [E]

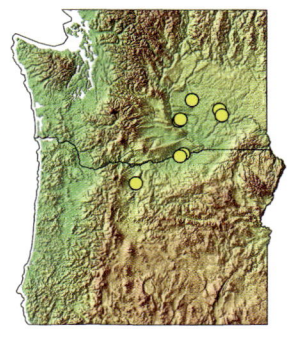

Strongly rhizomatous perennial, with shoots arising in tufts; **culms** 50–100 cm, 8–12 mm thick. Foliage glaucous; **blades** 8–20 mm wide; **ligules** 1.5–2.5 mm, membranous. **Spikes** 1–3 cm wide; 3–8 spikelets per node, forming a jagged profile due to thick clusters of spikelets at each node; 4–6 florets per spikelet. **Glumes** linear-lanceolate, keeled distally, **lemmas** 15–20 mm; hairy near the base, glabrous toward the tip, with awns 1.5–2.5 mm. **Anthers** about 5 mm. **Habitat**: Dry, sandy soils, sandy road cuts, SE WA and NE OR. Native to Eurasia. **Comments**: *Leymus racemosus* is a relatively short, stocky wildrye with a thick inflorescence that is often widest at the base. Introduced for sand stabilization in SE WA, it has recently been collected in sandy habitats along the Columbia and Deschutes Rivers in OR.

Leymus triticoides (Buckley) Pilg.
Beardless Wildrye

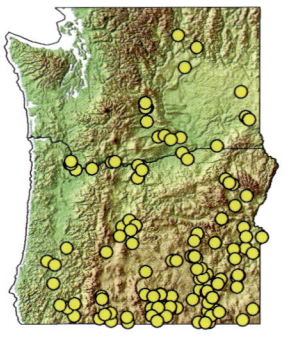

Strongly rhizomatous perennial; **culms** 45–125 cm, 1.8–3 mm thick. **Leaves** ± basally concentrated; **blades** 3.5–10 mm wide; **ligules** 0.2–1.3 mm; **auricles** to 1 mm. **Spikes** 0.5–1.5 cm wide, with 2 spikelets per node near middle of spike, 1–3 spikelets per node elsewhere. Spikelets sessile or sometimes on pedicels to 1.5 mm. **Glumes** awl-like, sometimes flattened, glabrous to scabrous distally; **lemmas** 5–12 mm, usually glabrous, or with sparse hairs to 0.3 mm, awns to 3 mm. **Anthers** 3–6 mm. **Habitat**: Moist, often alkaline meadows. **Comments**: *Leymus triticoides* is often abundant in the Great Basin. Note the very narrow glumes and the twisted rachilla that turns the florets about 90° within the spikelet. *Elymus lanceolatus* and *Pascopyrum smithii* are similar but have mostly 1 spikelet per node and wider glumes. The cultivar 'Shoshone', sold as *L. triticoides*, is actually Eurasian *L. multicaulis*. Commonly planted for forage, it has not yet been recorded in the PNW. *Leymus triticoides* hybridizes with other wheatgrasses including species of *Elymus*, *Hordeum*, and *Leymus*.

Lolium perenne L. ssp. *multiflorum* (Lam.) Husnot
Annual Ryegrass [E] (yellow dots)

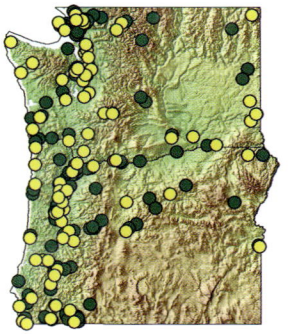

Tufted annual; **culms** to 150 cm; **blades** (2)3–8(13) mm wide; **ligules** membranous; **auricles** often present, claw-like. **Spikes** with 1 spikelet per node, arranged edgewise to the axis, sometimes sunken into the thickened axis. **Spikelets** with 10–22 florets per spikelet, with 1 glume on outside edge, terminal spikelet with 2 glumes. **Glumes** usually < mature spikelets; **lemmas** 4–8.2 mm, oblong-ovate, > 3× as long as wide, usually with awns to 15 mm. **Anthers** 2.5–4(5) mm. **Habitat**: Disturbed areas, prairies, meadows, savannas, roadsides. Native to Eurasia. **Comments**: Hybridizes with ssp. *perenne* with intermediate forms commonly occurring. Annual ryegrass is invasive in upland prairies and savannas.

Lolium perenne ssp. *perenne*
Perennial Ryegrass [E] (green dots)

Like ssp. *multiflorum*, but cespitose perennial; **culms** to 100 cm; **leaf blades** (1)2–4(6) mm wide; **spikelets** with (2)5–9(10) florets; **lemmas** 3.5–9 mm, awnless or with awns to 8 mm. **Anthers** 2–4.2 mm. **Habitat**: Disturbed open areas, meadows, roadsides, lawns. Native to Eurasia. **Comments**: Used for lawn seed, forage, and erosion control.

Lolium temulentum L. ssp. *temulentum*
Darnel or Bearded Ryegrass [E]

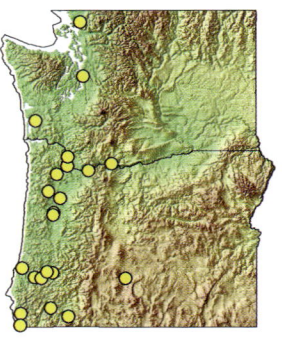

Annual; **culms** 120 cm. **Leaf blades** 1–12 mm wide; **ligules** membranous, to 2 mm; **auricles** prominent, claw-like. **Spikes** with 1 spikelet per node, arranged edgewise to the axis, not sunken into axis. **Spikelets** with 1 glume on outside edge, terminal spikelet with 2 glumes. **Longest glumes** about 1.5× longer than mature spikelets; **lemmas** 4.5–8.5 mm, ovate, ≤ 3× as long as wide, awnless or with awn to 23 mm. **Anthers** 1.5–4 mm. **Habitat**: Disturbed areas, roadsides, weed in grain fields. Native to Europe. **Comments**: Darnel was a serious contaminant in grain fields before methods were developed to separate its seeds from wheat.

Melica aristata Thurb. ex Bol.
Awned Melic

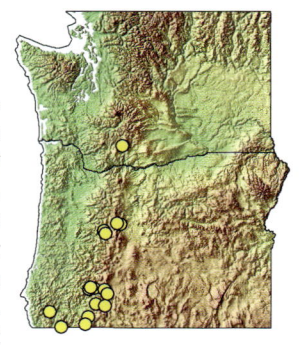

Cespitose perennial; **culms** 40–120 cm, bases not bulb-like. **Leaf sheaths** glabrous or scabrous, sometimes sparsely hairy, usually closed to the top; **blades** 2–6 mm wide, often sparsely hairy on both surfaces; **ligules** 2.5–5 mm, pubescent to the top, often closed in front. **Panicles** strongly contracted, the branches appressed to strongly ascending (spreading at flowering), lower branches usually paired. **Spikelets** with (2)3–5 fertile florets; **rachilla internodes** not swollen; disarticulation above glumes. **Fertile lemmas** 8–13 mm, glabrous or very short-hairy, except marginal veins with 0.3–0.6 mm hairs; bifid or notched; **awns** 5–12 mm. **Rudiments** 2.5–6 mm, similar to fertile florets. **Anthers** 2–3 mm. **Habitat**: Montane; dry open forest, clearcuts, rock outcrops, ridges; Cascades from S WA to SW OR. **Comments**: The long awns distinguish this species from all other *Melica* species except *M. smithii*, which has wider leaves.

Melica bulbosa Geyer ex Porter & J.M. Coulter
Oniongrass

Loosely cespitose, rhizomatous perennial; **culms** 30–100 cm, bases bulb-like, ± sessile on rhizomes. **Leaf sheaths** minutely scabrous, sometimes sparsely hairy, usually closed to the top; **blades** 1.5–5 mm wide, scabrous to pubescent; **ligules** 2–6 mm, usually open in front. **Panicles** strongly contracted; the branches erect, pedicels straight. **Spikelets** with 4–7 fertile florets; **rachilla internodes** not swollen; disarticulation above glumes; **glumes** > 50% of spikelet. **Fertile lemmas** (6)7–10(12) mm, glabrous, emarginate to acute, awnless. **Rudiments** 1.5–5 mm, truncate to tapering, sometimes similar to fertile florets. **Anthers** 1.5–4 mm. **Habitat**: Dry, rocky slopes, grasslands, sagebrush steppe, sometimes mesic sites with aspen. **Comments**: *Melica bulbosa* is distinguished by its sessile, bulb-like corms and acute lemmas. See *M. spectabilis* and *M. subulata*.

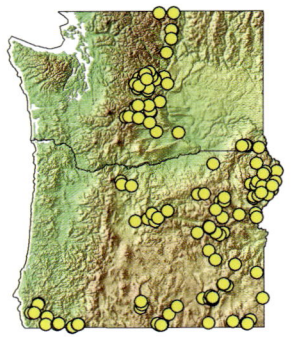

Melica fugax Bolander
Little Oniongrass, Little Melic

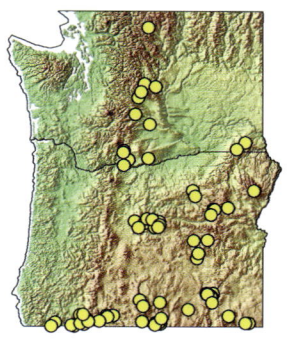

Cespitose perennial; **culms** 10–60 cm, **bases** bulb-like. **Leaf sheaths** glabrous to scabrous, usually split open 10–30 mm at the top; **blades** 1.2–5 mm wide, glabrous or hairy; **ligules** 0.5–2.6 mm, usually open in front. **Panicles** open, narrow; branches ascending when young, stiffly spreading at flowering; pedicels straight. **Spikelets** with 2–5 fertile florets; **rachilla internodes** swollen when fresh, wrinkled and orange when dry; disarticulation above glumes. **Fertile lemmas** 4–7 mm, glabrous to scabrous, rounded to acute, awnless. **Rudiments** similar to fertile florets. **Anthers** 1–2 mm. **Habitat**: Dry grasslands, pine savannas, sage steppe, sometimes on serpentine. **Comments**: The swollen rachilla internodes and spreading mature panicle branches distinguish *M. fugax* from other *Melica* species.

Melica geyeri Munro var. *geyeri*
Geyer's Oniongrass

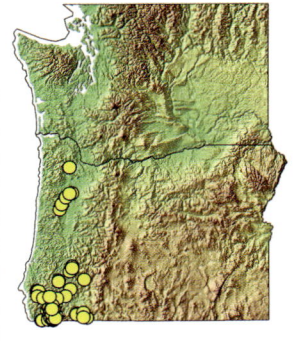

Cespitose, rhizomatous perennial; **culms** 65–200 cm, bases bulb-like, sessile on rhizomes. **Leaf sheaths** scabrous, sometimes sparsely hairy, closed to near top; **blades** 2–8 mm wide, scabrous below, hairy above; **ligules** 0.8–5 mm, usually open in front. **Panicles** open, branches spreading to drooping; pedicels straight. **Spikelets** with 4–7 fertile florets; **rachilla internodes** not swollen; disarticulation above glumes. **Fertile lemmas** 7.5–12.5 mm, glabrous to scabrous, acute to obtuse, awnless. **Rudiments** 3–7 mm, tapering, similar to fertile florets. **Anthers** 2.5–4 mm. **Habitat**: Dry grasslands, savannas, open woodlands, often on ultramafic substrates; W of the Cascades from the Willamette Valley to CA. **Comments**: The open panicle of *M. geyeri* distinguishes this large, attractive grass from other bulb-forming *Melica* species such as *M. subulata*. Most records of this grass from the Willamette Valley are more than 100 years old.

Melica harfordii Bolander
Harford's Melic

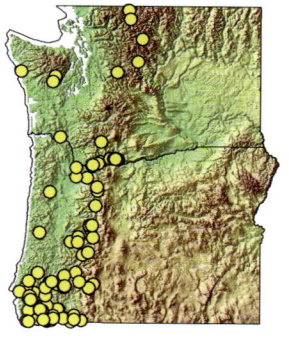

Cespitose perennial; **culms** 35–120 cm; **bases** not bulb-like, often reddish. **Leaf sheaths** glabrous or hairy, usually closed to the top; **blades** 1.5–4.5 mm wide, glabrous or hairy; **ligules** 0.5–1.5 mm, usually closed and projecting upward in front. **Panicles** strongly contracted, branches short, appressed; pedicels straight. **Spikelets** with 2–6 fertile florets; **rachilla internodes** not swollen; disarticulation above glumes. **Fertile lemmas** 6–16 mm, hairy at the base, hairs to 0.75 mm on back, 0.7–1.3 mm on margins, obtuse to rounded, sometimes with awns, 0.5–3 mm. **Rudiments** 2.5–6 mm, tapering, similar to fertile florets. **Anthers** 2.2–4 mm. **Habitat**: Sunny rocky slopes, dry open forest; in and W of the Cascades. **Comments**: *Melica harfordii* can be difficult to recognize as a *Melica* because it lacks bulb-like bases and has rudiments that resemble the fertile lemmas. Look for the closed leaf sheath and purplish lemmas with 9–11 veins.

Melica smithii (Porter ex A. Gray) Vasey
Smith's Melic

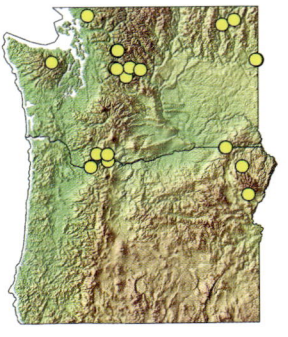

Loosely cespitose perennial; **culms** 60–160 cm, bases not bulb-like, sometimes ± thickened. **Leaf sheaths** glabrous, scabrous, or with stiff retrorse hairs, usually closed to the top; **blades** 5–13 mm wide, scabrous on both surfaces, sometimes hairy above; **ligules** 2–4 mm, glabrous except at the base, closed in front unless mechanically split. **Panicles** open, branches solitary, thin, widely spreading to reflexed, with **pulvini** in the axils. **Spikelets** with 3–5 fertile florets; **rachilla internodes** not swollen; disarticulation above glumes. **Fertile lemmas** 9.5–12 mm, glabrous or scabrous; long-acute with a bifid tip, awns 3–10 mm. **Rudiments** smaller than but similar to fertile florets. **Anthers** 1.3–2.5 mm. **Habitat**: Cool, moist, montane forests, riparian zones; rare. **Comments**: The wide, flat leaves of *M. smithii* are unique among PNW *Melica*. *Melica smithii* is often confused with *Bromus vulgaris*.

Melica spectabilis Scribn.
Purple Oniongrass

Loosely cespitose, rhizomatous perennial; **culms** 45–100 cm, bases bulb-like, attached to rhizome by a root-like structure 10–30 mm. **Leaf sheaths** glabrous, open 3–10 mm at the top; **blades** 2–5 mm wide, scabrous below, glabrous above; **ligules** 0.1–3 mm, open in front. **Panicles** contracted, branches usually appressed, sometimes divergent and flexuous. **Spikelets** with 3–7 fertile florets; **rachilla internodes** not swollen; disarticulation above the glumes; **glumes** < 50% spikelet length. **Fertile lemmas** 6–9 mm, glabrous or scabrous, rounded to acute, awnless. **Rudiments** 1.5–3.5 mm, acute, distinct from fertile florets. **Anthers** 1.5–3 mm. **Habitat**: Moist meadows, open forest. **Comments**: The "tail" connecting the bulb-like bases to the rhizome is distinctive in *M. spectabilis*. Similar *M. bulbosa* has bulb-like bases that are ± sessile on the rhizomes.

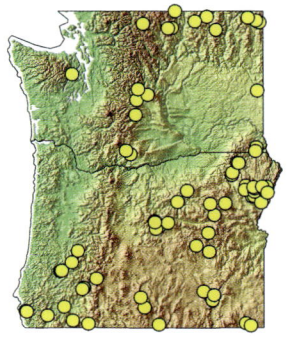

Melica stricta Bolander var. *stricta*
Rock Melic (yellow dots)

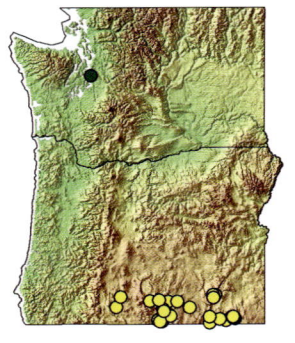

Densely cespitose perennial; **culms** 9–85 cm, bases not bulb-like. **Leaf sheaths** scabrous, rarely densely spreading-hairy, closed or open near the top; **blades** 1.5–5 mm wide, scabrous or hairy; **ligules** 2.5–5 mm, closed or open in front. **Inflorescence** a narrow, raceme-like panicle, branches appressed, pedicels sharply bent, spikelets dangling. **Spikelets** V-shaped at maturity, with 2–4 fertile florets; **rachilla internodes** not swollen; disarticulation below the glumes; **glumes** ≥ 80% spikelet length. **Fertile lemmas** 6–16 mm, glabrous or scabrous, acute, awnless. **Rudiments** 2–7 mm, acute to acuminate, similar to fertile florets. **Anthers** 1–2 mm. **Habitat**: Dry, rocky slopes, montane to alpine. **Comments**: *Melica stricta* is a small, attractive mountain grass with raceme-like panicles and dangling spikelets. It is rare in our area, occurring only in SE OR.

Melica ciliata L.
Silky-spike Melic, Hairy Melic [E] (green dot)

Like *M. stricta* but with 1 fertile floret, and lemmas with long, silky, white hairs 3.5–5 mm. **Comments**: Ornamental grass native to Europe, N Africa, SW Asia; naturalizing in Seattle, WA.

Melica subulata (Griseb.) Scribn.
Alaska Oniongrass

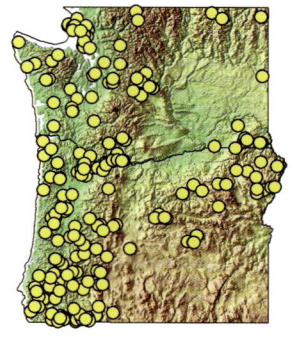

Cespitose, ± rhizomatous perennial; **culms** 55–125 cm, bases bulb-like, sessile on rhizomes. **Leaf sheaths** scabrous, glabrous, or hairy, closed nearly to top; **blades** 2–10 mm wide, glabrous or scabrous; **ligules** 0.4–5 mm, often closed or split open in front. **Panicles** usually ± contracted, branches appressed to ascending, sometimes spreading. **Spikelets** with 2–5 fertile florets; **rachilla internodes** not swollen; disarticulation above the glumes. **Fertile lemmas** 5.5–18 mm, usually with sparse long hairs at least on veins and margins near base, slenderly acuminate, awnless; **rudiments** 4–9 mm, tapering, similar to fertile florets. **Anthers** 1.5–2.5 mm. **Habitat**: Mesic forests in W OR and WA and mountains to E. **Comments**: Bulbous culm bases and slenderly acuminate lemmas distinguish it from other oniongrasses.

Miscanthus sinensis Andersson
Chinese Silvergrass [E]

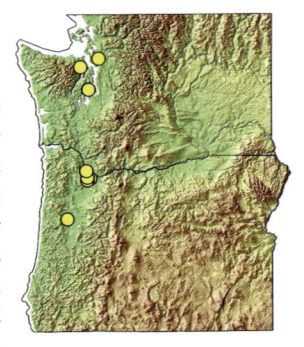

Cespitose perennial with short, thick rhizomes; **culms** 60–200 cm. **Leaves** mainly basal; **sheaths** long soft hairy at the throat; **blades** 20–70 cm, 6–20 mm wide, midribs whitish and conspicuous on the lower surface; **ligules** 1–2 mm, membranous, ciliate. **Panicles** dense to loose, usually with > 15 branches; rachises 6–16 cm, 33–67% as long as the inflorescence; branches 8–15(30) cm long, about 10 mm wide, mostly unbranched and spike-like, sometimes branched near the base. **Spikelets** 3.5–7 mm, lanceolate, borne in pairs, lower ones on pedicels 1.5–2.5 mm, upper ones on pedicels 3.5–6 mm; **callus hairs** of spikelets 6–12 mm, to 2× as long as the spikelets, white, straw-colored, or reddish. **Lower glumes** 3-veined; **upper glumes** 1-veined; **lower florets** sterile; **upper florets** bisexual; **upper lemmas** awned from between 2 teeth, awns 6–12 mm. **Habitat**: Escaping from ornamental plantings to roadsides, shorelines, disturbed areas. **Comments**: Chinese silvergrass is a striking, densely cespitose ornamental grass with fluffy inflorescences that look like feather dusters. Unfortunately, it escapes the garden and has been turning up on roadsides, riparian areas, and other disturbed sites in W OR and WA.

Molinia caerulea (L.) Moench
Purple Moorgrass [E]

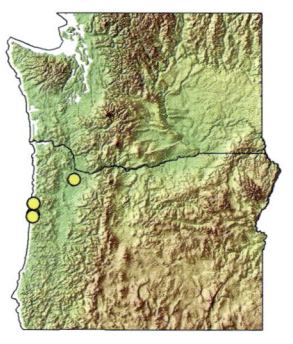

Cespitose perennial; **culms** 15–150(250) cm; basal internodes persistent, swollen, club-shaped. **Leaf sheaths** open, glabrous except margins sparsely hairy near top; **leaf blades** 2–10 mm wide, flat or convolute, eventually breaking from sheaths; **ligules** a fringe of hairs to 0.5 mm; **collars** marked by narrow line or ridge. **Panicles** 5–40 cm, contracted, usually purplish (to brown). **Spikelets** 4–9 mm, with (1)2–4(5) florets; first floret sessile, second floret raised on conspicuous rachilla internode, distal 1–3 florets sterile, reduced; **rachilla** prolonged beyond distal florets. **Lower glumes** 1.5–2.5 mm, obtuse; **upper glumes** 2–3 mm, acute; **lemmas** 2.5–4.5 mm, faintly 3–5-veined, acute to obtuse, awnless. **Anthers** 1.5–3 mm. **Habitat**: Deflation plains, moist riparian forest. Native to Eurasia. **Comments**: *Molinia caerulea* is a tall, tussock-forming grass with narrow panicles, commonly used in horticultural landscaping. In OR it has escaped ornamental plantings at the coast and near Portland.

Muhlenbergia andina (Nutt.) Hitchc.
Foxtail Muhly

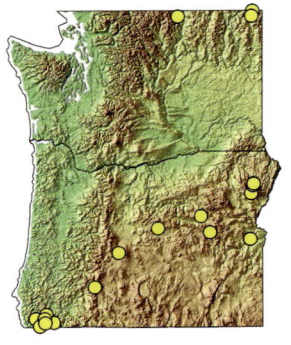

Rhizomatous perennial; **culms** 25–85 cm, hollow. **Leaf blades** 2–4 mm wide, flat, scabrous below, pubescent above; **ligules** 0.5–1.5 mm, membranous, lacerate to ciliate. **Panicles** dense, contracted, 0.5–2.8 cm wide, primary branches > 0.1 mm thick, appressed to ascending, or diverging to 30(40)° from axis at maturity. **Spikelets** 2–4 mm, with 1 floret; **glumes** ± ≥ floret, acuminate to awn-tipped; disarticulation above glumes. **Lemmas** 2–3.5 mm, 3-veined, hairy on callus and base with hairs 2–3.5 mm, tips acuminate with awns 1–10 mm. **Anthers** 0.4–1.5 mm. **Habitat**: Montane stream banks, marshes, lakeshores, moist meadows. **Comments**: This uncommon muhly has narrow inflorescences and awned lemmas with long-hairy bases and calluses. Similar *M. mexicana* and *M. glomerata* have shorter lemma hairs; *M. glomerata* is limited to NE WA in our area.

Muhlenbergia asperifolia (Nees & Meyen ex Trin.) Parodi
Scratchgrass or Alkali Muhly

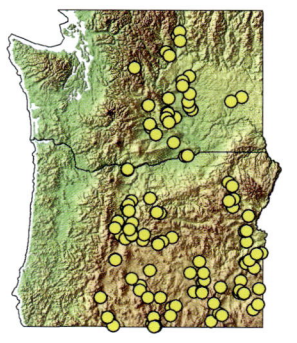

Rhizomatous perennial; **culms** 10–60 cm, solid. **Leaf blades** 1–3 mm wide, flat, smooth to scabrous; **ligules** 0.2–1 mm, ciliate. **Panicles** very open, 4–16 cm wide, primary branches 0.05–0.1 mm thick, spreading 30–90° from axis at maturity. **Spikelets** 1.2–2.1 mm, with 1(3) floret(s); **glumes** < floret, glabrous to scabrous on veins, purplish; disarticulation above glumes. **Lemmas** 1.2–2.1 mm, 3-veined, glabrous, awnless or with a short point to 0.3 mm. **Anthers** 1–1.3 mm. **Habitat**: Moist alkaline meadows, grassy swales, and roadside ditches receiving irrigation runoff; E of the Cascades. **Comments**: The light, airy panicles of scratchgrass float like mist over moist alkaline meadows. Its inflorescences resemble those of *Panicum capillare,* which has dorsiventrally compressed spikelets. The common name "Alkali sacaton" has been mistakenly used for *M. asperifolia,* but it refers to *Sporobolus airoides.* Spikelets are often infected with a smut (*Tilletia asperifolia*), which replaces the lemmas with black balls of spores.

Muhlenbergia filiformis (Thurb. ex S. Watson) Rydb.
Pull-up Muhly

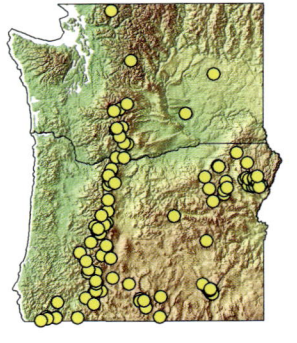

Tufted annual, sometimes appearing perennial; **culms** 5–20 cm, sometimes rooting at lower nodes, not nodulose, solid, glabrous. **Leaf blades** 0.6–1.6 mm wide, flat or involute, glabrous, scabrous, or pubescent; **ligules** 1–3.5 mm, membranous. **Panicles** narrowly spike-like, 0.2–0.5 cm wide, primary branches > 0.1 mm thick, appressed. **Spikelets** 1.5–3.2 mm, with 1 floret; **glumes** < floret, greenish-gray, glabrous, rounded to subacute. **Lemmas** (1.5)1.8–2.5(3.2) mm, 3-veined, pubescent on margins and midvein, awnless or with a short awn < 1 mm. **Anthers** 0.5–1.2 mm. **Habitat**: Moist montane meadows, lakeshores, stream banks. Common. **Comments**: *Muhlenbergia filiformis* is a tiny, delicate grass with a dark, narrow inflorescence. It can form dense turf-like stands. Similar *M. richardsonis* is rhizomatous and has nodulose culms and glabrous lemmas. See also *Agrostis variabilis* and *Podagrostis humilis*. The name "pull-up muhly" originated because cattle prefer not to eat roots and soil, so they spit them out, scattering grass tufts in their wake.

Muhlenbergia glomerata (Willd.) Trin.
Spike Muhly

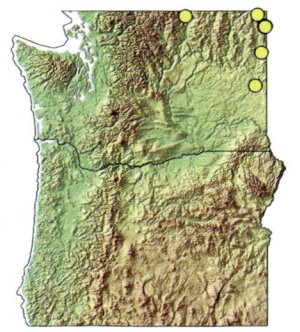

Rhizomatous perennial; **culms** 30–120 cm, hollow. **Leaf blades** 2–6 mm wide, flat, usually scabrous; **ligules** 0.2–0.6 mm, membranous, ciliolate. **Panicles** dense, contracted, lobed, 0.3–1.8 cm wide, primary branches > 0.1 mm thick, appressed. **Spikelets** 3–8 mm, with 1 floret; **glumes** 3–8 mm (including the awns), 1.3–2× lemma, acuminate, with awns to 5 mm. **Lemmas** 1.9–3.1 mm, 3-veined, hairy on callus, margins, and midvein, hairs to 1.2 mm, tips acuminate, awnless or with awn to 1 mm. **Anthers** 0.8–1.5 mm. **Habitat**: Meadows, bogs, fens, stream banks; NE WA. **Comments**: Spike muhly is widely distributed across the N US and S Canada, but it is rare in our area and is listed as a sensitive species in WA. Compare with *M. andina*, *M. mexicana*, and *Agrostis exarata*.

Muhlenbergia mexicana (L.) Trin. var. *filiformis* (Torrey) Scribn.
Wirestem Muhly (yellow dots)

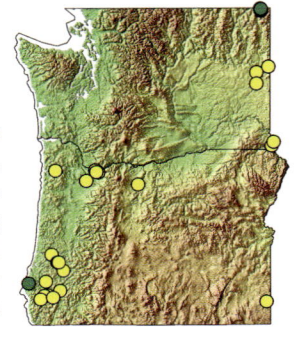

Rhizomatous perennial; **culms** 30–90 cm, hollow. **Leaf blades** 2–6 mm wide, flat, scabrous or smooth; **ligules** 0.4–1 mm, membranous, ciliolate. **Panicles** dense, contracted, 0.3–3 cm wide, primary branches > 0.1 mm thick, appressed or diverging to 30° from axis at maturity. **Spikelets** 1.5–3.8 mm, with 1 floret; **glumes** 1.5–3.7 mm, ± = floret, acuminate, awnless or with awns to 2 mm. **Lemmas** 1.5–3.8 mm, 3-veined, hairy on callus, margins, and midvein, hairs < 0.7 mm, tips acuminate, with awns 3–10 mm. **Anthers** 0.3–0.5 mm. **Habitat**: Rocky stream banks below high-water line. **Comments**: Wirestem muhly flowers on rocky riverbanks after water levels drop; scattered E and W of the Cascades in OR, and in E WA. See also *M. andina* and *M. glomerata*.

Muhlenbergia mexicana var. *mexicana* [N&E] (green dots)

Like var. *filiformis* but lemmas awnless or with awns < 3 mm. **Habitat**: Moist meadows, cranberry bogs. **Comments**: Rare in naturally occurring populations in NE WA; it was probably introduced in a coastal OR cranberry bog. Does not occur in Mexico.

Muhlenbergia minutissima (Steud.) Swallen
Least Muhly (yellow dots)

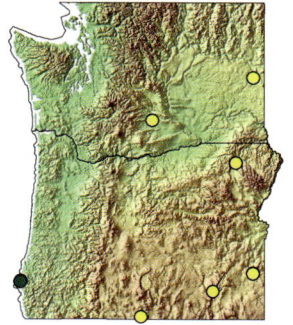

Tufted annual; **culms** 5–40 cm, solid. **Leaf blades** 0.8–2 mm wide, flat or involute, scabrous to short-hairy; **ligules** 1–2.6 mm, membranous. **Panicles** open, 1.5–6.5 cm wide, primary branches < 0.1 mm thick, spreading 25–80° from axis at maturity. **Spikelets** 0.8–1.5 mm, with 1 floret; **glumes** 0.5–0.9 mm, < floret, sparsely and minutely hairy, at least near the tips. **Lemmas** 0.8–1.5 mm, 3-veined, glabrous or hairy on midvein and margins, awnless. **Anthers** 0.2–0.7 mm. **Habitat**: Seeps over shallow soil on rock outcrops, lakeshores, and intermittent streams; SE OR, E WA. **Comments**: Rare in our area, least muhly is known from 4 sites in OR and 2 in WA; the N OR record and all WA records date from the late 1800s. See *M. asperifolia* and introduced *M. uniflora*, below.

Muhlenbergia uniflora (Muhl.) Fernald
Bog Muhly [E] (green dots)

Similar to *M. minutissima*, but plants perennial; **glumes** glabrous; **lemmas** 1.2–2 mm; **anthers** 0.6–0.9 mm. **Habitat**: Acidic bogs, lakeshores; native to E North America. **Comments**: Introduced in cranberry bogs near coast in SW OR, S BC; expected to spread.

Muhlenbergia richardsonis (Trin.) Rydb.
Mat Muhly

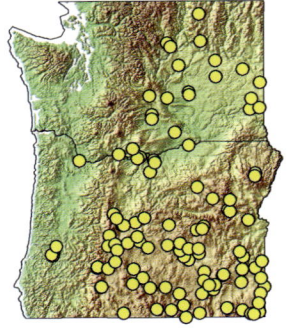

Rhizomatous, perennial, often mat forming; **culms** 5–30 cm, usually nodulose, solid, sometimes short-hairy below the nodes. **Leaf blades** 0.5–4.2 mm wide, flat or involute, glabrous below, minutely hairy above; **ligules** 0.8–3 mm, membranous, erose. **Panicles** narrowly spike-like, 0.1–1.7 cm wide, primary branches > 0.1 mm thick, appressed, rarely diverging up to 20° from the axis. **Spikelets** 1.7–3.1 mm, with 1(2) florets; **glumes** ≤ 50% floret, green, acute, sometimes mucronate. **Lemmas** 1.7–3.1 mm, 3-veined, glabrous, awnless or with a short point ≤ 0.5 mm. **Anthers** 0.9–1.6 mm. **Habitat**: Moist alkaline meadows, river margins; E of the Cascades and Douglas County, OR. **Comments**: Mat muhly can be a dominant in heavily grazed meadows, often with *Carex praegracilis* or *C. douglasii*. Raised nodules extend horizontally between vertical ribs on the culms, a unique feature among our muhlies. This feature is hard to see, especially on fresh plants. See *M. filiformis*.

Muhlenbergia rigens (Benth.) Hitchc.
Deergrass [E]

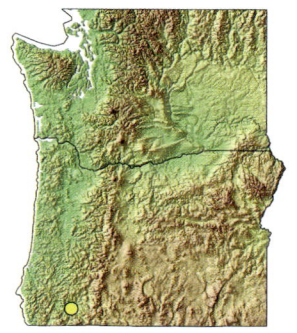

Densely cespitose perennial; **culms** (35)50–150 cm, to 5 mm thick near base. **Leaves** mainly basal, **blades** 1.5–6 mm wide, flat or involute, stiff, lower surface smooth, upper surface scabrous; **ligules** 0.5–2(3) mm, membranous, ciliolate. **Panicles** 15–60 cm, 0.5–1.2 cm wide, spike-like, dense, often interrupted below; branches 0.2–4 cm, appressed. **Spikelets** 2.4–4 mm; **glumes** 1.8–3.2 mm, almost as long as florets, 1(3)-veined, rounded to acuminate, rarely with small point to 0.6 mm; **lemmas** 2.4–4 mm, lanceolate, short-hairy near base, hairs to 0.4 mm, tips acute to obtuse, lacking awns, rarely short-pointed; **paleas** nearly = lemmas, acute. **Anthers** 1.3–1.8 mm. **Habitat**: Sandy or gravelly substrates in moist drainages, canyon bottoms. **Comments**: Deergrass is identified by its dense, hemispherical mound of long, drooping leaves and long, narrow inflorescence. It is planted ornamentally. Native from CA to Mexico, it was collected recently in SW OR.

Nardus stricta L.
Matgrass [E]

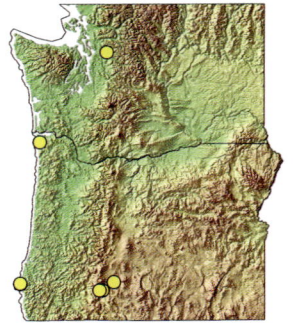

Densely cespitose perennial; **culms** (3)10–40(60) cm. **Leaf sheaths** open; **blades** 0.5–1 mm wide, tightly rolled, outer surface with spreading hairs, tips stiff, sharp; **ligules** 0.5–1(2) mm, membranous, rounded. **Inflorescences** 1-sided spikes, terminating in a bristle to 10 mm. **Spikelets** triangular in cross section, with 1 floret, purplish, base ± embedded in the spike axis. Disarticulation above the glume. **Lower glumes** reduced to cup-like rim; **upper glumes** absent or vestigial. **Lemmas** 5–10 mm, papery; **awns** 1–4.5 mm. **Anthers** 1–4 mm. **Habitat**: Sandy soils in coastal scrub and disturbed areas along the coast, irrigated pastures inland. **Comments**: The dense tuft of very narrow leaves of *Nardus stricta* resembles Idaho fescue. Matgrass is a Class A noxious weed in OR. It is spreading along the coast and in Klamath County. It is a serious threat to native ecosystems and also on pastures and rangelands because it is avoided by livestock and increases under grazing pressure.

Nassella lepida (Hitchc.) Barkworth
Foothills Needlegrass [N?]

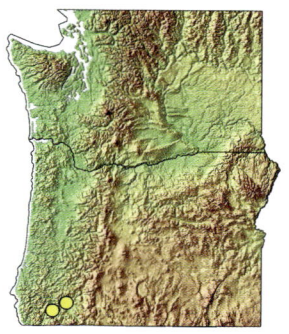

Cespitose perennial; **culms** 35–100 cm. **Leaves** 1–3.5 mm wide, flat to folded, lower surface minutely roughened; **ligules** 0.1–0.6 mm, truncate to rounded. **Panicles** 9–55 cm, open, branches 1–8 cm, ascending to spreading. **Spikelets** with 1 floret. **Glumes** 5.5–15 mm, narrow, long-acuminate, sometimes awned; **florets** 4–7 mm, terete, widest at or slightly above midlength; **calluses** 0.4–1.6 mm, sharp, with ± appressed hairs. **Lemmas** covered with minute bumps, at first evenly hairy, becoming glabrous between veins, tapering to an obscure crown; **crown** 0.25–0.3 × 0.15–0.2 mm, edged with hairs 0.3–0.6 mm; **awns** 12–55 mm, tardily deciduous, bent twice, 1st bend distinct, 2nd bend obscure, terminal segment wavy; **paleas** to 50% lemma length. **Anthers** 2–2.5 mm. **Habitat**: dry grasslands and hillsides. **Comments:** Foothills needlegrass is native from Mexico through CA and was recently found at two sites in SW OR.

Nassella pulchra (Hitchc.) Barkworth
Purple Needlegrass [N?]

Cespitose perennial, **culms** 35–100 cm. **Leaves** 0.8–3.5 mm wide, flat to folded, lower surface glabrous or sparsely hairy; **ligules** 0.3–1.2 mm, rounded. **Panicles** 18–60 cm, open; branches 3–9 cm, spreading, flexuous, often hairy in the axils. **Spikelets** with 1 floret. **Glumes** 12–20 mm; **florets** 7.5–11.5 mm, terete; **calluses** 1.8–3.5 mm, sharp, hairy; **lemmas**, with minute bumps, evenly hairy at maturity, tapering ± abruptly to the crown; **crown** 0.6–1.1 mm × 0.5–0.7 mm, edged with hairs 0.8–0.9 mm; **awns** 38–100 mm, tardily deciduous, distinctly twice geniculate, terminal segment straight; **paleas** to 50% lemma length. **Anthers** 3.5–5.5 mm. **Habitat**: Open oak woodlands and dry grasslands. **Comments**: Purple needlegrass is native from Mexico through CA, and was recently found in SW OR.

Nassella tenuissima (Trin.) Barkworth
Mexican Feathergrass, Mexican Needlegrass [E]

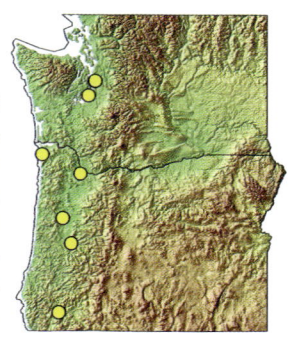

Densely cespitose perennial; **culms** 25–100 cm. **Leaf sheaths** open; **blades** 0.2–1.5 mm wide, convolute; **ligules** membranous, 1–5 mm. **Spikelets** with 1 floret. **Glumes** 5–13 mm. **Florets** somewhat laterally compressed; **callus** blunt with appressed hairs to 25–33% lemma length. **Lemmas** (1.5)2.5–3 × 0.5 mm, margins overlapping their entire length, covered with minute bumps, midveins pubescent on lower 50%; **crown** 0.1-0.2 mm, edged with hairs < 0.5 mm; **awns** 45–100 mm, twice-geniculate. **Paleas** 25–50% lemma length. **Anthers** 1.2–1.5 mm. **Habitat**: Disturbed urban areas, roadsides, dry grasslands. Native to SW US and N Mexico. **Comments**: *Nassella tenuissima* is a drought-tolerant bunchgrass with dense feathery inflorescences. At maturity, the awns twist and curl, creating matted "grass dreadlocks." Widely planted as an ornamental, it escapes cultivation in urban landscapes and in dry grasslands. It has potential to be a troublesome invasive in natural habitats.

Oryzopsis asperifolia Michx.
Roughleaf Ricegrass, Winter Grass

Cespitose perennial; **culms** 25–65 cm. **Leaf sheaths** open, bases bright purple below duff surface; **basal blades** 30–90 × 4–9 mm, twisted so dorsal surface is up, glossy-green dorsally, remaining green over winter; **uppermost flag leaf blade** reduced, 8–12 mm, narrower than sheath top; **ligules** 0.2–0.7 mm, membranous, ciliate. **Panicle** contracted; **spikelets** with 1 floret; **glumes** 5–7.5 mm, tips mucronate; **calluses** 0.8–2 mm, blunt, with ruff of dense, soft hairs; **lemmas** 5–7 mm, stout, pale, leathery or hard, pubescent, margins strongly overlapping, awns 7–15 mm, deciduous. **Anthers** 2–4 mm, usually penicillate. **Habitat**: Duff of cool coniferous forests. **Comments**: The basal leaves of *O. asperifolia* stay green through winter under the snow. This grass extends across Canada from British Columbia to Newfoundland, entering our region only in NE WA. It is easily recognized, even without flowers, by the glossy flat leaves lying on the duff, dorsal side up.

Panicum capillare L. ssp. *capillare*
Witchgrass

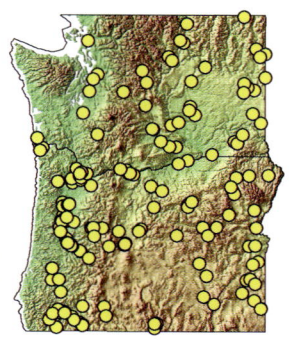

Annual; **culms** 15–130 cm; nodes hairy. **Leaves** not forming a basal rosette; sheaths and blades with dense, papillose-based hairs; **blades** 3–18 mm wide; **ligules** 0.5–1.5 mm, membranous, ciliate. **Panicles** diffuse, often with base enclosed in upper leaf sheath; branches spreading; at least some pedicels > 2 mm. **Spikelets** 1.9–4 mm, dorsiventrally compressed, elliptic to lanceolate; florets 2, the lower sterile, the upper fertile. **Lower glumes** 33–50% as long as spikelets; **upper glumes** 1.8–3.1 mm; **lower lemmas** ± = upper glume; **lower paleas** absent; **upper lemmas** 1.5–1.8 mm, fertile, hard, shiny, smooth, awnless. **Anthers** 0.5–1 mm. **Habitat**: Dry to seasonally wet, often disturbed sites. **Comments**: Witchgrass is a late-summer annual with delicate, diffuse panicles and spreading-hairy leaf sheaths. At maturity the panicle breaks off at the base and becomes a tumbleweed. Witchgrass is highly proficient in colonizing disturbed sites, which is somewhat unusual for a native species. It's good to have a few native weeds.

Panicum dichotomiflorum Michx. ssp. *dichotomiflorum*
Fall Panicgrass [E]

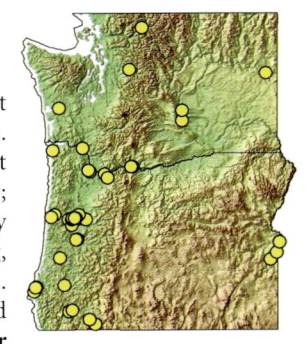

Annual or short-lived perennial; **culms** 5–200 cm, decumbent to erect, rooting at lower nodes when in water; nodes glabrous. **Leaves** not forming a basal rosette; **sheaths** glabrous except near base; **blades** 3–25 mm wide, glabrous or sparsely hairy; **ligules** 0.5–2 mm, membranous, ciliate. **Panicle bases** usually enclosed in upper sheath; primary branches spreading, secondary branches appressed; at least some pedicels > 2 mm. **Spikelets** 2.3–3.8 mm, dorsiventrally compressed, ellipsoid to ovoid; florets 2, the lower sterile, the upper fertile. **Lower glumes** 25–33% as long as spikelets; **upper glumes** 2.5–3.1 mm; **lower lemmas** similar to upper glumes; **lower paleas** vestigial to nearly = lower lemmas; **upper lemmas** 1.4–2.5 mm, fertile, hard, shiny, smooth, awnless. **Anthers** 1 mm. **Habitat**: Disturbed, usually moist sites, ditches, roadsides. Native to E North America. **Comments**: In the PNW, fall panicgrass is an annual that flowers in late summer; it is a perennial in the tropics. At anthesis, the upper florets sport bright orange anthers and magenta stigmas.

Panicum miliaceum L. ssp. *miliaceum*
Proso Millet, Common Millet, Broomcorn [E] (yellow dots)

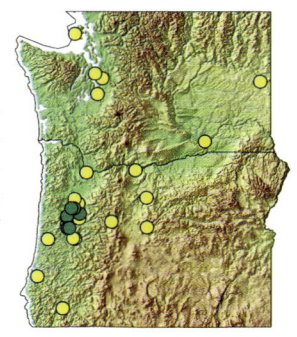

Annual. **Culms** 20–120 cm. **Leaves** not forming a basal rosette; **sheaths** open, with papillose-based hairs; **blades** 7–25 mm wide; **ligules** 1-3 mm, membranous, ciliate. **Panicles** dense, usually nodding, partially enclosed in upper sheaths, branches ascending to appressed. **Spikelets** 4–6 mm, dorsiventrally compressed, ovoid; florets 2, the lower sterile, the upper fertile. **Lower glumes** 50–75% as long as spikelets; **upper glumes** 4–5.1 mm; **lower lemmas** similar to upper glumes; **lower paleas** ≤ 50% as long as upper florets; **upper lemmas** 3–3.8 mm, fertile, straw-colored to orange, smooth or striate, shiny, awnless, not disarticulating. **Anthers** 1–2 mm. **Habitat**: Disturbed areas, roadsides, under bird feeders. Native to Asia.

Panicum miliaceum ssp. *ruderale* (Kitagawa) Tzvelev
Wild Millet [E] (green dots)

Like ssp. *miliaceum* but with **culms** 70–210 cm; **panicles** open, erect, usually fully exserted from sheaths, branches ascending to spreading. **Upper lemmas** smooth, glossy dark brown to blackish, disarticulating at maturity.

Panicum rigidulum Bosc ex Nees ssp. *rigidulum*
Redtop Panicgrass [E]

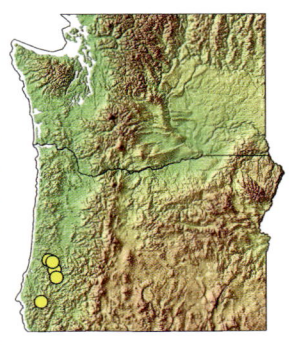

Loosely cespitose perennial; **culms** 40–150 cm, compressed. **Leaf sheaths** open, strongly compressed or keeled, often sparsely hairy near top; **blades** 5–12 mm wide, glabrous; **ligules** 0.3–3 mm, membranous, erose, or ciliate. **Panicles** terminal and axillary; ultimate branches 1-sided, ± appressed; pedicels < 2 mm. **Spikelets** 1.6–2.5 mm, dorsiventrally compressed, lanceolate, often purple; 2 florets, the lower sterile, the upper fertile. **Lower glumes** 67–75% as long as spikelets, 3-veined; **upper glumes** ± = spikelets. **Lower lemmas** like upper glumes; **upper lemmas** 1.4–2 mm, fertile, shiny, thick, hard, awnless but with tuft of short hairs at tip; tightly clasping paleas. **Anthers** 0.5–0.75 mm. **Habitat**: Sandy shorelines, disturbed areas in SW OR. Native from E North America to Central America. **Comments**: Redtop panicgrass differs from our other *Panicum* species in its 1-sided panicle branches and lanceolate spikelets. It is a warm-season grass with C4 metabolism that flowers in late summer. Formerly called *Coleataenia longifolia*.

Panicum virgatum L.
Switchgrass

Perennial with hard, scaly rhizomes; **culms** 40–300 cm; glabrous, often forming dense clumps. **Leaf** sheaths glabrous or hairy, usually with ciliate margins; **blades** 10–60 cm, 2–15 mm wide glabrous or hairy; **ligules** 2–6 mm, membranous, ciliate. **Panicles** open, 10–55 cm, 4–20 cm wide; base not enclosed in upper leaf sheath; branches ascending to spreading; pedicels 0.5–20 mm. **Spikelets** 2.5–8 mm, ± round in cross section to slightly laterally compressed, narrowly lanceoloid; florets 2, the lower sterile, the upper fertile. **Lower glumes** 50–80% as long as spikelets; **upper glumes** and **lower lemmas** extending 0.4–3 mm beyond upper florets; lower florets staminate; **lower paleas** 3–3.5 mm; **upper lemmas** 2.3–3 mm, fertile, shiny, smooth, glabrous, awnless. **Anthers** 1.3–2.3 mm. **Habitat**: Road shoulders, shorelines, disturbed areas, moist meadows. **Comments**: Switchgrass is native to tallgrass prairies of the Midwest from Canada to Mexico. It is widely grown as an ornamental and a forage grass. It readily self-sows around plantings and has escaped in a few places in WA.

Pappostipa speciosa (Trin. & Rupr.) Romasch.
Desert Needlegrass

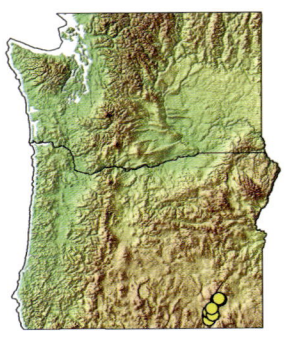

Cespitose perennial; **culms** 30–60 cm. **Leaf sheaths** open, mostly glabrous, densely hairy at throat, basal sheaths reddish-brown; **blades** 0.5–2 mm wide, usually rolled to < 1 mm diameter; **lower ligules** 0.3–1 mm, densely hairy; **upper ligules** to 2.5 mm, hyaline, glabrous to ± hairy. **Panicles** congested, often partly included in uppermost sheath. **Spikelets** with 1 floret; **lower glumes** 16–24 mm, **upper glumes** 13–19 mm; **calluses** sharp, 0.8–1.6(3) mm; **lemmas** (6)8–10 mm, densely short-hairy; awns 35–45(80) mm long, bent once; first segment with long, spreading hairs 3–8 mm, terminal segment glabrous, smooth; **paleas** about 50% as long as lemmas. **Anthers** 0.5–0.75 mm. **Habitat**: Rocky slopes in sagebrush steppe. **Comments**: Long-awned species of *Eriocoma* differ from *Pappostipa* in having the awn bent twice, and in lacking densely hairy ligules. Desert needlegrass is very rare in OR, more common in the arid SW US. See also *Eriocoma thurberiana*, which grows in the same locations but has longer ligules and awns bent twice.

Parapholis incurva (L.) C.E. Hubbard
Curved Sicklegrass [E]

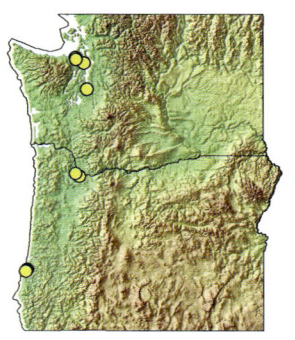

Short, tufted annual; **culms** 2–35 cm, erect to decumbent, branching at nodes, internodes hollow. **Upper leaf sheaths** strongly inflated, enclosing the lowest spikelets; **blades** 1–3 mm wide, scabrous; **ligules** membranous, to 1.5 mm. **Spikes** terminal and axillary, generally curved or twisted, with spikelets sunken into the thickened inflorescence axis. **Spikelets** 4–7.5 mm, cylindrical; **glumes** 2 per spikelet, stiff, lanceolate to acuminate, awnless, lying side by side opposite from the rachis, hiding the shorter, single floret; **lemmas** 3.5–5.5 mm, rounded, awnless. Disarticulation in the rachis. **Anthers** 0.5–1.3 mm. **Habitat**: Coastal salt marshes and disturbed, sandy sites at and above high tide. Native from W Europe to India. **Comments**: Curved sicklegrass is a rare introduction in coastal areas of OR and WA. It often has strongly incurving spikes and reddish coloration. Compare to *Hainardia cylindrica* with only 1 glume per spikelet, and the inland native species, *Deschampsia bolanderi*, with awned lemmas and inflorescence axes that do not disarticulate.

Pascopyrum smithii (Rydb.) Barkworth & D.R. Dewey
Western Wheatgrass

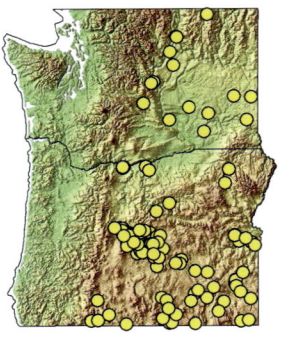

Rhizomatous perennial; **culms** 20–100 cm. **Leaves** mostly basal; **blades** 1–4.5 mm wide, flat to involute, upper surface with prominent veins; **ligules** about 0.1 mm, membranous; **auricles** 0.2–1 mm, often purple. **Spikes** 5–17 cm; spikelets 1 per node (occasionally 2 at lower nodes). **Spikelets** 12–26(30) mm; 1–3× > internodes, florets 2–12; **glumes** 5–15 mm, narrowly lanceolate, tapering to slender point from midlength or below, 3–5-veined, midveins usually twisting on their axis and curving to one side. **Lemmas** 6–14 mm, glabrous to moderately hairy, tips acute, awnless or with awns to 5 mm. **Anthers** 2.5–6 mm. **Habitat**: Mesic to moist meadows in sagebrush steppe, often in alkaline soils. **Comments**: *Pascopyrum smithii* originated as a hybrid of *Elymus lanceolatus* and *Leymus triticoides*. The former has leaves more evenly distributed along the culm; the latter usually has 2 spikelets per node.

Paspalum dilatatum Poir
Dallisgrass [E]

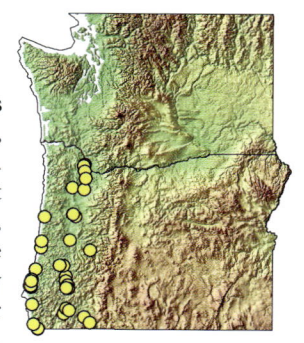

Loosely cespitose or short-rhizomatous perennial; **culms** 50–175 cm, decumbent at base. **Leaf blades** 2–16 mm wide, glabrous except near base; **ligules** 1.5–3.8 mm, membranous. **Panicles** with 2–7 spike-like branches arising at different points on the axis; branches 4–13 cm, spreading or dangling, 1-sided with overlapping spikelets forming 2 rows along the central axis of each branch. **Spikelets** 2.3–4 mm, dorsiventrally compressed, paired; florets 2, the lower sterile, the upper bisexual; **lower glumes** absent; **upper glumes** 2.3–4 mm, 5–7-veined, margins long-hairy; **lower lemmas** like upper glumes; **upper lemmas** 2–3 mm, hard, glabrous, awnless. **Anthers** 1–1.2 mm. **Habitat**: Disturbed moist sites, roadsides. Native to South America. **Comments**: *Paspalum dilatatum* panicle branches are well separated on the axis. In *Digitaria*, the panicle branches arise close together. Dallisgrass stays green after associated annual grasses have turned brown.

Paspalum distichum L.
Knotgrass [N?]

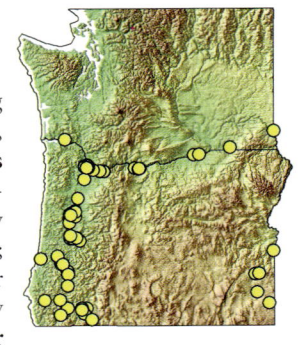

Strongly rhizomatous, stoloniferous perennial; often forming leafy tufts; **culms** 5–65 cm. **Leaf sheaths** mostly glabrous, sparsely long-hairy near top; **blades** 1.8–11.5 mm wide, **ligules** 1–2 mm, membranous. **Panicles** V-shaped, with 2(3) spike-like branches, 1.4–7 cm. **Spikelets** 2.4–3.2 mm, dorsiventrally flattened, usually not paired, in rows on 1 side of branch; florets 2, lower sterile, upper bisexual; **lower glumes** absent or to 1 mm; **upper glumes** 2.4–3.2 mm, 3-veined, sparsely hairy on back; **lower lemmas** = upper glumes, glabrous; **upper lemmas** 2.5–3 mm, hard, glabrous, awnless. **Anthers** 1–1.5 mm. **Habitat**: Margins of lakes, rivers, wet ditches. **Comments**: Rhizomes and stolons bear "knots" of densely overlapping leaves that serve as overwintering structures. There is uncertainty about whether knotgrass is native or exotic in the PNW.

Phalaris angusta Nees ex Trin.
Narrow Canarygrass [N?] (green dots)

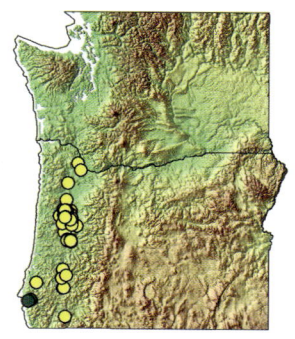

Annual; **culms** 10–170 cm. **Leaf blades** 2–12 mm wide; **ligules** 4–7 mm, membranous. **Panicles** 2–20 × 0.5–1.5 cm, cylindrical, usually > 10× longer than wide, often lobed and interrupted near base; spikelet borne singly. **Spikelets** with 3 florets, lower 2 florets sterile; **glumes** 2–5.5 mm, keels winged, wings ~ 0.4 mm wide; **sterile florets** 0.5–1.5 mm, linear, inconspicuously hairy; **fertile floret** 2–3.8 mm; **lemmas** pubescent, especially distally. **Anthers** 0.5–1.3 mm. **Habitat**: Low-elevation grasslands. Native in CA.

Phalaris aquatica L.
Harding Grass, Bulbous Canarygrass [E] (yellow dots)

Cespitose perennial; **culms** 50–200 cm; **bases** often bulbous. **Leaf blades** 0.5–10 mm wide; **ligules** 2–12 mm, membranous. **Panicles** 5–15(20) × 1–2.5 cm, densely cylindric to ovoid, spike-like; spikelets borne singly. **Spikelets** with 2–3(4) florets; lower floret(s) sterile; **glumes** 4.4–7.5 × 1.2–1.5 mm, keels winged distally, wings 0.2–0.4 mm wide; **sterile floret** 1–3 mm, usually solitary, hairy; second sterile floret, if present ≤ 0.7 mm; **fertile florets** 3–4.6 mm; **lemmas** pubescent. **Anthers** 3–3.6 mm. **Habitat**: Mesic to seasonally wet disturbed areas, roadsides, ditch banks. Invasive.

Phalaris arundinacea L.
Reed Canarygrass [E and N]

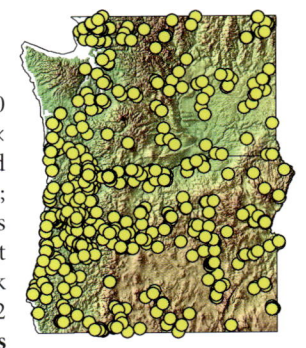

Rhizomatous perennial; **culms** 40–245 cm. **Leaf blades** 5–20 mm wide; **ligules** 4–11 mm, membranous. **Panicles** 5–40 × 1–4 cm, whitish or tinged with pink or purple, usually lobed near the base; branches appressed, spreading at anthesis; spikelets borne singly. **Spikelets** with 3 florets, lower 2 florets sterile; **glumes** 4–8.1 mm, keels smoothly curved, wing absent or to 0.2 mm wide; **sterile florets** 1.5–2 mm, hairy (they look like tufts of hairs rather than florets); **fertile florets** 2.5–4.2 mm; **lemmas** hairy on upper portion and margins. **Anthers** 2.5–3 mm. **Habitat**: Marshes, swales, shorelines, ditches, riparian forest, clearcuts; invades upland habitats in moister environments. **Comments**: Reed canarygrass is one of our most damaging invasive wetland grasses, forming dense, monocultural stands. Exotic and native forms occur in North America. Most PNW populations are descendants of European grasses planted for erosion control or forage. Once established, control is very difficult. A variegated horticultural variety, 'Picta,' has escaped cultivation. On depauperate plants the pinkish panicles can resemble those of *Holcus lanatus,* a cespitose grass with awned lemmas.

Phalaris californica Hook. & Arn.
California Canarygrass (green dots)

Rhizomatous perennial; **culms** 60–160 cm, bases bulbous. **Leaf blades** 3–12(18) mm wide; **ligules** 3–5(8) mm, membranous. **Panicles** 1.5–6 × 1–3 cm, dense, ovoid; spikelets borne singly. **Spikelets** with 3 florets; lower 2 florets sterile; **glumes** 5–8 mm, slender, wingless or nearly so; **sterile florets** 1.8–3.5 mm, densely hairy, usually > 50% length of fertile floret; **fertile floret** 3.5–5 mm, **lemmas** sparsely hairy. Anthers 3–3.5 mm. **Habitat**: Ravines and open, moist ground near the coast.

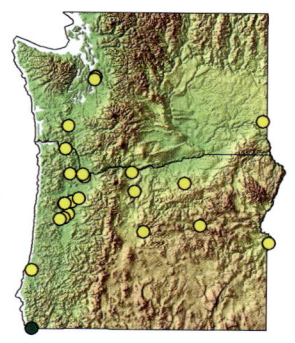

Phalaris canariensis L.
Annual Canarygrass [E] (yellow dots)

Annual; **culms** 30–100 cm. **Leaf blades** 2–10 mm wide; **ligules** 3–6 mm, membranous. **Panicles** 1.5–5 × 1.5–2 cm, dense, ovoid, with truncate bases; spikelets borne singly. **Spikelets** with 3 florets; lower 2 florets sterile; **glumes** 7–10 mm, keel wings to 0.6 mm wide, widest near the tip, so spikelet looks almost circular in outline; **sterile florets** 2–4.5 mm, ≥ 33% as long as the fertile floret, sparsely hairy; **fertile florets** 4.5–6.8 mm; **lemmas** densely hairy. **Anthers** 2–4 mm. **Habitat**: Disturbed areas, often near bird feeders.

Phalaris paradoxa L.
Hooded Canarygrass [E]

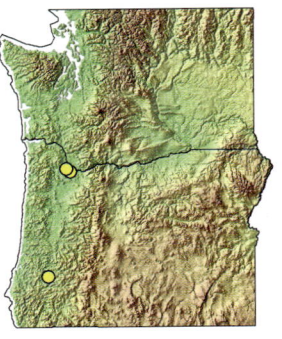

Annual; **culms** 20–100 cm. **Leaf blades** 2–5 mm wide; **ligules** 3–5 mm, membranous. **Panicles** 3–9 × about 2 cm, dense, cylindrical, tapering at base; spikelets borne in clusters, the central ones bisexual or pistillate, outer ones staminate or sterile; **glumes of fertile spikelets** 5–8 mm, with toothed wing 0.2–0.4 mm wide, tips acuminate; **glumes of sterile spikelets** similar, sometimes poorly developed; **sterile florets** knob-like projections on callus of fertile floret, each with 1 or 2 hairs; **fertile florets** 2.5–3.5 mm; **lemmas** hard, shiny, glabrous or short-hairy at tip. **Anthers** 1.5–2.5 mm. **Habitat**: Disturbed areas. Native to the Mediterranean region. **Comments**: The glumes of *P. paradoxa* have distinctive toothed wings. This grass is a rare weed in the PNW but is more common in CA. It may be mistaken for a solitary shoot of *P. aquatica*. Small plants may be mistaken for an odd *Alopecurus* or *Gastridium phleoides*.

Phleum alpinum L.
Alpine Timothy

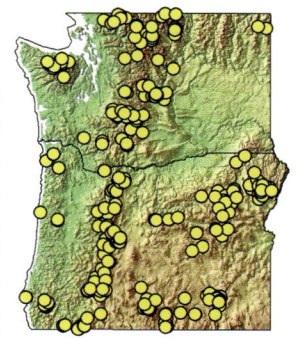

Cespitose to short-rhizomatous perennial; **culms** 15–50 cm; bases not bulbous. **Leaf sheaths** open, uppermost sheaths inflated; **blades** 4–7 mm wide, flat; **ligules** 1–4 mm, membranous, truncate. **Panicles** 1–6 cm × 5–12 mm, dense, spike-like, subglobose to cylindric, 1.5–3× as long as wide. **Spikelets** strongly laterally compressed, with 1 floret per spikelet; **glumes** 2.5–4.5 mm, 3-veined, scabrous, keels with spreading hairs, tips awned; **awns** 1.5–2.5(3.2) mm, straight, stiff; **lemmas** 1.7–2.5 mm, awnless. **Anthers** 1–2 mm. **Habitat**: Moist montane to alpine meadows and stream banks, also coastal headlands. **Comments**: Alpine timothy is a small tufted grass with dense, spike-like panicles and inflated upper sheaths. The stiff awns of the glumes lend a harsh feel to the inflorescence. *Phleum pratense* is a taller grass with longer panicles and shorter glume awns. It usually grows at lower elevations and in more mesic habitats.

Phleum pratense L.
Timothy [E]

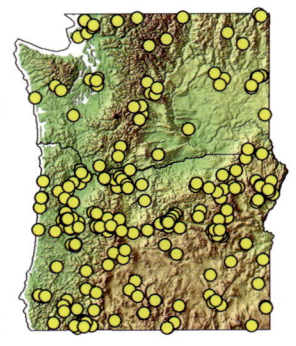

Cespitose perennial; **culms** (20)50–150 cm; bases often bulbous. **Leaf sheaths** open, not inflated; **blades** 4–8(10) mm wide, flat; **ligules** 2–4 mm, membranous, obtuse to acute. **Panicles** 5–10(16) cm × 5–7.5(10) mm, dense, spike-like, cylindric, 5–20× as long as wide. **Spikelets** strongly laterally compressed, with 1 floret per spikelet; **glumes** 3–4 mm, 3-veined, short-hairy, keels with spreading hairs, tips awned, awns 1–2 mm, straight, stiff; **lemmas** 1.2–2 mm, awnless. **Anthers** 1.6–2.3 mm. **Habitat**: Old fields, pastures, disturbed sites. **Comments**: *Phleum pratense* is a tall grass with bulbous culm bases and long, slender, cylindric panicles. The glume awns give the spikelets the look of a 2-tined fork. Superficially similar *Alopecurus pratensis*, a plant of generally wetter habitats, is strongly rhizomatous, flowers earlier in spring, and has soft, silky inflorescences and awnless glumes. Probably the first non-native grass deliberately introduced in the PNW, timothy is highly palatable to livestock and is commonly cultivated for hay.

Phragmites australis (Cav.) Trin. ex Steud. ssp. *americanus* Saltonstall et al.
American Common Reed (green dots)

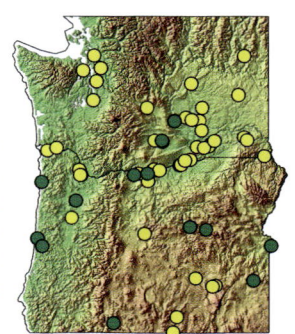

Rhizomatous perennial, rarely forming dense stands; fresh rhizomes yellow; **culms** 100–400 cm, smooth, shiny, red to chestnut. **Leaf sheaths** open, deciduous, exposing culms; blades 20–40 mm wide, flat, yellow-green; **ligules** membranous, ciliate, the membranous portion on middle culm leaves (0.2)0.4–0.9 mm. **Panicles** large, plumose. **Spikelets** with 2–10 florets; **rachillas** long-silky-hairy; lower **glumes** < upper; **lemmas** 8–15 mm, long-acuminate, glabrous, awnless. **Anthers** 1.5–2 mm. **Habitat**: Waterways, marshes, sloughs.

Phragmites australis ssp. *australis*
Old World Common Reed [E] (yellow dots)

Like ssp. *americanus* but forming dense stands. Fresh rhizomes white to light yellow, darkening with exposure to air; lower **culms** ridged, dull, yellowish or tan. **Leaf sheaths** persistent, hiding culms; **blades** dark green or glaucous, sometimes yellow-green near coast; membranous portion of **ligule** of middle leaf 0.1–0.4 mm. **Comments**: Subspecies *australis* is invasive and forms dense monocultures that exclude native wetland species. To identify the two subspecies, collect the distinctive lower culms.

Piptatheropsis exigua
(Thurb.) Romasch., P.M. Peterson & Soreng
Little Ricegrass

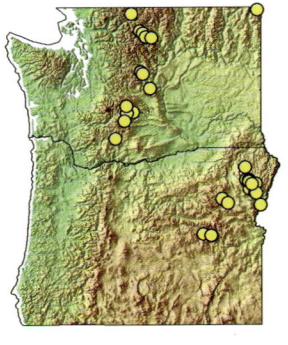

Densely cespitose perennial; **culms** 12–40 cm. **Leaf blades** usually folded and 0.4–0.8 mm wide, 0.6–1.4 mm wide when flat, **ligules** 1.2–3.5 mm, membranous, acute. **Panicles** narrow, branches appressed. **Spikelets** with 1 floret; **glumes** 3.5–6 mm; **florets** 3–6 mm, terete; **calluses** 0.2–0.5 mm, blunt, hairy; **lemmas** ± leathery, shiny, hairy especially in the upper half, tips with 2 thick lobes 0.3–0.4 mm, awned from between the lobes, awns 4–7 mm, persistent, scabrous, twisted and strongly bent once. **Anthers** 1.5–3 mm. **Habitat**: Rocky slopes and outcrops in upper montane to subalpine habitats. **Comments**: This small, densely tufted grass has stout, blunt, hairy lemmas and persistent awns. It is rare in OR and WA, but more common in the Rocky Mts. Other names for little ricegrass have included *Oryzopsis exigua*, *Piptatherum exiguum*, and *Stipa exigua*.

Pleuropogon oregonus Chase
Oregon Semaphoregrass

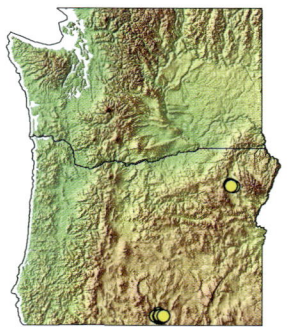

Rhizomatous perennial; **culms** 40–95 cm. **Leaf sheaths** closed 75% their length; **blades** 4–9 mm wide, flat, tips pointed or with awns to 4.5 mm; **ligules** 5–10 mm, membranous. **Racemes** 1-sided with 6–7 ascending, laterally compressed spikelets. **Spikelets** 20–40(50) mm, each with 7–14 florets; **lower glumes** 2–3 mm; **upper glumes** 2.5–4.5 mm; **lemmas** 4.5–7 mm, with 7 parallel veins, tips awned, awns 5–12 mm, straight; **palea keels** awned from lower half, awns 3–9 mm. **Anthers** ~ 4 mm. **Habitat**: Wet meadows, stream banks.

Comments: Endemic to OR. Rare. *Pleuropogon oregonus* has long spikelets, awned palea keels, and often awned leaf tips. *Pleuropogon refractus* differs with dangling spikelets, larger glumes, and awnless palea keels. *Pleuropogon oregonus* was declared extinct in 1975; then two small colonies were discovered in the 1980s. Efforts are underway to establish more populations on public land.

Pleuropogon refractus (A. Gray) Benth. ex Vasey
Nodding Semaphoregrass

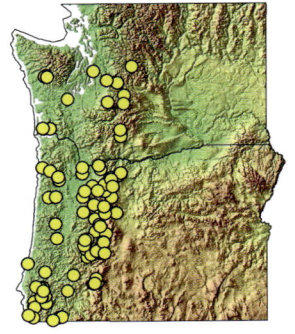

Rhizomatous perennial; **culms** (80)100–150 cm. **Leaf sheaths** closed at least near base, open above for 2–7 cm; **blades** 5–14 mm wide, flat, tips acute or acuminate with a short point, rarely awned to 0.3 mm; **flag leaf blades** often absent or reduced to a small bristle; **ligules** 2–7 mm, membranous. **Racemes** 1-sided with 6–14 reflexed, dangling spikelets. **Spikelets** (20)25–28 mm, each with 7–14 florets; **lower glumes** 3–6 mm; **upper glumes** 4–7(8.3) mm; **lemmas** 8–10 mm with 7(9) parallel veins, tips awned, awns 2–15(21) mm, straight; **palea keels** awnless but with a small triangular appendage 0.2–0.6 mm. **Anthers** 3.5–4 mm. **Habitat**: Wet meadows, streamsides, moist shaded woods; low elevations to midmontane. **Comments**: Nodding semaphoregrass has an attractive, 1-sided raceme of dangling spikelets. It is widespread but uncommon in moist habitats W of the Cascade Crest, extending as far E as the Metolius River and Odell Lake in OR.

Poa alpina L. ssp. *alpina*
Alpine Bluegrass

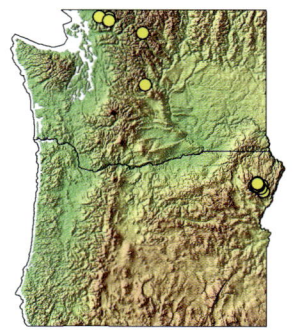

Small, densely cespitose perennial; not glaucous; **culms** 10–40 cm. **Leaf sheaths** closed 12–30% their length; **blades** 2–4.5 mm wide, flat, flag leaf blades much reduced; **ligules** to 4(5) mm, white, smooth, glabrous; **ligules** of sterile shoot leaves 1–2(3) mm. **Panicles** open to loosely contracted, branches ascending to spreading. **Spikelets** broadly ovoid, 1.5–2.5× as long as wide, with 3–7 florets, **rachilla internodes** 0.5–0.8 mm. **Calluses** glabrous; **lemmas** 3–5 mm, broadly lanceolate, keeled, silky-hairy on keel and veins, short-silky between veins. **Anthers** 1.3–2.3 mm. **Habitat**: Subalpine to alpine, meadows, ridges, talus, scree, unstable slopes, and gravelly streambeds, often on limestone (occasional lower elevation occurrences are probably ephemeral, destroyed by floods and re-established by seed washed down from higher elevations). **Comments**: Cespitose *P. alpina* has broadly ovoid spikelets and wide leaves. *Poa glauca* and *P. wallowensis* have proportionately narrower spikelets and narrower leaves. Also, *P. glauca* has longer rachilla internodes, and *P. wallowensis* has shorter anthers.

Poa annua L.
Annual Bluegrass [E]

Tufted annual (rarely biennial); **culms** 2–20(45) cm. **Leaf sheaths** closed about 33% their length; **blades** 1–3(6) mm wide, flat or loosely folded; **ligules** 0.5–3(5) mm, membranous. **Panicles** ± open, branches ascending, spreading, or reflexed. **Spikelets** with 2–6 florets. **Calluses** glabrous; **lemmas** 2.5–4 mm, keeled, crisp-puberulent to long-villous on keel, marginal veins, and usually lateral veins, glabrous between veins; **palea** keels pubescent. **Anthers** 0.6–1.1 mm, oblong. Habitat: Mesic disturbed areas, gravel parking lots, lawns, pavement cracks. Native to Eurasia. Comments: One of the world's most widespread weeds, annual bluegrass thrives in human-influenced habitats everywhere except the Arctic. Very similar to *P. infirma,* which has smaller lemmas and anthers, and lemmas with dense, silvery, shaggy-silky hairs on the keel and veins.

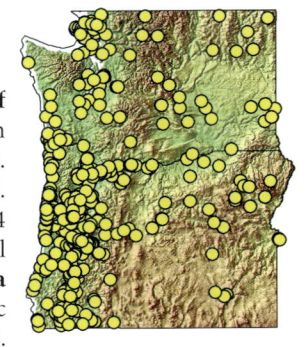

Poa arctica R. Br. ssp. *arctica*
Arctic Bluegrass

Rhizomatous perennial, usually purplish, shoots usually solitary; **culms** 7.5–60 cm. **Leaf sheaths** closed (16)20–40% their length; **blades** 1.5–3 mm wide, flat or folded; **ligules** (1)2–4 mm. **Panicles** open, branches ascending to widely spreading. **Spikelets** (3.5)4.5–6(7) mm, with (2)3–6 florets; **rachilla internodes** 0.4–1.1 mm. **Calluses** with cobwebby hairs; **lemmas** (2.7)3–4.5 mm, usually dark purple, keeled, keels hairy for most of their length, marginal and lateral veins hairy, short-hairy between veins at least near base, glabrous elsewhere. **Palea** usually at least sparsely short-hairy between keels. **Anthers** 1.4–2.5 mm, sometimes aborted late in development. **Habitat**: Steep, rocky alpine slopes. **Comments**: Similar to *P. pratensis* ssp. *alpigena*, which is glabrous between the palea keels. *Poa arctica* is widespread in N latitudes. It was reported in error from Mt. Rainier; it is present in BC, but has not been found in OR or WA.

Poa bolanderi Vasey
Bolander's Bluegrass

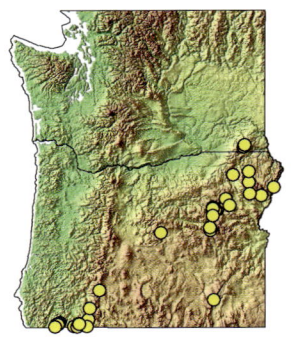

Tufted annual (rarely short-lived perennial), often glaucous; **culms** 20–60(70) cm. **Leaf sheaths** closed 50–75% their length; **blades** 1.5–5 mm wide, usually flat; **ligules** 2.5–7 mm, membranous. **Panicles** narrow at first, often becoming open, branches erect to often spreading or reflexed at maturity. **Spikelets** (3)4–7 mm, with 2–3(4) florets; **rachilla internodes** 1–1.2+ mm. **Calluses** of some or all florets with sparse cobwebby hairs; **lemmas** 2.5–4 mm, keeled, smooth or scabrous throughout, glabrous, tips purplish, acute; **palea keels** glabrous, sparsely scabrous. **Anthers** 0.5–1(1.8) mm. **Habitat**: Openings in dry, montane conifer forests to alpine slopes. Increases after wildfire. **Comments**: Similar *P. howellii* has hairy lemmas and usually grows at lower elevations. Recognized by its wide, relatively short leaves and narrow panicles that spread open at flowering time.

Poa bulbosa L.
Bulbous Bluegrass [E] (yellow dots)

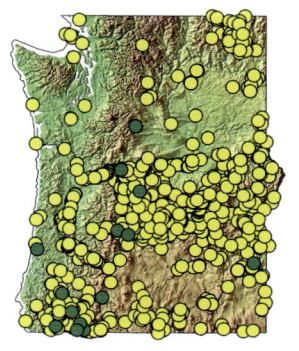

Cespitose perennial; **culms** 15–60 cm, bases bulbous. **Leaf sheaths** closed about 25% their length; **blades** 1–2.5 mm wide, flat, withering early, longest blades of basal tufts mostly > 4 cm; **ligules** of lowest leaves 1–3 mm, as long as or longer than wide, obtuse to acute, smooth (scabrous). **Panicles** ovoid, ± tightly contracted. **Spikelets** with 3–7 florets often modified into purplish bulblets, sometimes lowest florets or entire inflorescence producing flowers. **Calluses** with cobwebby hairs; **lemmas** 3–4 mm, keeled, glabrous, or keel and marginal veins short- to long-villous, glabrous or short-hairy between veins. **Anthers** well-developed and 1.4–2 mm, or aborted, or not developed. **Habitat**: Disturbed grasslands, roadsides, poor soils. Native to Europe. **Comments**: *Poa bulbosa* normally produces tiny plantlets with swollen bases instead of flowers, giving the panicle a shaggy appearance. Some plants produce only true flowers rather than bulblets. Invasive.

Poa iconia Azn. (Turkish bulbous bluegrass; green dots) is a similar species recently documented in OR and WA. It also produces bulblets, but it has shorter, scabrous or hairy ligules (0.25–1 mm) and hairy sheaths and leaves. Its full range is not known.

Poa chambersii Soreng
Chambers' Bluegrass

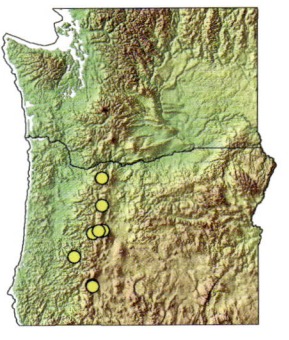

Loosely tufted, short-rhizomatous, dioecious perennial; **culms** 10–50 cm. **Leaf sheaths** closed 33–88% their length, glabrous; **culm blades** (3.2)4.2–5.7(7.3) × 2–5 mm wide, flat or folded, margins smooth, not scabrous; **ligules** (0.2)1.7–3.3(4.8) mm, membranous, smooth. **Panicles** ± contracted, branches 0.9–3.2(4) cm, erect to ascending. **Spikelets** 6–12 mm with 2–7 florets; **rachilla internodes** 0.8–1.5 mm. **Calluses** with sparse, cobwebby hairs (0.2)0.4–1.5(1.9) mm, rarely glabrous. **Lemmas** 5–7 mm, keeled, smooth or sparsely scabrous, keels and marginal veins with short, soft hairs on lower 25%. **Anthers** 1.8–3.7 mm on male plants; vestigial and 0.1–0.2 mm on female plants. **Habitat**: Openings in montane forest. **Comments**: *Poa chambersii* is rare and endemic to the Cascade Range in OR. *Poa rhizomata* has open panicles, and hairier lemma keels and marginal veins. *Poa cusickii* ssp. *purpurascens* is cespitose and always pistillate, and has minutely scabrous leaf margins. *Poa wheeleri* has retrorsely scabrous or hairy sheaths and scabrous ligules. See *P. mansfieldii*.

Poa compressa L.
Canada Bluegrass [E]

Extensively rhizomatous perennial, shoots solitary or loosely tufted; **culms** 15–60 cm; culms and nodes flattened and wiry; some lower nodes exserted. **Leaf sheaths** closed 10–20% their length; **blades** 1.5–4 mm wide, flat; **flag leaf blade** short, ascending to erect, squared off at base; **ligules** 1–3 mm. **Panicles** 2–10 cm, ± compact, proportionately small; branches erect to ascending, densely scabrous on angles. **Spikelets** (2.3)3.5–7 mm with (2)3–7(8) florets. **Calluses** usually with cobwebby hairs. **Lemmas** 2.3–3.5 mm, keeled, keels and marginal veins hairy, glabrous between veins. **Anthers** 1.3–1.8 mm. **Habitat**: Roadsides, pavement cracks, disturbed areas, moist meadows. Common. Native to Europe. **Comments**: *Poa compressa* is characterized by flattened culms and nodes and proportionately small inflorescences. Also look for the short, erect flag leaf blade with a squared-off base. This is a tough little grass that has nearly as many looks as *P. secunda*, depending on its growing conditions. It can be short and compact if repeatedly run over by tires, or sparse and upright if growing on severely compacted soil.

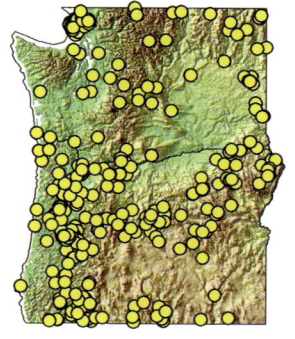

Poa confinis Vasey
Coastline Bluegrass

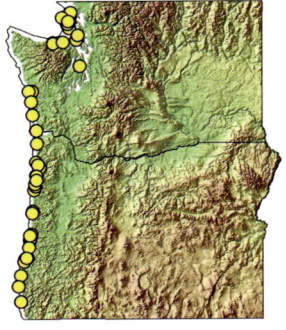

Tufted perennial from stolons and rhizomes; plants female or bisexual; **culms** 7–30(35) cm. **Leaves** firm, inrolled, sparsely scabrous; **sheaths** closed 33–67% their length; **cauline blades** 0.5–1(1.5) mm wide, upper surface scabrous; **ligules** 0.5–1.5(2.2) mm. **Panicles** 1–5(7) cm, compact, branches erect to ascending. **Spikelets** 3–5(8) mm with 2–5 florets. **Calluses** with sparse, diffuse cobwebby hairs 1–2 mm, rarely glabrous; **lemmas** 2.5–4(4.5) mm, keeled, scabrous, usually glabrous, sometimes with keels and marginal veins sparsely short-hairy near base. **Anthers** sterile, 0.1–0.2 mm, or normal, 1.5–2 mm. **Habitat**: Coastal sand dunes, stabilized sandy soil. **Comments**: *Poa confinis* populations have declined due to habitat loss from dune stabilization and development. Count yourself fortunate if you encounter this grass, and work to have it included in coastal habitat restoration projects. *Poa macrantha* grows in the same habitat but has much larger lemmas; *P. pratensis* has hairier lemmas and callus hairs that arise in tufts.

Poa curtifolia Scribn.
Little Mountain or Wenatchee Bluegrass

Densely cespitose perennial; **culms** (15)20–40 mm. **Leaf sheaths** closed 20–33% their length; **cauline blades** (1)1.5–3 mm wide, short, flat, thick, firm, with white margins and prominently prow-shaped tips; **ligules** (1.5)2–5 mm, conspicuous because blades spread so much. **Panicles** narrow, branches ± erect, with 1–4 spikelets per branch. **Spikelets** 7–9 mm, with (2)3–4 florets. **Calluses** glabrous; **lemmas** 4.5–6 mm, weakly keeled, glabrous, or the lower third sparsely short-hairy. **Anthers** 2.2–3.5 mm. **Habitat**: Subalpine to alpine serpentine slopes. **Comments**: Sometimes confused with *P. secunda*, *P. curtifolia* has shorter, firmer, flat leaves, conspicuous ligules, and fewer spikelets. Endemic to the Wenatchee Mts., WA.

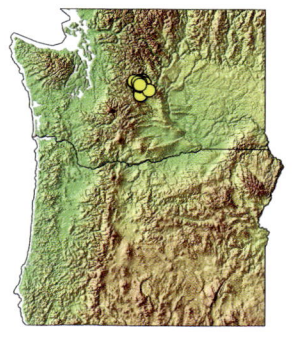

Poa cusickii Vasey
Cusick's Bluegrass

Cespitose perennial, rarely with short rhizomes; **culms** 10–60(70) cm. **Leaf sheaths** closed 25–75% their length; **blades** flat, folded, or involute; **ligules** 1–3(6) mm on culm leaves, 0.2–0.5(2.5) mm on tiller leaves. **Panicles** contracted, usually dense, with 10–100 spikelets. **Spikelets** (3)4–10 mm with 2–6 florets; **calluses** glabrous or with a few short hairs; **lemmas** 3–7 mm, keeled. **Anthers** vestigial, aborted, or well-developed. **Comments**: *Poa cusickii* is a common upland bluegrass with dense inflorescences and cespitose habit. See also *P. fendleriana*, *P. leibergii*, *P. pringlei*, and *P. wheeleri*. 4 subspecies.

Poa cusickii ssp. *cusickii*
Cusick's Bluegrass (yellow dots)

Plants 10–60(70) cm, gynodioecious and dioecious. **Leaf blades** usually < 1.5 mm wide. **Panicle branches** 1.7–4(5) cm, stout to slender, moderately to densely scabrous, with 2–15 spikelets per branch. **Calluses** and **lemmas** glabrous. **Anthers** normal, 2–3.5 mm, or sterile, 0.1–0.2 mm. **Habitat**: Mesic meadows in upper sagebrush zone, to alpine slopes and ridges. **Comments**: Produces functional anthers. Widespread.

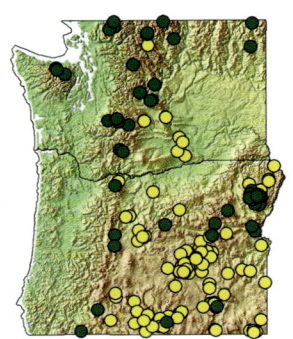

Poa cusickii ssp. *epilis* (Scribn.) W.A. Weber
Skyline Bluegrass (green dots)

Plants 20–45 cm, cespitose; pistillate. **Culm leaf blades** > 1.5 mm wide. **Panicle branches** 1–3 cm, smooth or slightly scabrous. **Calluses** glabrous; **lemmas** glabrous or scabrous. **Anthers** usually aborted late in development. **Habitat**: Subalpine to alpine meadows and rocky slopes.

Poa cusickii ssp. *pallida* Soreng
Pale Cusick's Bluegrass (green dots)

Plants cespitose 10–40(55) cm, all pistillate. **Leaf blades** usually < 1.5 mm wide. **Panicle branches** 0.5–1.7 cm, stout, moderately to densely scabrous, with 2–5 spikelets per branch. **Calluses** and lemmas glabrous. **Anthers** vestigial, 0.1–0.2 mm. **Habitat**: Montane to alpine forb-rich meadows. **Comments**: Similar ssp. *cusickii* has longer panicle branches.

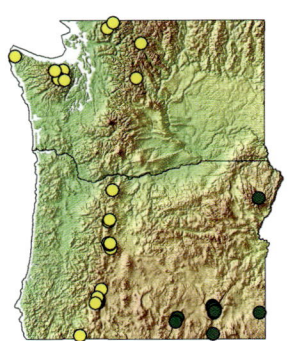

Poa cusickii ssp. *purpurascens* (Vasey) Soreng
Purple Cusick's Bluegrass (yellow dots)

Plants 25–50 cm, ± densely cespitose, basal shoots sometimes branching extravaginally, and resembling rhizomes; pistillate. **Culm leaf blades** > 1.5 mm wide, margins minutely scabrous. **Panicle branches** 1–3(4) cm, smooth to moderately scabrous. **Calluses** of lowest lemmas usually with sparse, short cobwebby hairs < 25% lemma length; those of other lemmas glabrous; **keels and marginal veins of lower lemmas** usually sparsely short-hairy near the base, sometimes glabrous, **upper lemmas** glabrous. **Anthers** aborted late in development. **Habitat**: Subalpine to alpine scree, meadows, thickets. **Comments**: Hairy lower lemmas distinguish this subspecies from the other *P. cusickii* subspecies.

Poa fendleriana (Steud.) Vasey
ssp. *longiligula* (Scribn. & T.A. Williams) Soreng
Vasey's Muttongrass

Densely to loosely cespitose perennial, with short, inconspicuous rhizomes; plants pistillate; **culms** 15–70 cm. **Leaf sheaths** closed about a third of their length; **blades** (0.5)1–3(4) mm wide; **upper cauline blades** greatly reduced or absent, usually < 1 cm; **ligules** (1.5)1.8–18 mm, membranous. **Panicles** 2–12(30) cm, contracted, branches erect. **Spikelets** 6–8(12) mm with 2–7(13) florets. **Calluses** glabrous; **lemmas** (3.5)4–6 mm, keeled, long-hairy on keels and marginal veins, usually glabrous between veins. **Anthers** 0.1–0.2 mm, sterile. **Habitat**: Subalpine to alpine grasslands, rocky slopes in sagebrush steppe. Uncommon. **Comments**: The flag leaves of *P. fendleriana* are reduced to inconspicuous points rarely as much as 1 cm. Short flag leaves may occur occasionally in *P. cusickii*, which has shorter ligules and lemmas with less hairy keels and marginal veins. Much more common in the Rocky Mts., muttongrass enters OR only in the SE corner, with its Basin and Range flora. Often erroneously reported in OR and WA.

Poa glauca Vahl ssp. *rupicola* (Nash) W.A. Weber
Timberline Bluegrass (yellow dots)

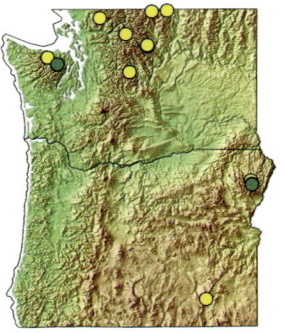

Densely cespitose, glaucous perennial; **culms** 5–15 cm; most shoots fertile at flowering time. **Leaf sheaths** closed 10–24% their length, **blades** 0.8–2 mm wide, flat or folded; **ligules** 1–4(5) mm, membranous. **Panicles** 1–5 cm, contracted to somewhat open, sparse, branches scabrous. **Spikelets** 3–7(9) mm with 2–5 florets; **calluses** glabrous; **lemmas** 2.5–4 mm, keeled, short-hairy on keel and veins, sometimes with shorter, crisped hairs between veins. **Anthers** (1)1.2–2.5 mm. **Habitat**: Dry alpine ridges and slopes. **Comments**: See *P. wallowensis*, *P. mansfieldii*, and *P. nemoralis*.

Poa glauca ssp. *glauca*
Glaucous Bluegrass (green dots)

Like ssp. *rupicola* but **calluses** usually with cobwebby hairs (sometimes glabrous); **lemmas** glabrous or hairy between the veins. **Comments**: Alpine habitats.

Poa howellii Vasey & Scribn.
Howell's Bluegrass

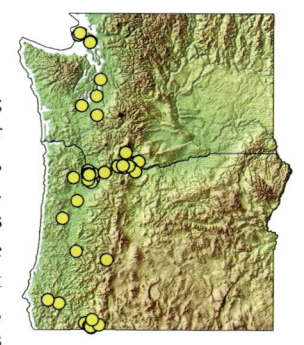

Densely tufted annual (rarely perennial), often glaucous; **culms** (10)25–80(120) cm. **Leaf sheaths** closed 50–90% their length; **blades** 1–7(10) mm wide, flat; **ligules** 1.5–5(10) mm, membranous. **Panicles** open, branches spreading to reflexed. **Spikelets** (2)4–6 mm with 2–5 florets; **rachilla internodes** about 1 mm; **calluses** sparsely cobwebby on at least some florets; **lemmas** 2.5–3.5 mm, keeled, evenly crisp-puberulent near base, scabrous distally; **palea keels** sparsely scabrous, glabrous, or with very short, soft hairs at midlength. **Anthers** 0.2–1 mm. **Habitat**: Moist openings in mixed conifer or oak woodlands, often on northerly aspects, often with moss or boulders; usually below 3,500 feet, up to 5,000 feet in SW OR. **Comments**: Similar *P. bolanderi* has glabrous to scabrous lemmas, longer rachilla internodes, and more contracted panicles, and usually grows at higher elevations. This delicate annual grass doesn't appear to compete well against robust perennials.

Poa infirma Kunth
Weak Bluegrass [E]

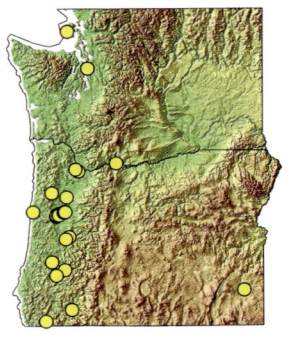

Tufted annual; **culms** 2–15 cm. **Leaf sheaths** closed about 33% their length; **blades** 1–3(4) mm wide, flat; **ligules** 0.5–3 mm. **Panicles** ± narrow, branches usually ascending. **Spikelets** with 2–6 florets. **Calluses** glabrous; **lemmas** 2–2.5 mm, keeled, usually shaggy-silky on keel, marginal veins, and lateral veins, glabrous between veins; **palea keels** pubescent. **Anthers** 0.1–0.5(0.6) mm, spherical to short-elliptical. **Habitat**: Seasonally moist, disturbed areas, gravel parking lots, pavement cracks. Native to Europe. **Comments**: *Poa infirma* is very similar to *P. annua* but it has shorter lemmas with silvery, shaggy-silky hairs on the keel and veins, and tiny, ± spherical anthers that look like grains of sand. This spindly, inconspicuous weed often grows with *P. annua*, often in slightly wetter microsites. It sometimes forms a "bathtub ring" at the edges of mud puddles between the puddle and neighboring *P. annua* plants. Rarely documented but probably widespread, especially W of the Cascades. *Poa annua* is a tetraploid thought to have been derived from hybrids between *P. infirma* (small anthers) and *P. supina* (large anthers). A cultivar of *P. supina* used in lawn seed mixes is expected to escape.

Poa laxiflora Buckley
Looseflower Bluegrass

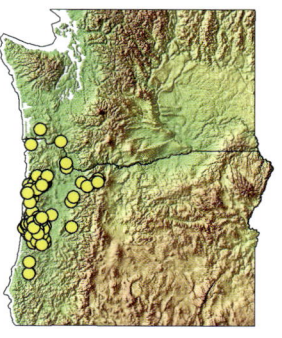

Long-rhizomatous perennial; shoots loosely tufted or solitary; **culms** 50–120 cm, retrorsely scabrous. **Leaf sheaths** closed 50–75% their length, retrorsely scabrous; **blades** 3–8 mm wide, flat, lax; **ligules** 2–3.5 mm, membranous. **Panicles** open, sparse, often nodding, branches widely spreading to drooping, angles scabrous. **Spikelets** 4–8 mm with 2–4 florets; **glumes** 3-veined. **Calluses** with cobwebbly hairs; **lemmas** 3.2–6 mm, keels hairy 67–75% their length, marginal veins with sparse soft hairs, lacking hairs between veins. **Anthers** 0.5–1.1 mm.

Habitat: Moist, shady riparian terraces in coniferous forest, Coast Range and W slope of Cascades. **Comments**: *Poa laxiflora* has a very open panicle and retrorsely scabrous culms. This shade-loving grass is rare and potentially threatened by logging. See also *P. leptocoma*, *P. marcida*, and *P. nervosa*.

Poa leibergii Scribn.
Leiberg's Bluegrass

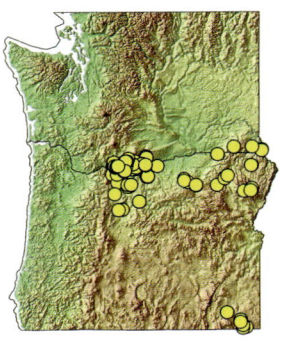

Densely cespitose perennial; gynodioecious; **culms** 5–35 cm. **Leaf sheaths** closed 40–80% their length; **culm blades** 0.5–1(1.5) mm wide, folded or involute, usually soon withering; **ligules** (1)2–4 mm on all leaves, membranous, smooth. **Panicles** open to loosely contracted, often nodding, with (1)6–17(22) spikelets, branches erect to spreading, smooth or sparsely (rarely moderately) scabrous. **Spikelets** 4–8 mm, with 2–8 florets; **calluses** glabrous; **lemmas** 3.5–7 mm, keeled, glabrous. **Anthers** vestigial, 0.1–0.2 mm, or normal, 1.3–3 mm.
Habitat: Wet rock ledges, edges of vernal pools, and moist swales in the Columbia Plateau; also snow pockets on ridges of the Wallowa Mts. and the Owyhee-Malheur divide. **Comments**: *Poa leibergii* is a small, early-flowering grass with glabrous calluses and lemmas. It flowers as early as March in the Columbia Gorge, often before *P. secunda*, which has rounded lemmas and more cigar-shaped spikelets. Both species senesce by early summer into withered brown tufts.

Poa leptocoma Trin.
Western Bog Bluegrass

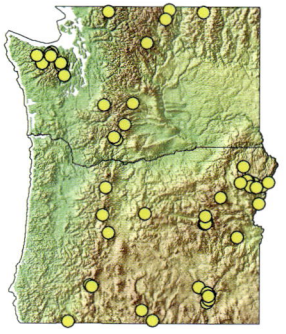

Slender, ± cespitose perennial, often purplish; **culms** 15–100 cm. **Leaf sheaths** closed 25–60% their length; **blades** 1–4 mm wide, flat; **ligules** 1.5–4(6) mm, membranous. **Panicles** sparse, purplish, nodding, branches usually moderately scabrous, ± spreading to reflexed. **Spikelets** 4–8 mm with 2–5 florets; **lower glume** 1-veined; **calluses** with sparse, cobwebby hairs; **lemmas** 3–4 mm, often purple distally, keeled, keels hairy on lower 25–67%, marginal veins hairy, lateral veins and areas between veins glabrous, lemma tips usually bronze-colored. **Palea** keels smooth, scabrous, or with short curved hairs. **Anthers** 0.2–0.6(1.1) mm. **Habitat**: Montane to alpine streamsides, lakeshores, wet meadows. **Comments**: The sparse stems of this slender grass coming up through the moss suggest it should have rhizomes, but it rarely does. Similar *P. reflexa* grows in drier habitats and has smoother panicle branches and longer anthers; *P. paucispicula* also has smoother panicle branches, fewer spikelets, and broader glumes.

Poa lettermanii Vasey
Letterman's Bluegrass

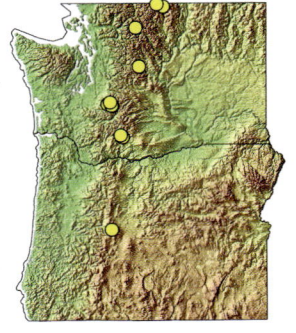

Small, densely cespitose perennial; **culms** 1–12 cm. **Leaf sheaths** closed 15–25% their length; **blades** 0.5–2 mm wide, flat or folded; **ligules** 1–3 mm, membranous. **Panicles** 1–3 cm, contracted, branches erect to strongly ascending. **Spikelets** 3–4 mm with 2–3 florets; **rachilla internodes** < 1 mm; **glumes** 2.6–3.4(4) mm, often ≥ the florets; **calluses** glabrous; **lemmas** 2.5–3 mm, glabrous, keeled, keels and marginal veins rarely sparsely hairy. **Anthers** 0.2–0.8 mm. **Habitat**: Shelter of rocks or in mesic to wet sites on rocky slopes of the highest alpine peaks and ridges. **Comments**: Tiny *P. lettermanii* resembles *P. suksdorfii* but is smaller in all its parts. It grows on South Sister in the central OR Cascades, and on peaks of the WA Cascades, also N into BC and in the Rocky Mts.

Poa macrantha Vasey
Seashore Bluegrass, Dune Bluegrass

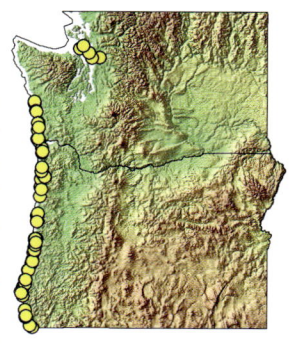

Dioecious perennial; shoots arising in tufts from rhizomes to 4 m long; **culms** 15–60 cm. **Leaf sheaths** closed about half their length; **culm blades** 2–4 mm wide, thick, firm, involute; **ligules** 1–5 mm. **Panicles** congested, branches erect. **Spikelets** 9–17 mm with 3–6(10) florets; **calluses** usually with sparse crown of short hairs, sometimes glabrous or with short, cobwebby hairs. **Lemmas** (6)7.5–11 mm, keeled, keels, marginal veins, and sometimes lateral veins with short, soft hairs, areas between veins glabrous, scabrous, or short soft-hairy. **Anthers** (2)3–4(5) mm or vestigial and 0.1–0.2 mm. **Habitat**: Coastal sand dunes. **Comments**: *Poa macrantha* is a giant among bluegrasses with broad inflorescences and unusually long lemmas. Populations have declined sharply from habitat loss due to dune stabilization, European beachgrass invasion, and coastal development.

Poa mansfieldii Otting & B.L. Wilson
Steens Mountain Bluegrass

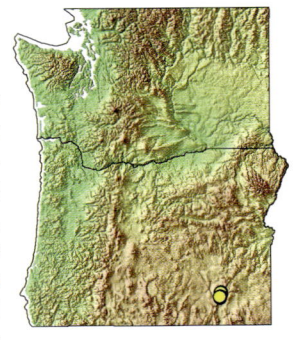

Rhizomatous, gynodioecious perennial; shoots arising singly or in tufts; **culms** 8–33 cm. **Leaf sheaths** closed 35–60% their length; **upper culm blades** (0.75)1.5–3.1(4.7) cm, (0.8)2–3(3.4) mm wide, flat or folded; **culm leaf ligules** (1)2–4.5(6) mm, membranous. **Panicles** tightly to loosely contracted, purplish, branches erect to slightly spreading. **Spikelets** 3–8.4(10.5) mm with (1)2–5(6) florets; **rachilla internodes** (0.3)0.8–1.6 mm; **calluses** glabrous; **lemmas** 3.8–5.2(6.7) mm, glabrous or sparsely scabrous, keeled, keels and veins glabrous. **Anthers** 1.8–3.4 mm, or vestigial and 0.1–0.2 mm. **Habitat**: Alpine slopes and meadows, where snowbanks persist. **Comments**: *Poa mansfieldii* is a recently described bluegrass endemic to Steens Mt. in SE OR. Similar *P. chambersii* is endemic to the Cascades. It is dioecious and has hairy calluses, lemma keels, and marginal veins, and longer leaf blades. Also see *P. cusickii*, *P. pratensis*, and *P. wheeleri*.

Poa marcida Hitchc.
Weeping Bluegrass, Withered Bluegrass

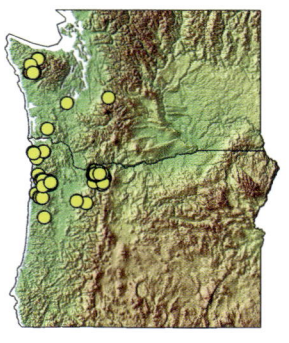

Cespitose, dioecious perennial; **culms** 20–80 cm. **Leaf sheaths** closed ≥ 90% their length; **blades** 1.5–5 mm wide, flat, lax; **ligules** 0.5–2 mm, membranous. **Panicles** 6–22 cm, nodding, branches ascending, angles scabrous. **Spikelets** 3.5–7 mm, with (1)2(4) florets. **Lower glumes** 1-veined; **calluses** with sparse cobwebby hairs; **lemmas** 3.2–5 mm, keeled, glabrous, narrow, tips acuminate. **Anthers** 0.5–1.2 mm. **Habitat**: This uncommon PNW endemic grows in moist, late-successional forests in the Coast Range and in the foothills of the W Cascades. It has a narrow, nodding inflorescence with few branches that are ± parallel with the main axis, and lemmas that are long-tapered. Rhizomatous *P. laxiflora* has widely spreading panicle branches and lemmas with tips acute to rounded.

Poa × *multnomae* Piper
Multnomah Bluegrass

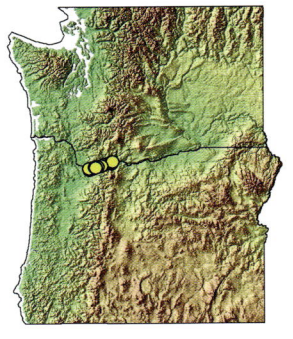

Cespitose perennial; **culms** 15–45 cm. **Leaf sheaths** closed to 33% their length, glabrous; **blades** 0.5–1.5 mm wide, flat or folded; **ligules** 0.3–1(2.8) mm, rounded to truncate, margin finely ciliate. **Panicles** open, branches spreading. **Spikelets** 4.5–8 mm, cylindrical to laterally compressed, with 2–4 florets; **upper glumes** 2.5–3.8 mm; **calluses** with short, ± curled hairs; **lemmas** rounded to weakly keeled on the back, keels short-hairy on basal 50–67%, scabrous or smooth distally, marginal veins short-hairy on the basal 50%, areas between veins scabrous or hairy. **Anthers** 1.5–2 mm, well-formed but often with defective pollen. **Habitat**: Waterfalls, streamsides, moist slopes. **Comments**: *Poa* × *multnomae* is a hybrid between *P. nervosa* and *P. secunda*. It grows in the Columbia Gorge and near Portland.

Poa nemoralis L.
Woodland Bluegrass [E]

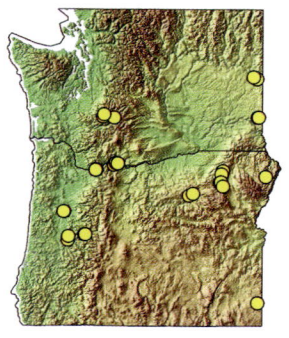

Densely cespitose perennial; **culms** 30–80 cm, smooth, glabrous below lower nodes, top node at or above middle of culm. **Leaf sheaths** closed 10–20% their length; **blades** 0.8–3 mm wide, lax; **ligules** 0.2–0.8(1) mm. **Panicles** 7–16(20) cm, erect or lax, delicate, somewhat sparse, branches ascending to widely spreading, scabrous. **Spikelets** 3–8 mm with (1)2–5 florets; **lower glumes** 3-veined, 6.4–11× as long as wide; **calluses** with sparse, often short, cobwebby hairs; **lemmas** 2.4–4 mm, usually partially bronze-colored, keeled, keels and marginal veins short-hairy, lateral veins and areas between veins glabrous. **Anthers** (1)1.2–1.5(1.9) mm. **Habitat**: Low-elevation forests, streamsides, meadows; native to N Eurasia. **Comments**: Both *P. palustris* and *P. nemoralis* have nearly all shoots fertile at flowering time; new shoots form in late summer and flower the following year. *Poa palustris* differs in having well-developed callus hairs, longer ligules, and culms often branched above the base, giving it a more spreading habit.

Poa nervosa (Hook.) Vasey
Veiny Bluegrass

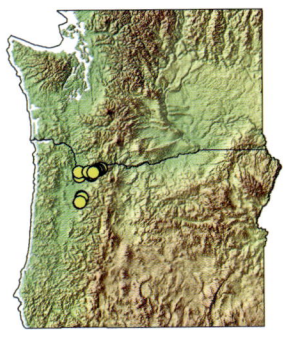

Rhizomatous perennial; gynomonoecious (having both pistillate and bisexual flowers on the same plant); **culms** 20–65 cm, loosely tufted or solitary. **Leaf sheaths** closed 67–90% their length, smooth or sparsely scabrous, sometimes hairy; **collars** of lower leaves hairy, the hairs longer than sheath hairs; **blades** 2–4.5 mm wide, usually flat; **ligules** 0.5–1.5 mm. **Panicles** open or loosely contracted, sparse, branches ascending to spreading. **Spikelets** 4–7 mm with 3–8 florets; **calluses** usually glabrous, rarely minutely webbed; **lemmas** 3–4.5 mm, keeled, keels and marginal veins usually glabrous, rarely sparsely soft-short-hairy, glabrous or hairy between veins. **Anthers** 2.5–4 mm, sometimes sterile and 0.1–0.2 mm. **Habitat**: Low-elevation moist forests, waterfalls, seeps. **Comments**: *Poa nervosa* is endemic to SW WA and NW OR. *Poa wheeleri*, formerly considered a variety of *P. nervosa*, has densely scabrous or hairy lower leaf sheaths, ± glabrous collars, and vestigial anthers. It is widespread in a variety of habitats and also grows at higher elevations. *Poa pratensis* and *P. palustris* have conspicuously cobwebby calluses, a feature lacking in *P. nervosa*. See also *P.* × *multnomae* and *P. chambersii*.

Poa palustris L.
Fowl Bluegrass [E (and N?)]

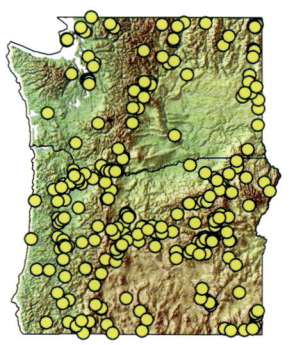

Loosely to densely cespitose perennial, often stoloniferous; **culms** 20–120 cm, retrorsely scabrous or hairy below lower nodes, sometimes branched above base, top node at or above middle of culm. **Leaf sheaths** closed 10–20% their length; **blades** 1.5–8 mm wide, flat; **ligules** (1)1.5–6 mm, membranous. **Panicles** (9)13–30(41) cm, open; branches widely spreading at anthesis, scabrous. **Spikelets** 3–5 mm with (1)2–5 florets; **lower glumes** 6.4–10× as long as wide; **calluses** with cobwebby hairs ≥ (50)67% as long as lemma; **lemmas** 2–3 mm, keeled, keels and marginal veins hairy, lateral veins and areas between veins glabrous. **Anthers** 1.3–1.8 mm. **Habitat**: Moist meadows, shorelines, forests, disturbed areas. Native to North America and Eurasia; most PNW populations are considered to be introduced, but some may be native. **Comments**: Nearly all *P. palustris* shoots are fertile at flowering time; post-flowering shoots that form in late summer remain vegetative until the following year. In OR and WA these traits are shared only by *P. glauca* and *P. nemoralis*. Strongly rhizomatous *P. pratensis* occurs in many of the same habitats but has larger spikelets and many vegetative shoots at flowering time.

Poa paucispicula Scribn. & Merr.
Few-flowered Bluegrass

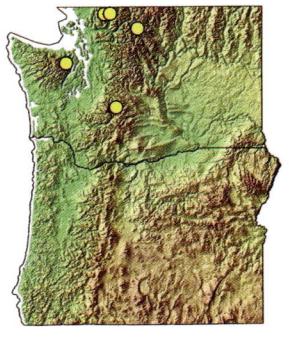

Loosely cespitose perennial; **culms** 10–30 cm. **Leaf sheaths** closed 25–60% their length; **blades** 1–3 mm wide, flat, thin, soft; **ligules** 1–2 mm. **Panicles** open, sparse, lowest branches usually ascending to spreading, smooth (rarely scabrous), with 1–3(5) spikelets. **Spikelets** 4–6 mm, usually dark purple, with 3–5 florets; **calluses** sparsely webbed; **lemmas** 3–4 mm, keeled, keels and marginal veins hairy on lower 50–67%, lateral veins and areas between veins glabrous; **palea keels** sparsely scabrous at midlength. **Anthers** 0.4–1 mm. **Habitat**: Mesic rocky alpine slopes; high mountains of WA. Rare. **Comments**: *Poa paucispicula* is a delicate alpine grass with a sparse, open, purple inflorescence. *Poa reflexa*, another high elevation *Poa*, also has hairs on the lemma veins but in our area is found only in the Wallowa Mts. *Poa leptocoma* is a larger, loosely tufted grass of wet sites.

Poa piperi Hitchc.
Piper's Bluegrass

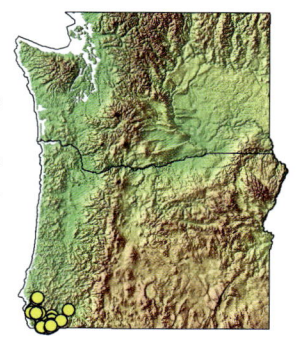

Loosely cespitose, rhizomatous perennial; dioecious; foliage usually glaucous; **culms** 20–55 cm. **Leaf sheaths** closed 33–67% their length; **collars** usually sparsely short-hairy; **cauline blades** 1–3 mm, involute, firm; **ligules** 1–2 mm. **Panicles** loosely contracted, branches ascending. **Spikelets** 6–9(11) mm with 2–5(7) florets; **calluses** with diffuse cobwebby hairs about half as long as lemmas; **lemmas** 4–6(7) mm, glabrous, smooth or scabrous, keeled, keels scabrous. **Anthers** 2–3 mm or vestigial and 0.1–0.2 mm. **Habitat**: Low- to mid-elevation forest openings and Jeffrey pine savanna on ultramafic soils. Rare. **Comments**: *Poa piperi* is a rare dioecious bluegrass endemic to serpentine substrates in SW OR and NW CA. Similar *P. rhizomata* has a more open panicle, silky-hairy lemma keels, and leaves that are softer and mostly flat.

Poa pratensis L.
Kentucky Bluegrass [E & N]

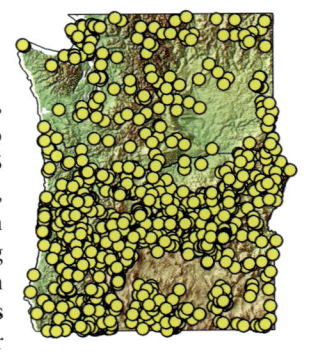

Extensively rhizomatous perennial; **culms** 5–70(100) cm, smooth, loosely to densely tufted. **Leaf sheaths** closed 25–50% their length, ± smooth; **collars** glabrous; **cauline blades** 0.4–4.5 mm wide, flat or folded; **ligules** 0.9–2(3.1) mm, membranous, truncate to obtuse. **Panicles** open to contracted, often nodding, 1-sided and tawny at maturity, branches ascending to spreading, (1)2–9 cm, with 4–30 spikelets ± crowded on distal half. **Spikelets** 3.5–6(7) mm with 2–5 florets; **calluses** with abundant cobwebby hairs from half as long as to longer than lemmas; **lemmas** 3–4.3(6) mm, keeled, keels and marginal veins long-hairy. **Anthers** 1.2–2 mm. **Habitat**: Sea level to alpine, mesic or seasonally moist grasslands. **Comments**: *Poa pratensis* is a common bluegrass with nodding inflorescences and copious cobwebby callus hairs. Five subspecies, both native and introduced, are believed to occur in OR and WA; if you're feeling ambitious, consult the *Flora of North America* to identify them. This species is an important lawn and pasture grass, but it can be invasive in mesic native grasslands. Hybrids with *P. secunda* ssp. *juncifolia* are called *P.* × *limosa*. See *P. confinis, P. palustris, P. rhizomata, P. trivialis,* and *P. wheeleri*.

Poa pringlei Scribn.
Pringle's Bluegrass

Densely cespitose perennial; plants dioecious; **culms** 5–35 cm, bases long-decumbent, resembling rhizomes. **Leaf sheaths** closed 15–33% their length; **blades** 1.5–3 mm wide, involute; **ligule** of culm leaf 1–6 mm, membranous. **Panicles** 1–6 cm, moderately congested, branches 0.5–1.5(2) mm, erect, with 1–3 spikelets. **Spikelets** 6–8(12) mm with 2–5 florets. **Glumes** 3.5–7.5 mm, shiny, broadly hyaline except near veins (pull glume away from lemma to check); **calluses** glabrous; **lemmas** 5–8 mm, keeled, smooth or scabrous, glabrous. **Anthers** 2–4 mm, or vestigial and 0.1–0.2 mm. **Habitat**: Rocky subalpine and alpine slopes. **Comments**: *Poa pringlei* reaches its N extent in the mountains of SW OR. It has relatively long, nearly transparent glumes; its long, decumbent plant bases resemble rhizomes. It hybridizes with *P. cusickii*, which lacks the broadly hyaline glumes and has sheaths usually closed for a greater portion of their length.

Poa reflexa Vasey & Scribn.
Nodding Bluegrass

Cespitose perennial; **culms** 10–60 cm, in small tufts or sometimes solitary. **Leaf sheaths** closed 33–67% their length; **blades** 1.5–4 mm wide, flat, thin, soft; **ligules** 1.5–3.5 mm. **Panicles** open, nodding, branches spreading to reflexed, flexuous, smooth, longest ones with (3)6–18 spikelets. **Spikelets** 4–6 mm, often purplish, with 3–5 florets; **calluses** with cobwebby hairs; **lemmas** 2–3.5 mm, keeled, keels and marginal veins short- to long-hairy, keels hairy for 67–80% their length, lateral veins usually sparsely short-hairy on at least 1 side, area between veins glabrous; **palea keels** scabrous, usually with sparse, tiny hairs at midlength. **Anthers** 0.6–1 mm. **Habitat**: Subalpine and alpine meadows, streamsides, talus. Rare. **Comments**: In our area, *P. reflexa* is known only from mesic meadows in the Wallowa Mts. Similar *P. leptocoma* grows in wetter sites, lacks soft hairs on lateral lemma veins and palea keels, and has shorter anthers. See also *P. paucispicula*.

Poa rhizomata Hitchcock
Rhizome Bluegrass

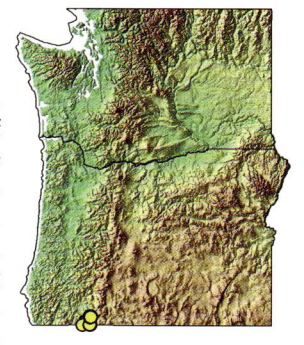

Short-rhizomatous perennial; most plants all male or all female, some plants bisexual; **culms** 20–65 cm, tufted or solitary. **Leaf sheaths** closed 50–67% their length; **collars** glabrous; **cauline blades** 1–3.5 mm, mostly flat, sometimes folded, soft; **ligules** 2–8 mm, acute to acuminate. **Panicles** nodding, open, sparse, branches ascending to spreading. **Spikelets** (4)6–9(12) mm with 3–8 florets; **calluses** with cobwebby hairs > half as long as lemmas; **lemmas** 4–6.5 mm, keeled, keels and marginal veins short- to long-silky-hairy, areas between veins scabrous or glabrous. **Anthers** 2.5–4 mm or vestigial and 0.1–0.2 mm. **Habitat**: Rocky gabbro or peridotite soils in montane mixed conifer forests. **Comments**: *Poa rhizomata* is a rare endemic grass usually on ultramafic substrates in the Klamath Mts. of SW OR and NW CA. It is most likely to be confused with *P. pratensis* or *P. wheeleri*; *P. pratensis* has a more pyramidal panicle, bigger tufts of cobwebby hairs on the callus, and shorter, truncate to rounded ligules. *Poa wheeleri* also has truncate ligules and glabrous calluses. See also *P. chambersii* and *P. piperi*.

Poa secunda J. Presl

Densely cespitose perennial, sometimes glaucous; **culms** (10)15–120 cm. **Leaf sheaths** closed 10–25% their length; **blades** 0.4–3(5) mm wide, flat, folded, or involute; **ligules** 0.5–6(10) mm. **Panicles** usually contracted, ± open in flower, often nodding, branches usually erect or ascending with (1)2–20(60+) spikelets per branch. **Spikelets** (4)5–10 mm, narrowly lanceolate, ± cylindrical to slightly laterally compressed with (2)3–5(10) florets; **lemmas** 3.5–6 mm, ± rounded over the back or weakly keeled, the keel not extending to the base, keels and marginal veins glabrous or hairy, area between veins glabrous or hairy over the basal 67%, hairs of keel and veins similar in length to those between the veins. **Anthers** 1.5–3 mm. **Comments**: Widespread and abundant in W North America, *P. secunda* can be excellent early spring forage. It grows from desert to the subapline zone, often one of the last native grasses on disturbed sites.

Key to subspecies of *Poa secunda*

1a Leaves ± firm, retaining their shape; lemmas glabrous to scabrous, rarely sparsely short-hairy on lower 25% of keel and marginal veins; ligules of vegetative shoot leaves and often of culm leaves < 2(6) mm, ± truncate, scabrous; inflorescence branches appressed; habitat moist meadows, riparian areas........***P. s.* ssp. *juncifolia***

1b Leaves usually ± soft, soon withering; lemmas ± evenly short-hairy on lower 67%; ligules of vegetative shoot leaves and culm leaves > 2(6) mm, acute or acuminate, smooth or sparsely scabrous; inflorescence branches appressed or spreading; drier habitats: sagebrush steppe, upland meadows, dry forest, rock outcrops and ridges .... ***P. s.* ssp. *secunda***

Poa secunda ssp. *juncifolia* (Scribn.) Soreng
Alkali Bluegrass, Big Bluegrass, Nevada Bluegrass

Culms 30–120 cm; **ligules** of culm leaves 0.5–6 mm, those of vegetative leaves 0.5–2 mm, truncate to obtuse; **blades** 1–3(5) mm, ± thick, not early withering. **Calluses** glabrous; **lemmas** scabrous, usually glabrous, sometimes short-hairy on lower 25% of keels and marginal veins. **Habitat**: Moist meadows, riparian terraces, sometimes in alkaline soils. **Comments**: Subspecies *juncifolia* is usually more robust than ssp. *secunda* and usually grows on deeper soils and in more mesic habitats. Seeded as Big Bluegrass to revegetate disturbed sites.

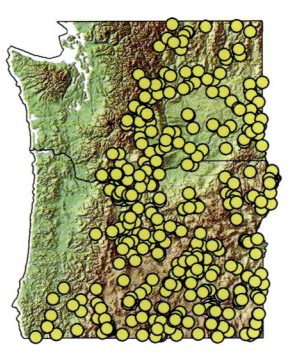

Poa secunda ssp. *secunda*
Sandberg, Canby, Pacific, or Pine Bluegrass

Culms (10)15–100 cm; **ligules** of culm leaves 2–6(10) mm, those of vegetative leaves 2–6 mm, obtuse to acuminate; **blades** 0.4–3 mm, thin, early withering. **Calluses** glabrous or with crown of hairs 0.1–0.5(2) mm; **lemma** keels and marginal veins hairy to 67% their length, area between veins usually hairy. **Habitat**: Lowlands to low alpine, sagebrush steppe, upland prairies, savannas, dry forests. **Comments**: Subspecies *secunda* usually grows on shallow soils or in drier habitats than ssp. *juncifolia*. It has early-withering leaves, longer ligules, and hairy lemmas.

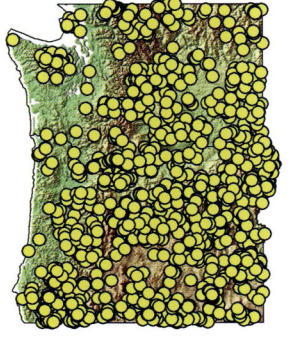

Poa secunda encompasses a complex of common, highly variable upland bunchgrasses distinguished by nearly cylindrical spikelets and weakly keeled lemmas. Many variants have historically been recognized as species, but recent treatments gather these forms into two variable subspecies. Plants may outcross but often set seed without sex (apomixis). Apomixis is often associated with odd numbers or large numbers of chromosomes; unbalanced numbers prevent pairing for meiosis, thus limiting the grass to asexual reproduction. *Poa secunda* ssp. *juncifolia* (2n=63) probably produces all of its seeds without fertilization. Thus, thousands of plants, each grown from seed, may constitute a single clone. For restoration purposes it may be useful to recognize ecotypes that may correspond to the historically recognized taxa within the subspecies as follows. Recently, many of these historical taxa have been recognized as varieties of the two subspecies; the variety names are listed below in parentheses.

Poa secunda ssp. **juncifolia** is more robust, with longer leaves remaining through the growing season. Lemmas usually glabrous with keels and marginal veins occasionally sparsely puberulent near the base. Ligules of vegetative shoots to 2 mm (except to 6 mm in Nevada Bluegrass).

Big Bluegrass (*P. secunda* ssp. *juncifolia* var. *ampla*) has short ligules, glabrous leaf sheaths, and flat leaf blades; available as a cultivar for restoration. **Habitat**: Upland forest and montane meadows with deep soils.

Alkali Bluegrass (*P. secunda* ssp. *juncifolia* var. *juncifolia*) has short ligules, glabrous leaf sheaths, and rolled leaf blades. **Habitat**: Riparian areas and wet meadows, often in alkaline or saline soils.

Nevada Bluegrass (*P. secunda* ssp. *juncifolia* var. *nevadensis*) has longer, decurrent ligules and slightly scabrous leaf sheaths. **Habitat**: Open forest, dry grasslands, and sagebrush steppe.

Poa secunda ssp. **secunda** is usually a smaller plant with lax leaves that wither relatively early in the season. Lemmas are sparsely to densely short-hairy on the basal two-thirds. Ligules of vegetative shoots are usually more than 2 mm long. It grows on drier sites than ssp. *juncifolia*.

Sandberg's Bluegrass and Canby Bluegrass (*P. secunda* ssp. *secunda* var. *secunda*) have been combined in variety *secunda,* which has loosely to tightly contracted panicles. Sandberg's bluegrass is < 30 cm tall and has short leaves. Canby bluegrass is usually > 30 cm tall, with pubescence covering the lower half of the lemma, and sheaths that are smooth or sparsely scabrous. Variety *secunda* is common and widespread in W North America. **Habitat**: Open forest and sagebrush to alpine areas; in well-drained soils. **Comments**: Available as cultivar 'Canbar.' The common name Sandberg's bluegrass comes from the old name *P. sandbergii*.

Pacific Bluegrass (*P. secunda* ssp. *secunda* var. *gracillima*) is a slender, sparse plant with open panicles, usually growing in mesic, shady habitats. **Habitat**: Open forest, rocky areas, and subalpine to alpine meadows; in soils that are moist through the growing season. Often confused with *P. stenantha*.

Pine Bluegrass (*P. secunda* ssp. *secunda* var. *scabrella*) has scabrous or short-hairy leaf sheaths that are notably reddish at the base. **Habitat**: Open forest, scrub, and grassland; in well-drained soils; mainly W of the Cascades.

Poa stenantha Trin. var. stenantha
Northern or Narrow-flower Bluegrass

Densely to loosely cespitose perennial; **culms** 20–60(100) cm. **Leaf sheaths** closed 10–20(25)% their length; **blades** 1.5–4(5) mm wide, flat or folded; ligules 2–6 mm, membranous. **Panicles** open to loosely contracted, branches ascending to spreading. **Spikelets** 6–10 mm, laterally compressed, 3–3.6× as long as wide, with 3–4(7) florets; **upper glumes** (3.7)4.1–6.5 mm; **calluses** usually with crown of hairs 0.2–2 mm, sometimes glabrous; **lemmas** 4–6 mm, keeled, keels short- to long-hairy 75% their length, marginal and sometimes lateral veins hairy, area between veins glabrous or sparsely short-hairy basally, hairs distinctly shorter than those of the keel and veins. **Anthers** 1.2–2 mm. **Habitat**: Montane to low alpine, forests, meadows, rocky slopes, cliffs, coastal bluffs. **Comments**: *Poa stenantha* can be difficult to distinguish from *P. secunda* ssp. *secunda*, which has nearly cylindrical spikelets, lemmas that are rounded or weakly keeled, and hairs between the veins as long as those on the keel and marginal veins.

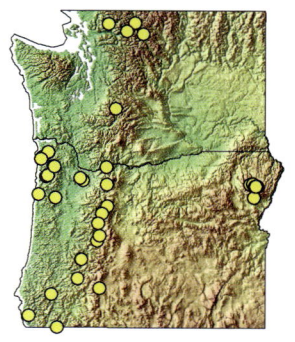

Poa suksdorfii (Beal) Vasey ex Piper
Suksdorf's Bluegrass

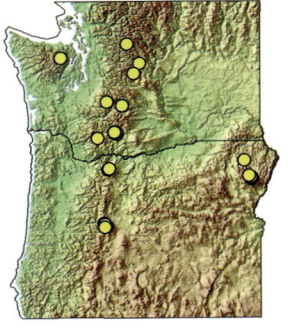

Densely cespitose perennial; **culms** 7–25 cm. **Leaf sheaths** closed 15–25(33)% their length; **blades** 1–2 mm wide, folded to involute, ± thick and firm; **ligules** of cauline leaves 1–3 mm, membranous. **Panicles** contracted, narrow, branches erect. **Spikelets** 4.2–7 mm, often purplish, with 2–4 florets; **rachilla internodes** 1–1.5 mm; **glumes** (3.3)4.3–5.7 mm, **lower glumes** ≤ lowest lemmas; **upper glumes** not > upper florets; **calluses** glabrous; **lemmas** 4.1–5.8 mm, distinctly keeled, glabrous. **Anthers** 0.8–1.2(1.7) mm, often aborted in top floret. **Habitat**: dry, rocky alpine slopes. **Comments**: *Poa suksdorfii* is a small, densely tufted, alpine bluegrass that grows in the Olympic, Cascade, and Wallowa Mts. *Poa lettermanii* is finer-textured and has shorter glumes and lemmas.

Poa trivialis L. ssp. trivialis
Rough Bluegrass [E]

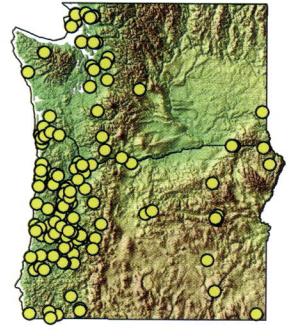

Loosely to densely cespitose, short-lived perennial, weakly stoloniferous; **culms** 25–120 cm, sometimes trailing and rooting at nodes. **Leaf sheaths** closed 33–50% their length, usually densely scabrous; **blades** 1–5 mm wide, flat, lax, scabrous; **ligules** 3–10 mm, membranous, acute to acuminate, scabrous on the exterior. **Panicles** 8–25 cm, open, often erect, branches spreading. **Spikelets** 2.5–4(5) mm with 2–4 florets; lower glumes sickle-shaped, 1-veined; **calluses** with cobwebby hairs > 67% lemma length; **lemmas** 2.3–3.5 mm, keeled, keels usually sparsely short-hairy on lower 60%, marginal veins usually glabrous, rarely short-hairy near base, areas between veins glabrous. **Anthers** 1.3–2 mm. **Habitat**: Moist deciduous forest, riparian areas, pastures, lawns. **Comments**: Run your fingers up the sheaths to feel the roughness. The scabrous foliage and 1-veined, sickle-shaped lower glumes distinguish *P. trivialis* from rhizomatous *P. pratensis*, which has smoother foliage and broader, 3-veined glumes. Invasive in riparian zones and moist forest understories and a weed in lawns.

Poa unilateralis Scribn. ex Vasey
Seacliff Bluegrass

Densely cespitose, glaucous perennial; **culms** 5–40 cm. **Leaf sheaths** closed 10(20)% their length, **cauline blades** 2–5 mm wide, flat, folded, or involute; **ligules** 2–6 mm, membranous. **Panicles** dense, nearly cylindrical, branches erect. **Spikelets** 4.5–7 mm with 3–5 florets; **calluses** glabrous or with a crown of tiny hairs; **lemmas** 3–4.5 mm, keeled, keels and marginal veins glabrous or short-hairy, areas between veins glabrous or short-hairy near base. **Anthers** 1.5–3 mm. **Habitat**: Grassy coastal bluffs and cliffs. **Comments**: *Poa unilateralis* is a cespitose grass with dense, contracted panicles that grows on north-facing rocky bluffs of coastal headlands and offshore islands. Both subspecies of *P. unilateralis* are rare in OR and WA. *Poa unilateralis* was apparently more common 100 years ago, but grazing and trampling of the headlands and grassy bluffs has eliminated it except on inaccessible ledges of steep cliffs and rocky islands. *Poa secunda* is similar but has weakly keeled lemmas and does not grow at the immediate coast, except in the San Juan Islands.

Key to subspecies of *Poa unilateralis*

1a Panicles contracted, ± irregular in outline, tapering to a pointed tip, longest branches (1.2)2.5–4 cm; lemma keels and marginal veins hairy ≥ 33% their length; N OR to S WA... ***P. u.* ssp. *pachypholis***

1b Panicles densely contracted, ovoid, ± smooth in outline, rounded at the top; longest branches 0.5–1.5 mm; lemma keels and marginal veins hairy ≤ 20% their length; S OR to CA.. ***P. u.* ssp. *unilateralis***

Poa unilateralis ssp. *pachypholis* (Piper) D.D. Keck ex Soreng
Seacliff Bluegrass (green dots)

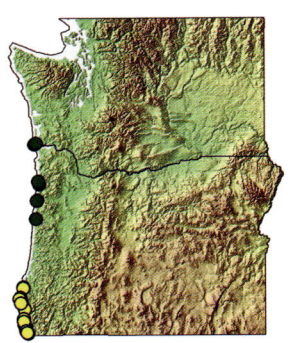

Cauline blades similar to innovation blades. **Panicles** tapering to ± pointed tip, longest branches (1.2)2.5–4 cm. **Calluses** with short crown of hairs 0.1–0.2 mm, sometimes with a few longer hairs to 0.5 mm, rarely glabrous; **lemmas** hairy on keels and marginal veins 33–50% their length, area between veins glabrous or with short hairs near base. **Comments**: Range N OR to S WA. Subspecies *pachypholis* differs from ssp. *unilateralis* in having less dense, more irregularly shaped inflorescences tapered to the tip, more extensive lemma hairiness, and calluses with short hairs.

Poa unilateralis ssp. *unilateralis*
San Francisco Bluegrass (yellow dots)

Cauline blades sometimes wider and thicker than innovation blades. **Panicles** very dense, ovoid, usually 1-sided, ± smooth in outline, rounded at the top, longest branches 0.5–1.5 cm. **Calluses** glabrous, sometimes with crown of minute hairs to 0.1 mm; **lemmas** glabrous or keels and marginal veins hairy on lower 20%, area between veins glabrous. **Comments**: Range S OR to CA. Its compact, smooth-sided, ovoid inflorescences and less hairy lemmas distinguish ssp. *unilateralis* from ssp. *pachypholis*.

Poa wallowensis Soreng
Wallowa Bluegrass

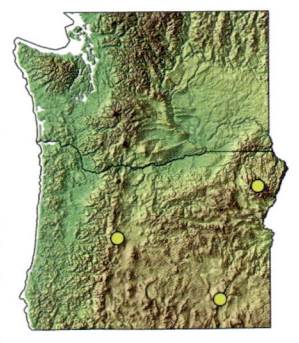

Small cespitose perennial; **culms** 8–14 cm. **Leaf sheaths** closed 20–33% their length; **blades** 1.5–2 mm wide, folded, ± firm; **ligules** 1.25–2 mm, membranous, blunt. **Panicles** narrowly contracted, sparse to moderately dense, branches 0.8–1.5(3) cm, steeply ascending. **Spikelets** 4–6 mm with 3–4 florets; **rachilla internodes** < 1 mm; **upper glumes** subequal to lowest lemmas; **calluses** with cobwebby hairs ≥ 50% lemma length; **lemmas** 2.8–3.5 mm, keeled, keels short-hairy in basal 50% or more, marginal veins hairy in basal 40%, areas between veins glabrous. **Anthers** 0.6–1.0 mm. **Habitat**: Rocky alpine slopes and ridges. **Comments**: *Poa wallowensis* was described from the Wallowa Mts. and subsequently discovered on Steens Mt. and in the central Cascades, OR. *Poa suksdorfii* has longer, completely glabrous lemmas and involute leaves; *P. leptocoma* and *P. reflexa* have panicles that are more open.

Poa wheeleri Vasey
Wheeler's Bluegrass

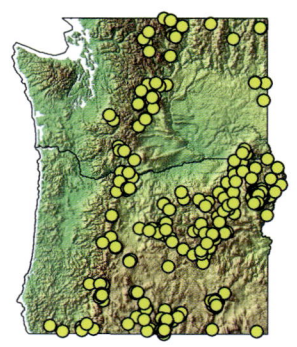

Short-rhizomatous perennial; plants nearly all functionally female; **culms** 35–80 cm, loosely tufted or solitary. **Leaf sheaths** closed 33–75% their length, retrorsely hairy or scabrous, at least on some lower leaves; **collars** of lower leaves glabrous or with hairs = sheath hairs; **cauline blades** 2–3.5 mm wide, flat or folded; **ligules** 0.5–2 mm, membranous, lacerate. **Panicles** erect or nodding, loosely contracted or open, branches ascending, spreading, or reflexed. **Spikelets** 5.5–10 mm with 2–7 florets; **calluses** glabrous; **lemmas** 3–6 mm, keeled, keels and marginal veins glabrous or rarely short-hairy, area between veins glabrous or short-hairy. **Anthers** vestigial and 0.1–0.2 mm or aborted late in development and up to 2 mm, rarely normal. **Habitat**: Montane, mesic conifer forests, subalpine meadows. **Comments**: *Poa wheeleri* is common and widespread E of the Cascade Crest and in SW OR. It was formerly considered a variety of *P. nervosa*, which is restricted to wetter habitats in the Columbia Gorge and W Cascades. *Poa wheeleri* hybridizes with *P. pratensis* and in the field is sometimes mistaken for that species. The callus of *P. wheeleri* lacks the cobwebby hairs of *P. pratensis*. See also *P. chambersii*.

Podagrostis aequivalvis (Trin.) Scribn. & Merr.
Arctic Bentgrass

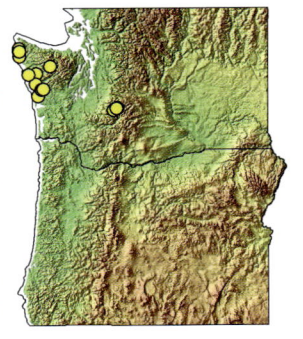

Rhizomatous perennial; **culms** 25–90 cm. **Leaf blades** 1–2.5 mm wide, flat; **ligules** 0.4–4 mm. **Panicles** 5–15 cm × 2–10 cm, often drooping, sparsely branched, branches erect to ascending or spreading; spikelets usually restricted to outer half of branches. **Spikelets** usually purplish-bronze, sometimes greenish-purple; **glumes** 2.3–4.3 mm, ± equal; **lemmas** 2.5–3.5 mm, awnless; **paleas** 2–3 mm; **rachillas** prolonged 0.5–1.9 mm beyond the base of the floret, bristle-like, with a tuft of hairs < 0.3 mm at the tip. **Anthers** 0.8–1.3 mm. **Habitat**: Bogs, fens, lake margins, streamsides, more common in coastal regions, but to 5,500 ft on Mt. Rainier. **Comments**: *Podagrostis aequivalvis* generally has longer glumes and longer prolonged rachillas than *P. thurberiana,* but the characters overlap, particularly where the ranges approach each other in the Cascade Range.

Podagrostis humilis (Vasey) Bjorkman
Alpine Bentgrass

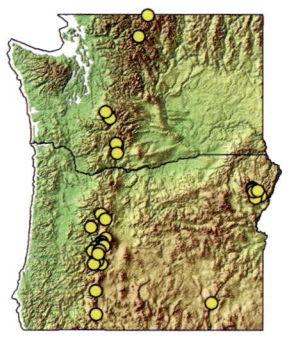

Cespitose perennial, rarely rhizomatous; **culms** 3–15(26) cm. **Leaf blades** 0.3–0.8(1.5) mm wide, usually folded; **ligules** (0.2)0.5–2(2.7) mm. **Panicles** 1.5–6 cm × 0.2–1.5 cm, contracted, linear to narrowly oblong, dark red to purple; branches erect to appressed, ascending at anthesis. **Spikelets** purple; **glumes** 1.6–2 mm, ± equal; **lemmas** 1.5–2 mm, awnless or occasionally awned from the tip with awns to 1.3 mm; **paleas** 65–75% as long as lemmas; **rachillas** prolonged 0–0.1 mm beyond the base of the floret. **Anthers** 0.4–1 mm.
Habitat: Subalpine to alpine wet meadows. **Comments**: This delicate little grass is most similar to *Agrostis variabilis*, which has paleas a third the length of the lemmas. The prolonged rachilla is so small as to be nearly invisible, and indeed it is sometimes absent in spikelets in the lower part of the panicle.

Podagrostis thurberiana (Hitchc.) Hultén
Thurber's Bentgrass

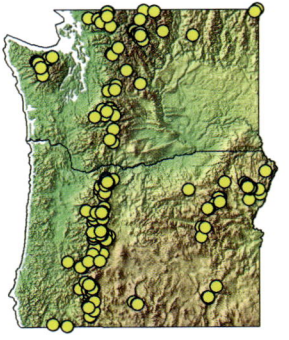

Cespitose perennial; tufted on short rhizomes; **culms** 10–40(50) cm. **Leaf blades** 1–3 mm wide, flat or folded; **ligules** 0.5–2 mm. **Panicles** (2.5)5–12 cm × 0.5–0.7(2.5) cm, linear to narrowly oblong; branches erect to ascending. **Spikelets** purple or greenish; **glumes** 2–2.5 mm, usually slightly unequal; **lemmas** 2–2.5 mm, usually awnless or occasionally awned from the tip with awns to 1.3 mm; **paleas** 75–100% as long as lemmas; **rachillas** prolonged 0.1–0.5 mm beyond the base of the floret, with or without a tuft of short hairs at the tip. **Anthers** 0.4–0.8 mm. **Habitat**: Montane to alpine wet meadows. **Comments**: *Podagrostis thurberiana* is distinguished from similar-looking *Agrostis* by paleas that are ≥ 75% as long as the lemmas, prolonged rachillas, and opaque, distinctly veined lemmas. Small plants can be distinguished from *P. humilis* by the longer rachillas, which are minute or absent in *P. humilis*, and wider leaf blades. Thurber's bentgrass is fairly common in mountains of OR and WA.

Polypogon fugax Nees ex Steud.
Asian Beardgrass [E]

Annual; **culms** 15–60 cm, often decumbent at base and rooting at nodes. **Leaf blades** 2–11 mm wide; **ligules** 2–8 mm, membranous. **Panicles** oblong to cylindrical, usually lobed, yellowish or pale green. **Pedicels** 0–0.5 mm; **stipes** 0.2–1.3 mm; **glumes** 1.8–2.4 mm, scabrous, tips acute to rounded, with lobes 0.1–0.2 mm; **glume awns** 0.6–3 mm; **lemmas** 0.9–1.2 mm, smooth, awnless or with short awns to 2 mm; **paleas** 0.7–1.2 mm, 75–100% as long as lemmas. **Anthers** 0.3–0.6 mm. **Habitat**: Disturbed places, salt marshes. Native to Asia.

Comments: *Polypogon interruptus* glumes have similar awns but shorter lobes than *P. fugax*. Historic records of *P. fugax* along the West Coast appear to have been waifs, but the species was recently collected in Seattle on a disturbed roadside.

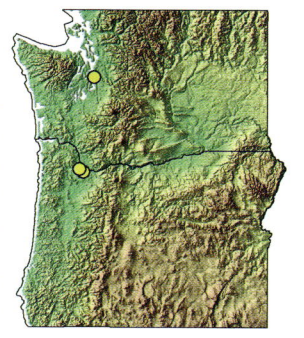

Polypogon interruptus Kunth
Ditch Beardgrass [E]

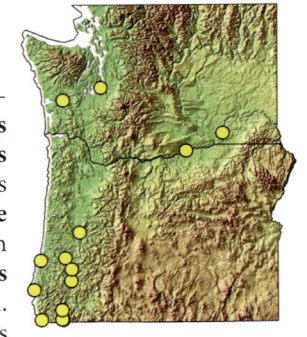

Cespitose perennial, often flowering the first year; **culms** 20–80 cm, ± decumbent. **Leaf blades** 1–6 mm wide, flat; **ligules** 2–6 mm, membranous. **Panicles** lobed to interrupted. **Pedicels** lacking; **stipes** 0.2–0.7 mm; **glumes** 2–3 mm, scabrous, tips acute to truncate, unlobed or with lobes ≤ 0.1 mm; **glume awns** 1.5–3.2 mm; **lemmas** 0.8–1.5 mm, smooth, shiny, with awns 1–3.2 mm; **paleas** about 75% as long as lemmas. **Anthers** 0.5–0.7 mm. **Habitat**: Disturbed wet sites. Uncommon. Native to S America. **Comments**: *Polypogon interruptus* has a lobed inflorescence somewhat like a sparse *P. monspeliensis,* but with shorter awns. *Polypogon fugax* is an annual with slightly longer glume tip lobes.

400   SPECIES ACCOUNTS

Polypogon maritimus Willd.
Mediterranean Beardgrass [E]

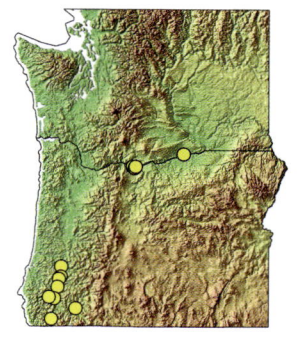

Annual; **culms** (5)20–40(50) cm. **Leaf blades** 0.5–5 mm wide; **ligules** to 7 mm, membranous. **Panicles** narrowly ellipsoid, sometimes lobed. **Pedicels** to 0.5 mm; **stipes** 0.1–1.2 mm; **glumes** 1.8–3.2 mm, with large, thick scabers or bristles on the lower portion, ciliate lobes 0.3–1.2 mm, > 15% as long as glume body; **glume awns** (4)7–12 mm; **lemmas** 0.5–1.5 mm, awnless or with awns < 1 mm; **paleas** subequal to lemmas. **Anthers** 0.4–0.5 mm. **Habitat**: Disturbed wet sites; stream banks, rock quarries. SW OR and Columbia Basin, to be expected elsewhere; native to the Mediterranean region. **Comments**: *Polypogon maritimus* looks like a depauperate *P. monspeliensis* with narrower panicles. The bizarre glumes have coarse scabers on the lower part and long, fringed lobes that look like rabbit ears. It may be overlooked in the PNW due to its similarity to *P. monspeliensis*.

Polypogon monspeliensis (L.) Desf.
Rabbitsfoot Grass, Annual Beardgrass [E]

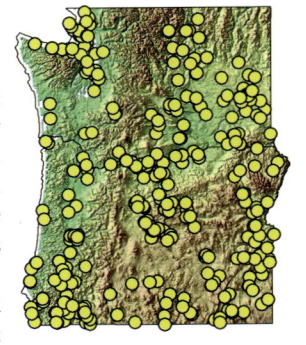

Tufted annual; **culms** 5–65 cm. **Leaf blades** 1–7 mm wide; **ligules** 2.5–16 mm, membranous, lacerate. **Panicles** narrowly ellipsoid, soft, often lobed, lobes often obscured by long awns. **Pedicels** 0–0.2 mm; **stipes** 0.1–0.2 mm; **glumes** 1–2.7 mm, minutely hairy, lobes 0.1–0.2 mm, ≤ 10% as long as glume body; **glume awns** 4–10 mm; **lemmas** 0.5–1.5 mm, with awns 0.5–1(4.5) mm; **paleas** subequal to lemmas. **Anthers** 0.2–1 mm. **Habitat**: Disturbed wetlands, ditches, roadsides, often on alkaline soils. Native to S Europe and Turkey. **Comments**: Compare to other *Polypogon* species and *Gastridium phleoides*. Hybridizes with *Agrostis stolonifera* to form ×*Agropogon lutosus*, which has more tapered, short-awned glumes and subterminal lemma awns.

Polypogon viridis (Gouan) Breistr.
Water Beardgrass, Beardless Rabbitsfoot Grass [E]

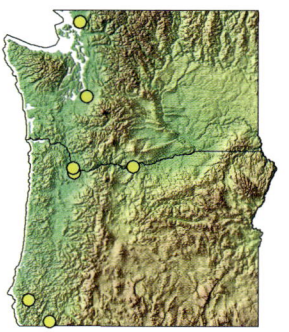

Cespitose perennial, often flowering first year. **Culms** 10–90 cm. **Leaf blades** 1–6 mm wide; **ligules** to 5 mm, membranous. **Panicles** ovoid-oblong to pyramidal, interrupted. **Pedicels** lacking; **stipes** 0.1–0.6 mm; **glumes** scabrous on back and keel, tips obtuse to truncate, not lobed, awnless; **lemmas** about 1 mm, awnless; **paleas** subequal to lemmas. **Anthers** 0.3–0.5 mm. **Habitat**: Stream banks, shorelines, ditches. Native from S Europe to Pakistan. **Comments**: With its awnless glumes and lemmas, *P. viridis* could easily be mistaken for an *Agrostis,* but its spikelets disarticulate below the glumes, and the paleas are nearly as long as the lemmas.

Pseudoroegneria spicata (Pursh) Á. Löve
Bluebunch Wheatgrass

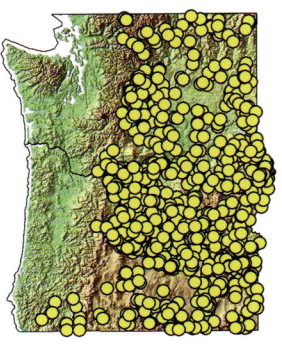

Cespitose perennial, sometimes shortly rhizomatous, especially in moist conditions; **culms** 30–100 cm tall. **Leaf sheaths** open; **blades** 2–6 mm wide, flat to involute, smooth and glabrous below, scabrous to hairy above; **ligules** truncate, 0.1–0.4 mm, **auricles** clasping, often reddish. **Spikes** 3–8(10) mm wide excluding awns, with 1 spikelet per node, the spikelets scarcely overlapping on the same side of the rachis. **Glumes** 0.9–2.2 mm wide, half as long as the spikelet, glabrous, sometimes scabrous over the veins, lanceolate to oblanceolate, not especially hard, (3)4–5(7)-veined. **Lemmas** awnless or with divergent awns up to 25 mm long. **Anthers** 4–6 mm. **Habitat**: Arid and semiarid grasslands, shrub steppe, open forest. **Comments**: Prior to the introduction of domestic livestock in the West, bluebunch wheatgrass was a dominant bunchgrass in many habitats; now it is found in abundance primarily in places inaccessible to cattle or far from water sources. Excellent forage. Compare to *Elymus wawawaiensis*.

Puccinellia distans (Jacq.) Parl.
European Alkali Grass [E]

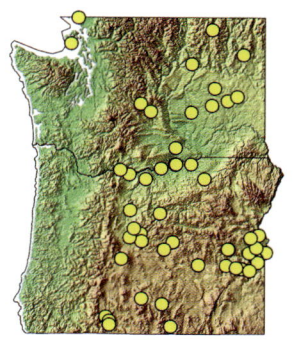

Cespitose perennial; **culms** 5–60 cm, erect to decumbent. **Leaf blades** 1–7 mm wide; **ligules** 0.8–1.2 mm, membranous, acute. **Panicles** open at maturity, lower branches horizontal to descending with spikelets borne on the outer 67%. **Spikelets** with 2–7 florets. **Lower glumes** 0.4–1.3 mm; **upper glumes** 0.9–1.8 mm. **Lemmas** 1.5–2(2.2) mm, apical margins hyaline and often yellowish, with minutely jagged edges, tips widely obtuse to truncate. **Anthers** 0.4–0.8 mm. **Habitat**: Seasonally moist, alkaline or saline substrates, roadsides. **Comments**: Introduced from Europe. Usually inland (rarely coastal on pilings and logs in NW WA). Spreads along salted roads.

Puccinellia lemmonii (Vasey) Scribn.
Lemmon's Alkali Grass

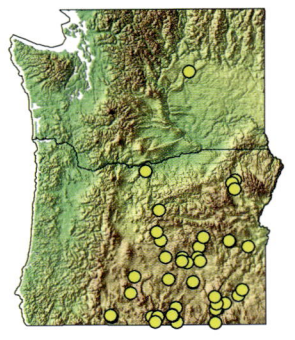

Cespitose perennial; **culms** 5–40 cm. **Leaves** mostly basal, involute, 1.2–1.9 mm wide when flattened; **ligules** 0.8–2.2 mm, membranous, acute. **Panicles** compact to open at maturity, lower branches ascending to descending, usually spikelet-bearing to near the base. **Spikelets** with 2–6 florets. **Lower glumes** 0.7–1.5 mm; **upper glumes** 1.8–3 mm. **Lemmas** 2.4–4 mm, mostly smooth, glabrous or with a few hairs near the base, midveins often slightly scabrous on the distal half, often extending to apical margins; apical margins smooth to slightly jagged; tips acute. **Anthers** 1–2 mm. **Habitat**: Alkaline flats E of the Cascades.

Puccinellia maritima (Huds.) Parl.
Common Saltmarsh Grass [E]

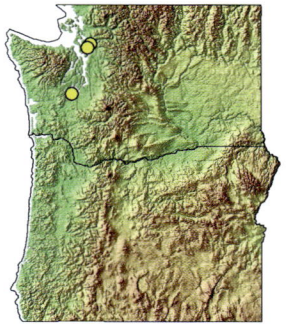

Cespitose perennial, often stoloniferous; **culms** 20–100 cm. **Leaf blades** 2–4.4 mm wide; **ligule** 1–3.5 mm. **Panicles** compact to diffuse at maturity, lower branches erect to descending, spikelet-bearing from about midlength; **pedicels** usually densely scabrous. **Spikelets** with 4–9 florets. **Lower glumes** 2–3.4 mm; **upper glumes** 3–4.5 mm. **Lemmas** 3–5 mm, slightly to very leathery throughout, sparsely to densely hairy in the lower half, mainly on veins, apical margins usually smooth (rarely with minutely jagged edge). **Anthers** 1.5–2.6 mm. **Habitat**: Coastal salt marsh. **Comments**: Introduced from Europe to a few locations around the Puget Sound.

Puccinellia nutkaensis (J. Presl) Fernald & Weath.
Pacific Alkali Grass

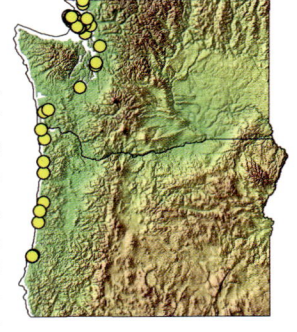

Cespitose perennial, sometimes rooting at nodes of buried stems; **culms** 10–90 cm. **Leaves** usually distributed evenly along culms; **blades** sometimes involute, 1.5–6 mm wide when flat; **ligules** 1–3 mm, membranous. **Panicles** compact to diffuse at maturity, lower branches usually erect to ascending; **pedicels** scabrous and usually papillate or with slightly swollen cell surfaces distally. **Spikelets** with 3–7 florets. **Lower glumes** 1–1.6 mm; **upper glumes** 2–3 mm. **Lemmas** (2.2)3–4.5(5) mm, glabrous or sparsely hairy on the proximal half, mainly on the veins, apical margins with minutely jagged edges, tips usually acute to obtuse, sometimes acuminate. **Anthers** 0.5–1.4 mm. **Habitat**: Coastal marshes. **Comments**: Similar *P. nuttalliana* does not have papillae near pedicel apices. Most coastal plants that key to the *P. nuttalliana/nutkaensis* pair are *P. nutkaensis*.

Puccinellia nuttalliana (Schult.) Hitchc.
Nuttall's Alkali Grass

Cespitose perennial; **culms** 10–100 cm. **Leaves** mostly basal or distributed along culms. **Leaf blades** 1–4 mm wide, flat or involute; **ligules** 1–3 mm, membranous, acute. **Panicles** compact to diffuse at maturity, lower branches erect to divergent, occasionally descending; **pedicels** scabrous but lacking papillae or swollen cell surfaces. **Spikelets** with 2–7 florets. **Lower glumes** 0.5–1.5 mm; **upper glumes** 1–2.8 mm. **Lemmas** (2)2.2–3(3.5) mm, glabrous or sparsely hairy on the proximal half, mainly on the veins, apical margins minutely jagged, tips obtuse to acute. **Anthers** 0.6–2 mm. **Habitat**: Mostly inland alkaline wetlands, rarely coastal salt marshes and rocky shores. **Comments**: See similar *P. nutkaensis*.

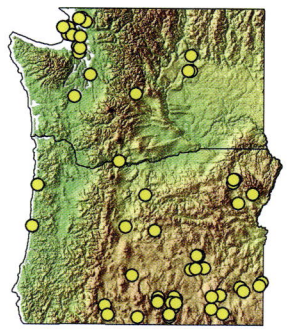

Puccinellia pumila (Vasey) Hitchc.
Smooth Alkali Grass

Cespitose perennial, sometimes rooting at nodes of buried stems; **culms** 8–40 cm, erect to decumbent. **Leaf blades** 1–3 mm wide; **ligules** 0.8–2.5 mm, membranous, acute. **Panicles** dense to open at maturity, lower branches ascending to descending, branches and **pedicels** smooth or with a few scattered scabrules. **Spikelets** with 3–7 florets. **Lower glumes** 1.4–2(4) mm; **upper glumes** 2–3 mm. **Lemmas** 2.5–4.6 mm, glabrous or with a few hairs on veins near the base, apical margins usually smooth, occasionally with a few scattered scabrules, acute to obtuse. **Anthers** 0.5–1.2 mm. **Habitat**: Coastal mudflats, salt marshes, sandy or rocky shores. **Comments**: Smooth pedicels and panicle branches help distinguish *P. pumila* from our other coastal alkali grasses.

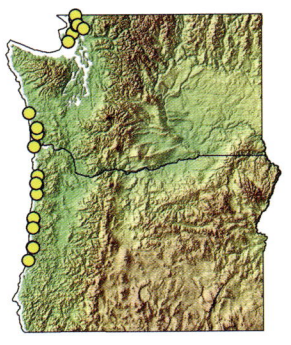

Puccinellia simplex Scribn.
Western Alkali Grass [E]

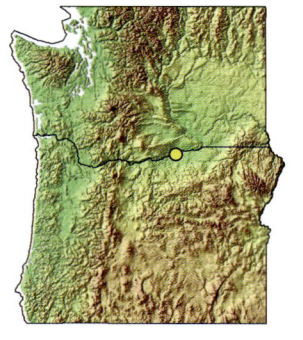

Annual; **culms** 2–25 cm, erect, not mat-forming. **Leaf blades** 0.7–2 mm wide; **ligules** 1–3 mm. **Panicles** 1–18 cm, compact, mostly linear at maturity. **Spikelets** 3.5–8 mm with 2–7 florets. **Lower glumes** 1.3–2 mm; **upper glumes** 2.3–3 mm; **lemmas** 2.5–4 mm, evenly short-hairy with longer, twisted and tangled hairs along the veins and near the base; apical margins smooth or with a few scattered scabrules, tips acute. **Anthers** 0.2–0.5 mm. **Habitat**: Seasonally moist, saline or alkaline soils. Introduced from the Central Valley of CA. **Comments**: *Puccinellia simplex* is a small alkali grass of saline soils in CA, introduced to the Great Basin. It was recently collected in N OR. Like *Crypsis*, *P. simplex* appears to be carried to new locations by waterfowl. Does that make it introduced? You decide.

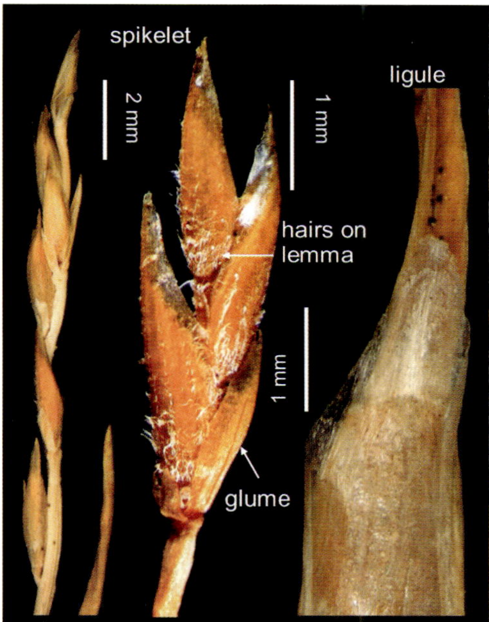

Redfieldia flexuosa (Thurb. ex A. Gray) Vasey
Blowout Grass [E]

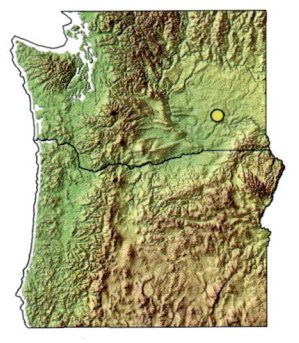

Strongly rhizomatous perennial, with deep horizontal or vertical rhizomes; **culms** 50–130 cm, solid, bases often buried in sand. **Leaf blades** 2–8 mm wide; **ligules** to 1.5 mm, membranous, with a fringe of hairs longer than the membranous portion. **Panicles** diffuse, 20–50 cm × 8–25 cm, branches thin and divergent, with spikelets borne at tips of branchlets. **Spikelets** ovate to obovate, with 2–6 florets. **Glumes** 3–4.5 mm, glabrous. **Lemmas** 4.5–6 mm, glabrous or short-hairy; **callus hairs** to 1.5 mm. **Anthers** 2–3.6 mm.
Habitat: Interior sand dunes, bare, wind-blown soil. **Comments**: Native to the Great Plains, introduced in E WA for erosion control on sand. The diffuse open panicle resembles *Eriocoma hymenoides*, which differs in having a singled awned floret per spikelet, glumes longer than the floret, and long-hairy lemmas.

Rytidosperma penicillatum (Labill.) Connor & Edgar
Hairy Wallaby Grass, Hairy Oatgrass, Poverty Grass [E] (yellow dots)

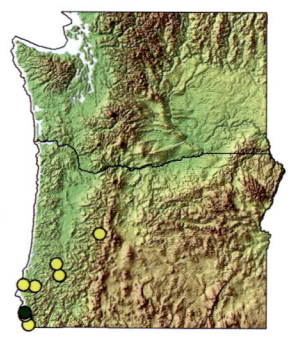

Loosely cespitose to short-rhizomatous perennial; **culms** 30–90 cm. **Leaves** mostly basal, the basal tuft much < culms; **flag leaves** usually not reaching inflorescence; **sheaths** open, with tufts of hairs at top; **blades** to 5 mm wide; **ligules** a fringe of hairs 0.1–1 mm. **Inflorescences** panicles or racemes, narrow, dense, curiously inconspicuous. **Spikelets** 9–15(18) mm, with 5–10 florets; **glumes** 8–14(17.5) mm, acute; **calluses** longer than wide with marginal tufts of hairs; **lemma bodies** 2–4 mm, 9-veined, thick, leathery, with 2 transverse rows of tufted hairs; lower row continuous or with weak central tufts; upper row usually consisting of just 2 marginal tufts; **lemma tips** with 2 awn-like apical **lobes** 5–13 mm and a central awn arising from between the lobes, the awn (7)9–16 mm, bent, twisted. **Anthers** 0.4–2.5 mm. **Habitat**: Grasslands, lawns, nutrient-poor sites. Native to Australia.

Rytidosperma biannulare (Zotov) Connor & Edgar
Wallaby Grass [E] (green dot)

Similar to *R. penicillatum* but **lemma bodies** 1.8–2.4(2.8) mm, 7–9-veined; upper row of lemma hair tufts ± continuous; **lemma lobes** 3.5–5(8.5) mm. Introduced from New Zealand.

Schedonorus arundinaceus (Schreb.) Dumort.
Tall Fescue [E]

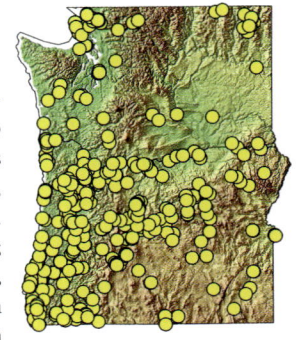

Coarse, cespitose perennial, rarely rhizomatous; **culms** 50–200 cm. **Leaf sheaths** open; **blades** 4–12 mm wide; **ligules** to 1 mm; **auricles** flange-like, sometimes claw-like, ciliate (hairs sometimes wearing off with age). **Panicles** erect to drooping, branches erect to spreading. **Spikelets** with 3–6(9) florets. **Lemmas** (4)5–9(11.5) mm, awnless or awned from the tip; **awns** to 4 mm. **Anthers** 2.5–4 mm. **Habitat**: Mesic grasslands, pastures, roadsides, lawns. **Comments**: Tall fescue has been widely introduced for forage, erosion control, and as a lawn grass. It is abundant and invasive in mesic grasslands and irrigated pastures. The cilia on the flange-like auricles are diagnostic, but they wear off with age; even one hair clinches the determination that the plant is tall fescue.

Schedonorus pratensis (Huds.) P. Beauv.
Meadow Fescue [E]

Rhizomatous perennial; **culms** to 130 cm, loosely clustered or solitary. **Leaf sheaths** open; **blades** 2–7 mm wide; **ligules** to 0.5 mm; **auricles** clasping, glabrous. **Panicles** ± erect, branches ascending to spreading. **Spikelets** with (2)4–10(12) florets. **Lemmas** 5–8 mm, awnless or with very short point to 0.5 mm. **Anthers** (1.5)2–4.6 mm. **Habitat**: Moist meadows, pastures, ditches. **Comments**: *Schedonorus pratensis* is a Eurasian pasture grass introduced for forage but seldom planted now. It differs from *S. arundinaceus* in its rhizomatous habit, nonciliate auricles, awnless lemmas, and generally more open, erect panicle.

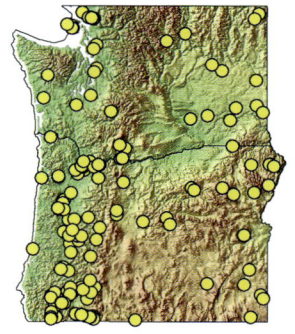

Schizachyrium scoparium (Michx.) Nash var. *scoparium*
Little Bluestem

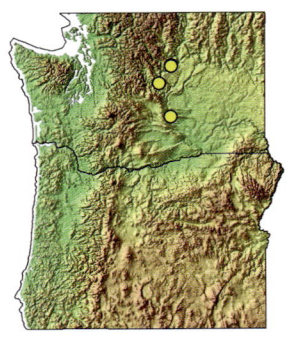

Cespitose to short-rhizomatous perennials; **culms** 30–150 cm, often purplish near the nodes. **Leaf sheaths** open; **blades** 1.5–9 mm wide; **ligules** 0.5–2 mm, membranous. **Inflorescences** with hairy, spike-like branches attached directly to the axis or on a peduncle to 10 cm long. **Spikelets** in sessile-pedicellate pairs. **Sessile spikelets** fertile, narrow, 6–11 mm, with awn 2.5–17 mm. **Glumes** enfold one another so that it is easy to mistake one glume for a lemma. **Pedicellate spikelet** sterile, 1–6 mm long, often resembling a continuation of the pedicel. **Anthers** 2.4–3.8 mm. **Habitat**: High water line of the Columbia River shore in central WA, growing in sand, silt, cobble, and gravel, often in high-quality riparian plant communities dominated by native bunchgrasses. Rare. **Comments**: Little bluestem is a community dominant in the tallgrass prairie of the Great Plains. It is also grown ornamentally.

Sclerochloa dura (L.) P. Beauv.
Fairgrounds Grass, Hardgrass [E]

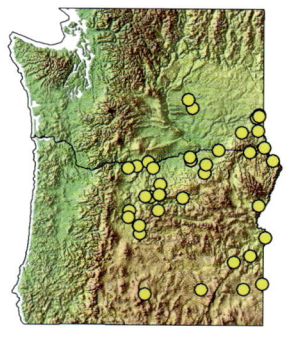

Matted annuals, **culms** 2–15(30) cm. **Leaf sheaths** open or closed to 50% their length, **blades** 1–4 mm wide, **ligules** 0.7–2(3.5) mm, membranous. **Inflorescences** dense, 1-sided racemes, usually partially enclosed by the enlarged upper sheaths; each spikelet with a pedicel attached directly to the inflorescence axis. **Spikelets** (3.5)5–12 mm long with (2)3–4(7) florets. **Glumes** and **lemmas** blunt, rounded, awnless, glabrous to slightly scabrous. **Anthers** 0.8–1.5 mm. **Habitat**: Disturbed, compacted areas, roadsides, gravel parking lots; E of the Cascades. **Comments**: Native to S Europe and W Asia. Fairgrounds grass really does grow in fairground parking lots! It is often overlooked and is very similar to *Poa annua*, which has panicles, and obtuse to acute lemmas with hairy veins.

Scolochloa festucacea (Willd.) Link
Common Rivergrass, Whitetop Grass

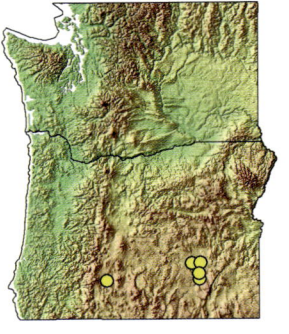

Strongly rhizomatous perennial; rhizomes succulent; **culms** 70–200 cm, 6–8 mm thick at the base. **Leaf sheaths** open; **blades** 4–12 mm wide; **ligules** 3–7(9) mm, membranous. **Panicles** open, primary branches ascending to spreading, with spikelets appressed to the branches. **Spikelets** 7–11 mm with 3–4 florets. **Glumes** acute to acuminate, awnless; **lower glume** shorter than the uppermost floret; **upper glume** usually > or = the uppermost floret. **Lemmas** 4–9 mm, with 3–9 convergent veins, minutely 3-lobed, awnless; **calluses** with hairs 1–1.5 mm. **Anthers** 2–4 mm. **Habitat**: Freshwater to moderately alkaline wetlands, ponds, marshes, seasonally flooded areas, shores of lakes and streams. Rare. **Comments**: Provides waterfowl nesting habitat. Native to N Great Plains, rare and disjunct in E OR.

Secale cereale L.
Rye, Cereal Rye [E]

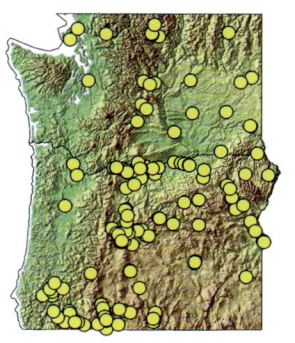

Tall, light green annual or biennial; **culms** (35)50–120(300) cm. **Leaf sheaths** open; **blades** (3)4–12 mm wide; **ligules** to 2 mm, membranous, obtuse; **auricles** usually present. **Spikes** laterally compressed, angled to nodding when mature. **Spikelets** 1 per node with 2(3) florets. **Glumes** 8–20 mm with awns 1–3 mm. **Lemmas** 14–18 mm, strongly compressed, marginal veins keeled and coarsely ciliate; **awns** 7–50 mm. **Anthers** about 7 mm. **Habitat**: Disturbed grasslands, roadsides, cultivated fields. **Comments**: Used for grain, whiskey, forage, and erosion control. Rye is a staple crop in climates too cold for wheat. Populations planted for erosion control often persist for years. Rye can be invasive in dry, disturbed grasslands E of the Cascades. Triticosecale (or triticale), a hybrid of rye and wheat, is grown for grain and forage and is particularly valuable for its high protein content.

Setaria faberi R.A.W. Herrm.
Giant Foxtail, Japanese Bristlegrass [E]

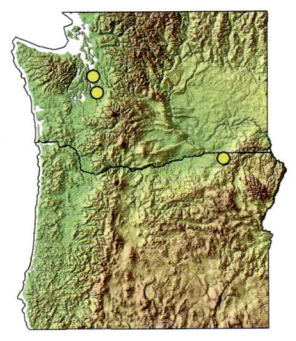

Annual; **culms** 50–200 cm. **Leaf blades** 10–20 mm wide, upper surface soft-hairy; **ligules** 2 mm, consisting of a fringe of hairs. **Panicles** 6–20 cm, dense and spike-like, arching and nodding. **Spikelets** 2.5–3 mm, dorsiventrally compressed, subtended by (1)3(6) antrorsely scabrous bristles. **Lower glumes** about 1 mm, acute, 3-veined; **upper glumes** about 2.2 mm, obtuse, 5-veined. **Upper lemmas** finely, distinctly, transversely rugose. **Anthers** 0.5–1.1 mm. **Habitat**: Disturbed areas, railroad ballast, roadsides, cultivated fields. **Comments**: Introduced from Asia. So far there are only a few records in OR and WA, but it is expected to spread. Giant foxtail is a major weed in the Midwest, especially in cornfields.

Setaria pumila (Poir.) Roem. & Schult.
Yellow Foxtail, Yellow Bristlegrass, Cattail Grass [E]

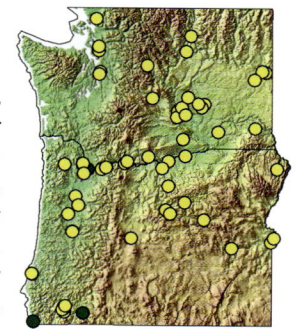

Annual; **culms** (5)30–130 cm. **Leaf blades** 4–10 mm wide, with papillose-based hairs toward the base of the upper surface; **ligules** membranous, ciliate. **Panicles** 3–15 × 1 cm, dense, spike-like and cylindrical. **Spikelets** 2–3.4 mm, dorsiventrally compressed, subtended by 4–12 antrorsely scabrous bristles. **Lower glumes** about 33% spikelet length, 3-veined; **upper glumes** 50% spikelet length, 5-veined. **Upper lemmas** exposed, strongly transversely rugose. **Anthers** 0.4 mm. **Habitat**: Disturbed soil, roadsides, lawns, and cultivated fields. **Comments**: Similar to *S. viridis*, but *S. pumila* has a smaller, paler inflorescence, and more bristles subtending each spikelet.

Setaria pumila ssp. *pallide-fusca* (Schumach.) B.K. Simon
Cattail Grass [E] (green dots) (illustrations below)

Leaf blades dark green. **Bristles** reddish to purple. **Spikelets** 2–2.5 mm long. **Comments**: *Setaria pumila* ssp. *pallide-fusca* is known from SW OR and historically from ballast dumps in Portland. Introduced from southern Africa.

Setaria pumila ssp. *pumila*
Yellow Foxtail, Yellow Bristlegrass [E] (yellow dots) (illustrations p. 422)

Leaf blades yellowish-green. **Bristles** yellowish, occasionally rusty. **Spikelets** 3–3.4 mm. **Comments**: Widely distributed in the PNW. Introduced from Europe.

Setaria verticillata (L.) P. Beauv.
Hooked Bristlegrass [E]

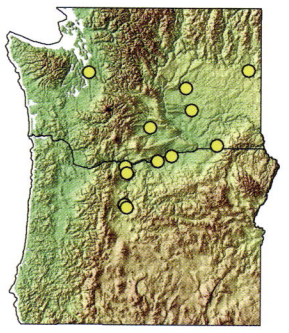

Annual; **culms** 30–100 cm. **Leaf blades** 5–15 mm wide, upper surfaces scabrous; **ligules** to 1 mm, membranous and densely ciliate. **Panicles** 5–15 × 0.8 cm, dense and spike-like, tapering at the tip. **Spikelets** 2–2.3 mm, dorsiventrally compressed, subtended by a solitary, retrorsely scabrous bristle. **Lower glumes** about 33% spikelet length, obtuse; **upper glumes** nearly as long as spikelets. **Upper lemmas** mostly hidden, finely and transversely rugose. **Anthers** 0.6–0.9 mm. **Habitat**: Disturbed ground, agricultural fields. **Comments**: Introduced from Europe. The inflorescence tends to cling to clothing (among other objects) because of its retrorsely scabrous bristles (small barbs point toward the base). This is our only *Setaria* with retrorsely scabrous bristles. A serious weed in CA vineyards.

Setaria viridis (L.) P. Beauv. var. *viridis*
Green Foxtail [E] (yellow dots)

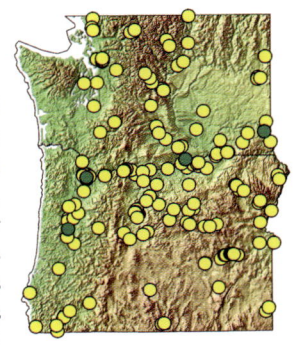

Annual; **culms** 20–100 cm, with 6–7 nodes. **Leaf blades** 4–12 mm wide, scabrous or smooth; **ligules** 1–2 mm, membranous and ciliate. **Panicles** 3–8(15) cm, dense and spike-like, erect or nodding only near the tip, unlobed. **Spikelets** 1.8–2.2 mm, subtended by 1–3 antrorsely scabrous, green bristles. **Lower glumes** 33% as long as the spikelet, 3-veined; **upper glumes** nearly as long as the spikelet, 5–6-veined. **Upper lemmas** mostly hidden, pale green, finely, transversely rugose. **Anthers** 0.4–0.8 mm. **Habitat**: Roadsides, disturbed areas, agricultural fields. **Comments**: Introduced from Eurasia. *Setaria viridis* var. *viridis* is our most common foxtail. *Setaria italica*, an occasional waif in the PNW, has thicker, arching, often lobed inflorescences, longer spikelets, and smooth upper lemmas.

Setaria viridis var. *major* (Gaudin) Peterm.
Giant Green Foxtail [E] (green dots)

Like var. *viridis*, but much larger, with **culms** 100–250 cm, with 7–12 nodes. **Leaf blades** 10–25 mm wide. **Panicles** 10–20 cm and often somewhat lobed at the base. **Comments**: Uncommon in the PNW, but a serious weed in the Midwest.

Sorghum bicolor (L.) Moench
Milo, Sorghum, Shattercane [E]

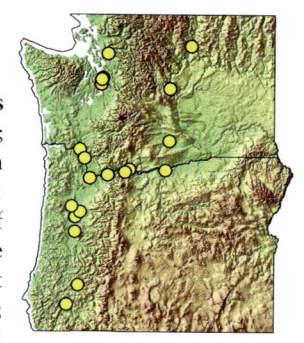

Annual; **culms** solitary or clustered, 50–500 cm. **Leaf blades** 5–100 mm wide, glabrous to sparsely hairy, margin serrate; **ligules** 1–4 mm, membranous, ciliate. **Panicles** 5–60 cm, open to very dense. **Spikelets** arranged in pairs or triplets with 1 sessile spikelet and 1 or 2 pedicellate spikelets; **calluses** of spikelets with inconspicuous hairs much < the spikelets. **Sessile spikelets** with one bisexual floret and a lower sterile floret (lemma only), 3–9 mm; **glumes** leathery or membranous; **lemmas** of bisexual florets awnless or with a bent, twisted awn to 30 mm. **Seeds** exposed at maturity, not falling when ripe, or falling tardily. **Pedicellate spikelets** staminate or sterile, 3–6 mm, awnless. **Anthers** 2–2.8 mm. **Habitat**: Crop fields and under bird feeders. **Comments**: Sorghum was domesticated in Africa 3,000 years ago. It is the fifth most important grain crop in the world and is also grown for forage and sweeteners. Shattercane, a form of *S. bicolor* whose seeds fall early, is a troublesome agricultural weed. With its wide leaves, vegetative *S. bicolor* resembles domestic corn, which differs in having smooth leaf margins.

Sorghum halepense (L.) Pers.
Johnson Grass [E]

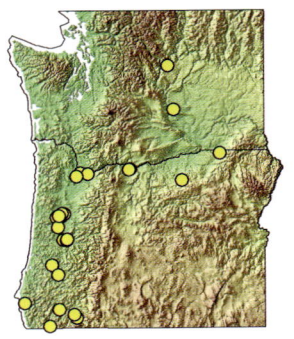

Strongly rhizomatous perennial; **culms** to 200 cm. **Leaf blades** (8)10–20(40) mm wide, glabrous to sparsely hairy; **ligules** 2–6 mm, membranous, conspicuously ciliate. **Panicles** 10–50 cm, open. **Spikelets** arranged in pairs or triplets with 1 sessile spikelet and 1 or 2 pedicellate spikelets; **calluses** of spikelets with inconspicuous hairs much < the spikelets. **Sessile spikelets** with one bisexual floret and a lower sterile floret (lemma only), 3.8–6.5 mm, fully enclosed by two glumes; **glumes** leathery or hard, usually dark brown to blackish when mature, **lemmas** of bisexual florets awnless or with a bent, twisted awn to 13 mm. **Seeds** not exposed, falling at maturity. **Pedicellate spikelets** staminate, 3.6–5.6 mm, awnless. **Anthers** 1.9–2.7 mm. **Habitat**: Roadsides, ditches, orchards, disturbed seasonally moist sites. **Comments**: Native to the Mediterranean region. *Sorghum halepense* is invasive, particularly in warmer areas, and is on the OR and WA noxious weed lists.

Spartina alterniflora Loisel.
Smooth Cordgrass [E]

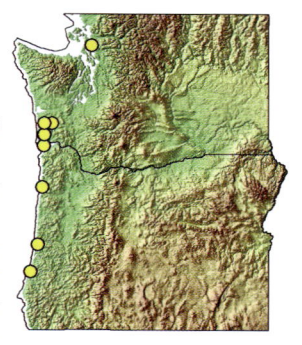

Rhizomes long, white, with scales not or slightly overlapping; **culms** 60–250 cm, soft; fresh culms with sulfur odor. **Leaf blades** 3–25 mm wide, flat at base, inrolled toward tip, diverging 15–18° from stem; margins smooth to slightly scabrous; **ligules** 1–2 mm. **Panicles** with 3–25 branches; **branches** 5–15 cm, erect, with 20–30 spikelets per branch; **branch axes** prolonged beyond spikelets. **Spikelets** 8–16.5 mm; **glumes** usually glabrous on sides, sometimes hairy near base; **lower glumes** 4–10 mm; **upper glumes** 7–12 mm; **lemmas** 8–11 mm. **Anthers** 3–6 mm. **Habitat**: Intertidal mudflats, salt marshes, esturaries. Native to E North and South America. **Comments**: Smooth cordgrass is a serious threat to salt marshes and estuaries on the West Coast. It forms dense stands that crowd out native vegetation. It seldom flowers but can be identified vegetatively with rhizome, culm, and leaf characters.

Spartina anglica C.E. Hubb.
English Cordgrass [E] (yellow dots)

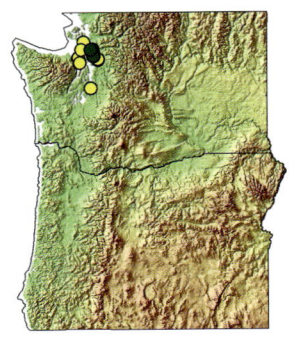

Rhizomes long, thick, fleshy, whitish, with overlapping scales; **culms** 30–130 cm, relatively hard; fresh culms not smelling of sulfur. **Leaf blades** (6)11–13(15) mm wide, flat at least at base, diverging 30–60° from the stem; margins smooth or slightly scabrous; **ligules** 2–3 mm. **Panicles** with 2–12 branches; **branches** 14–25 cm, erect or somewhat divergent; **branch axes** prolonged to 5 cm beyond spikelets. **Spikelets** 14–21 mm; **glumes** appressed-hairy on sides, keels ciliate; **lower glumes** 10–14 mm; **upper glumes** to 17 mm. **Anthers** 5–13 mm, well-developed, dehiscent at maturity. **Habitat**: Coastal, low intertidal to high marsh zone. Native to Great Britain. **Comments**: This fertile species originated by chromosome duplication in the sterile hybrid *S.* × *townsendii* (not pictured, green dots on map), which has narrower leaf blades, shorter ligules, and shorter, nondehiscent anthers. Both taxa have been introduced in the PNW and are threats to native tidal habitats.

Spartina densiflora Brongn.
Dense-flowered Cordgrass [E]

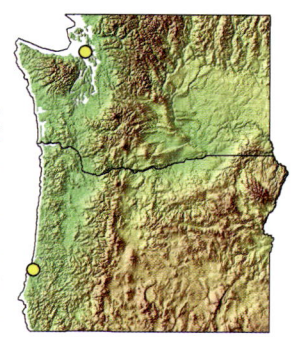

Cespitose, forming dense clumps; **rhizomes**, if present, short, thin, wiry, to 10 mm thick; **culms** 27–150 cm, firm. **Leaf blades** 3–8 mm wide, involute when fresh; margins scabrous; **ligules** 1–2 mm. **Panicles** with 2–15 branches; **branches** 1–11 cm, tightly appressed, overlapping, comb-like, with 10–30 spikelets per branch; **branch axes** not prolonged beyond the spikelets. **Spikelets** 8–14 mm; **glumes** glabrous or sparsely short-hairy; **lower glumes** 4–7 mm; **upper glumes** 8–14 mm. **Anthers** 3–5 mm. **Habitat**: Coastal, mid to high salt marshes. Native to South America. **Comments**: *Spartina densiflora* is the only cespitose PNW cordgrass. *Spartina patens* is similar with narrow culms and leaves, but it is strongly rhizomatous, and its panicle branches are spreading and widely separated. *Spartina densiflora* grows at 1 site each in OR and WA; it is a serious invasive on the N CA coast.

Spartina gracilis Trin.
Alkali Cordgrass

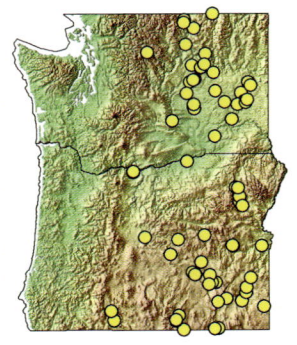

Strongly rhizomatous; **rhizomes** 1.5–5 mm thick, white, scales closely overlapping; **culms** 40–100 cm, 2–3.5 mm thick, firm. **Leaf blades** 2.5–8 mm wide, flat, becoming involute; margins scabrous; **ligules** 0.5–1 mm, consisting of a fringe of dense silky white hairs. **Panicles** with 3–12 branches; **branches** 1.5–8 cm, comb-like, usually appressed, rarely spreading, with 10–30 spikelets per branch; **branch axes** not prolonged beyond the spikelets. **Spikelets** 6–11 mm; **glumes** acute or short-pointed, not awned; **lower glumes** 3–5 mm; **upper glumes** 6–10 mm, ± = the floret. **Lemmas** 6–7 mm. **Anthers** 2.5–5 mm. **Habitat**: Margins of lakes and streams, meadows, in alkaline soils. **Comments**: Both of our native cordgrasses occur E of the Cascades. *Spartina gracilis* is the more delicate of the two; *S. pectinata* has thicker, darker rhizomes, thicker culms, and distinctly awned glumes.

Spartina patens (Aiton) Muhl.
Saltmeadow Cordgrass [E]

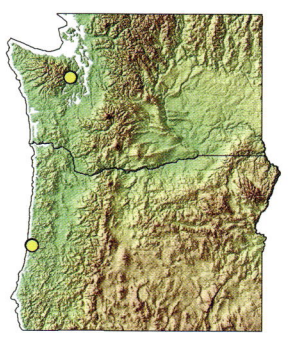

Strongly rhizomatous; **rhizomes** 1–6 mm thick, thin, wiry, whitish, scales not overlapping; **culms** 15–80 cm, firm. **Leaf blades** 0.5–4(7) mm wide, involute except sometimes near base, margins strongly scabrous; **ligules** 0.5–1 mm. **Panicles** with 2–15 branches; **branches** 1–8 mm, comb-like, appressed to strongly divergent, with 10–30 spikelets per branch; **branch axes** not prolonged beyond the spikelets. **Spikelets** 7–12 mm; **glumes** glabrous or sparsely short-hairy; **lower glumes** 3–8 mm; **upper glumes** 7–12 mm; **lemmas** 7–8 mm. **Anthers** 3–5 mm. **Habitat**: Coastal high salt marsh. Native to the East Coast of North, Central, and South America. **Comments**: Similar *S. densiflora* is cespitose and has tightly appressed panicle branches. Like *S. alterniflora, S. patens* forms dense stands that degrade coastal salt marsh habitats on the West Coast. In E North America *S. patens* is expanding its range inland along roads that are salted in winter.

Spartina pectinata Link
Prairie Cordgrass

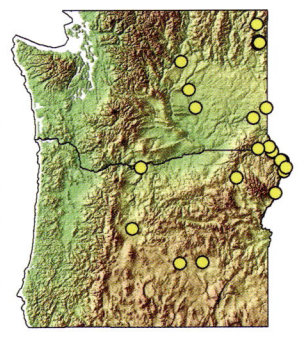

Strongly rhizomatous; **rhizomes** (2)3–8 mm thick, purplish brown or light brown but drying white, scales closely overlapping; **culms** to 250 cm, 2.5–11 mm thick. **Leaf blades** 5–15 mm, flat when fresh, involute when dry, margins strongly scabrous; **ligules** 1–3 mm, a narrow membranous band topped with a fringe of dense silky hairs. **Panicles** with 5–50 branches; **branches** 1.5–10 cm, comb-like, usually appressed to somewhat spreading, with 10–80 spikelets per branch; **branch axes** not prolonged beyond the spikelets. **Spikelets** 10–25 mm; **glumes** tapering to awned tips; **lower glume** 5–10 mm; **upper glume** 10–25 mm including awn; awn of upper glume 3–8 mm. **Lemmas** about 10 mm. **Anthers** 4–6 mm. **Habitat**: Marshes, wetlands, along streams and rivers, sometimes in alkaline soils. **Comments**: *Spartina pectinata* is the larger of our native inland cordgrasses (see *S. gracilis*). It is rare in our area, mainly due to destruction of its habitat; most collections are more than 50 years old; widespread in the Great Plains. This species would be a good selection for wetland and riparian habitat restoration.

Sphenopholis intermedia (Rydb.) Rydb.
Slender Wedgegrass

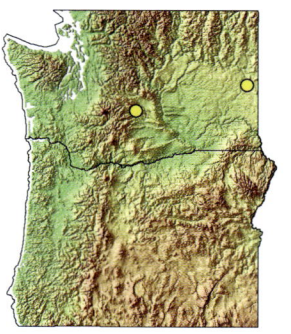

Cespitose perennial or winter annual; **culms** (5)30–120 cm. **Leaf sheaths** open; **blades** 2–6 mm wide; **ligules** 1.5–2.5 mm, membranous, erose-ciliate. **Panicles** narrow but not spike-like, often nodding. **Spikelets** 2.1–4 mm, with 2 florets; disarticulation below glumes; **lower glumes** < 33% as wide as upper glumes; **upper glumes** 1.9–2.9 mm, oblanceolate or obovate, not hooded, 23–35% as wide as long, tips acute, rounded, or almost truncate; **lemmas** 2.1–3 mm, awnless. **Anthers** 0.2–0.8 mm. **Habitat**: Wet sites, shores, stream banks, often with vernally wet clay soils that dry out in summer. **Comments**: Rare in the PNW. *Sphenopholis intermedia* differs from *S. obtusata* in its more open panicle and narrower upper glumes. *Koeleria macrantha* has a more spike-like panicle and spikelets that disarticulate above the glumes.

Sphenopholis obtusata (Michx.) Scribn.
Prairie Wedgegrass

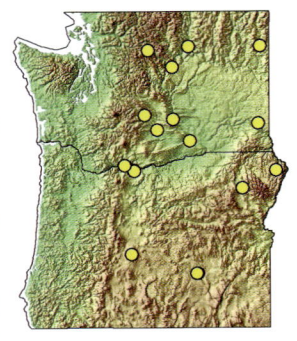

Cespitose perennial or winter annual; **culms** (9)20–130 cm. **Leaf sheaths** open; **blades** 2–8 mm wide; **ligules** (1)1.5–2.5 mm, membranous, erose-ciliate. **Panicles** dense, spike-like, usually erect. **Spikelets** 2.2–3.6 mm, with 2 florets; disarticulation below glumes; **lower glumes** < 33% as wide as upper glumes; **upper glumes** 1.5–2.5 mm, obovate, 30–50% as wide as long, tips rounded to truncate, hooded; **lemmas** 1.9–2.8 mm, awnless. **Anthers**: 0.2–1 mm. **Habitat**: Wet meadows, prairies, marshes, waste places, sometimes in alkaline soils. **Comments**: The shiny spike-like panicles of *S. obtusata* resemble those of *Koeleria macrantha*, but *Koeleria* grows in drier habitats and has acute glumes and spikelets that disarticulate above the glumes. See also *Agrostis exarata*. Nearly all PNW records of *S. obtusata* are more than 50 years old, and the species is apparently rare or overlooked in our area.

Sporobolus airoides (Torr.) Torr.
Alkali Sacaton

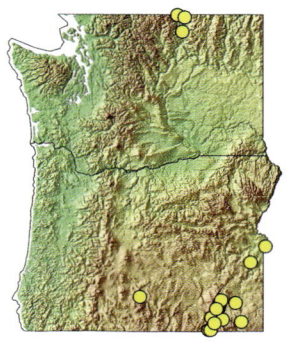

Cespitose perennial; **culms** 35–120(150) cm. **Leaf sheath** tops glabrous or sparsely hairy; **blades** 2–5 mm wide, tough, flat to involute; **ligules** 0.1–0.3 mm, membranous, ciliate with a row of long stiff hairs on the blade just above the ligule and extending to the sides of the upper sheath; **upper culm leaves** ascending, nearly parallel to the culm. **Panicles** (10)15–45 cm, diffuse, open, light brown, often partly enclosed in sheath, branches spreading. **Spikelets** 1.3–2.8 mm; **lower glumes** 0.5–1.8 mm; **upper glumes** 1.1–2.8 mm; **lemmas** 1.2–2.5 mm; glabrous. **Anthers** 1.1–1.8 mm. Seed coat gelatinous when wet. **Habitat**: Dry, alkaline, sandy soils. **Comments**: Alkali sacaton is rare in our area, occurring in SE OR, and disjunct in Okanagan County, WA. Immature spikelets of *Panicum* resemble *S. airoides* but have strongly veined glumes and lemmas. The name "alkali sacaton" is sometimes used for *Muhlenbergia asperifolia*, which has a whitish panicle and membranous ligule. The name "sacaton" comes from the American Spanish word *zacatón*, grass, which originated from the Nahuatl word *zacatl*, also meaning grass.

Sporobolus compositus (Poir.) Merr. var. *compositus*
Rough Dropseed

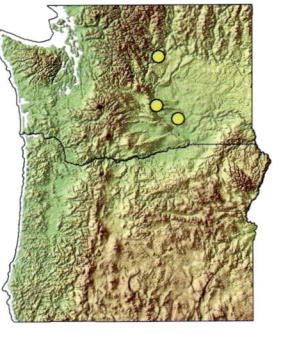

Cespitose perennial; **culms** (20)30–130(150) cm. **Leaf sheath** tops sparsely hairy; **blades** 1.5–10 mm wide, flat, folded, or involute; **ligules** 0.1–0.5 mm, membranous, ciliate. **Panicles** 5–30 cm, terminal and axillary, usually spike-like, often partly enclosed in sheaths; branches appressed. **Spikelets** 4–6(10) mm; **glumes** subequal, (1.2)2–5(6) mm. **Lemmas** (2.2)3–6(10) mm, 1(3)-veined; glabrous. **Anthers** 0.2–3.2 mm. **Habitat**: Sandy soils, river shorelines, roadsides. **Comments**: Rare, known from a few sites along the Columbia River in central WA. *Sporobolus* are unusual among grasses because the ovary wall does not adhere to the seed. The seed coat becomes sticky (gelatinous) when wet, helping disperse the seeds. In some cases, the drying ovary wall forcibly ejects the seed.

Sporobolus cryptandrus (Torr.) A. Gray
Sand Dropseed

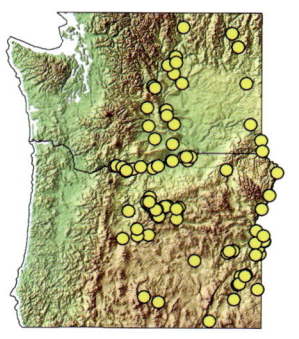

Cespitose perennial; **culms** 30–100(120) cm. **Leaf sheath** tops with conspicuous tuft of hairs; **blades** 2–6 mm wide, flat to involute; **ligules** 0.5–1 mm, membranous, ciliate, mostly obscured by copious long hairs; **upper leaf blades** nearly perpendicular to culms. **Panicles** 15–40 cm; contracted, spike-like when young, becoming narrowly pyramid-shaped as primary branches spread; secondary branches appressed to the primary branches; panicle base often enclosed in upper leaf sheath. **Spikelets** 1.5–2.7 mm; **lower glumes** 0.6–1.1 mm, **upper glumes** 1.5–2.7 mm. **Lemmas** 1.4–2.5(2.7) mm; glabrous. **Anthers** 0.5–1 mm. Seed coat gelatinous when wet. **Habitat**: Sandy soils, rocky slopes, calcareous ridges, roadsides. **Comments**: Sand dropseed is a common late-flowering grass of arid habitats. It is a warm-season grass, with C4 metabolism. Its seeds are among the tiniest of our grasses, with over 5 million seeds per pound. *Sporobolus* has an unusual feature; the palea splits down the back at maturity, and the seed falls out between the two halves.

Sporobolus neglectus Nash
Puffsheath Dropseed

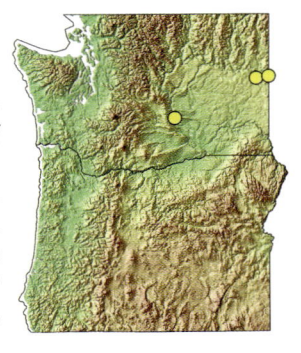

Delicate tufted annual; **culms** 10–45 cm, wiry, decumbent to erect. **Leaf sheaths** inflated, glabrous except tops with a tuft of hairs to 3 mm; **blades** 0.6–2 mm wide, flat to loosely inrolled; **ligules** 0.1–0.3 mm, membranous, ciliate. **Panicles** 2–5 cm; contracted, narrowly cylindrical, partially or fully enclosed in upper leaf sheath. **Spikelets** 1.6–3 mm; **lower glumes** 1.5–2.4 mm, **upper glumes** 1.7–2.7 mm. **Lemmas** 1.6–2.9 mm, glabrous. **Anthers** 1.1–1.6 mm. **Habitat**: Sandy soils, shorelines, disturbed areas. **Comments**: Easily overlooked, *S. neglectus* sprawls its straggly stems sparsely on shoreline sands and silts just above highwater line. Last collected in NE WA in the early 1900s, it was recently found along the Columbia River in Grant County, WA. Similar *S. vaginiflorus* differs in having larger, short-hairy lemmas, and larger spikelets and anthers.

Sporobolus vaginiflorus (Torr. ex A. Gray) Alph. Wood var. *vaginiflorus*
Poverty Grass [E]

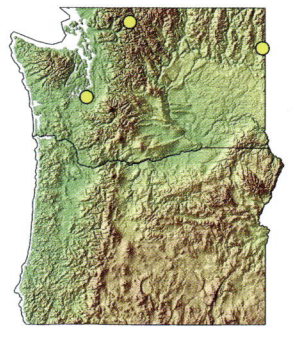

Delicate tufted annual; **culms** 15–60(70) cm, wiry, decumbent to erect. **Leaf sheaths** often inflated, usually glabrous, tops sometimes with a tuft of hairs to 3 mm; **blades** 0.6–2 mm wide, flat to loosely inrolled; **ligules** 0.1–0.3 mm, membranous, ciliate. **Panicles** 1–5 cm, contracted, narrowly cylindrical, partially or fully enclosed in upper leaf sheath. **Spikelets** 3–6 mm; **lower glumes** (2.2)2.8–4.7 mm; **upper glumes** (2.4)3–5 mm. **Lemmas** 3–5.4 mm, with short, appressed hairs. **Anthers** 1.2–3.2 mm. **Habitat**: Sandy or gravelly roadsides, disturbed areas. **Comments**: Poverty grass spreads along roadsides and has been introduced to a few sites in WA. It is present in ID and to be looked for in OR. *Sporobolus neglectus* has smaller, glabrous lemmas, and smaller spikelets and anthers.

Taeniatherum caput-medusae (L.) Nevski
Medusahead [E]

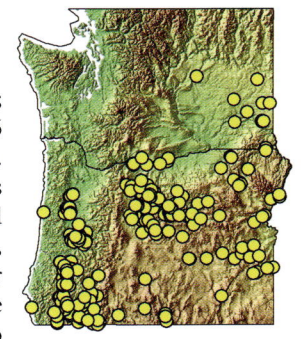

Tufted annuals; **culms** (5)10–55(70) cm. **Leaf sheaths** open; **blades** (0.2)0.7–2.5 mm wide, flat to involute; **ligules** 0.2–0.6 mm. **Spikes** terminal, 1.2–6 cm. **Spikelets** 2(4) per node. **Glumes** (5)7–80 mm, awn-like. **Lemmas** 5.5–8 mm, awns (20)30–110 mm, longer than spike axis, becoming twisted and tangled. **Anthers** 0.8–1 mm. **Habitat**: Disturbed grasslands, dry slopes, roadsides. **Comments**: Medusahead, named for the Greek monster with snakes for hair, is native from the Mediterranean to Central Asia. It is similar in appearance to *Hordeum*, which has 3 spikelets per node and relatively shorter awns. Highly invasive, it forms a dense thatch that decays more slowly than other grasses due to its high silica content. It outcompetes cheatgrass on heavier clay soils that are seasonally saturated.

Thinopyrum intermedium (Host) Barkworth & D.R. Dewey ssp. *intermedium*
Intermediate Wheatgrass [E] (yellow dots)

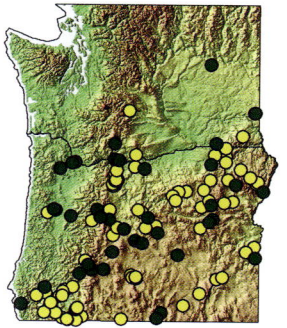

Strongly rhizomatous perennial, often glaucous; **culms** 50–115 cm. **Leaf sheath margins** often ciliate; **blades** flat, 2–8 mm wide; **ligules** 0.1–0.8 mm; **auricles** 0.5–1.8 mm. **Spikes** 8–21 cm, with 1 spikelet per node. **Spikelets** 11–18 mm, with 3–10 florets. **Glumes** thick, stiff, and hard, glabrous, midvein longer than lateral veins, tips acute to obtuse or obliquely truncate, awnless. **Lemmas** 7.5–10 mm, usually glabrous (sometimes hairy on margins), awnless or with a short awn. **Anthers** 5–7 mm. **Habitat**: Grasslands, roadsides, open forest, pastures. **Comments**: Introduced from Eurasia. Commonly planted for erosion control. Forms large clones. The 2 subspecies of intermediate wheatgrass overlap considerably.

Thinopyrum intermedium ssp. *barbulatum* (Schur) Barkworth & D.R. Dewey
Pubescent Wheatgrass [E] (green dots)

Like ssp. *intermedium* but the glumes and lemmas hairy.

Thinopyrum obtusiflorum (DC.) Banfi
Tall Wheatgrass [E]

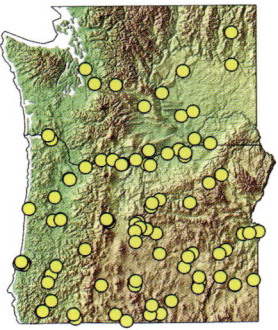

Densely cespitose perennial; **culms** 50–200 cm. **Leaf sheath margins** ciliate; **blades** 2–5.6 mm wide, usually convolute; **ligules** 0.3–1.5 mm; **auricles** 0.2–1.5 mm. **Spikes** 10–42 cm, with 1 spikelet per node. **Spikelets** 13–30 mm, often curving away from the culm. **Glumes** thick, stiff, and hard, 6.5–10 mm, with all veins about equal, tips truncate, awnless. **Lemmas** 9–12 mm, glabrous, awnless. **Anthers** 4–6 mm. **Habitat**: Roadsides, grasslands, pastures, often on alkaline soils. **Comments**: Introduced from Eurasia. Once you get an eye for it, this is a 60-mile-an-hour grass that you can identify through a car windshield. It is sometimes confused with *Leymus cinereus*, which has 2–7 spikelets per node and narrowly acute glumes. The glumes of *T. obtusiflorum* look like they have been chopped off with a pair of scissors. It has been the victim of multiple name changes over the past few decades, having been placed in *Agropyron, Elymus,* and *Elytrigia,* and has had several specific epithets as well.

Torreyochloa erecta (Hitchc.) G.L. Church
Spiked False Mannagrass

Rhizomatous perennial; **culms** 20–62 cm 1–1.4 mm thick at the base. **Leaf sheaths** open; largest culm **leaf blades** 3.4–6(7.2) mm wide; **ligules** of largest leaves 2.6–6.5 mm, membranous, lacerate. **Panicles** linear to narrowly elliptic, 5.5–19× as long as wide, ≤ 1 cm wide, branches usually appressed. **Spikelets** 4.2–6.4 mm with 4–6 florets; **lower glumes** 0.8–1.6 mm; **upper glumes** 1–2.1 mm; **lemmas** 2.3–3.1 mm, with (5)7–9 prominent, ± parallel veins, awnless. (Lemma veins become more prominent with maturity.) **Anthers** 0.6–0.8 mm.

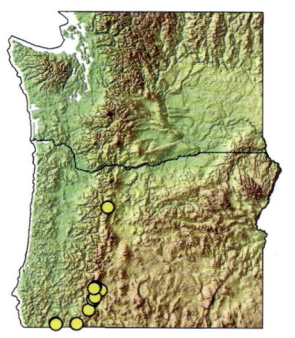

Habitat: Margins of subalpine lakes, streams, bogs, marshes; not in alkaline habitats.
Comments: *Torreyochloa erecta* is a rare grass of high-elevation shores and bogs in the S Cascades and Sierra Nevada. It has also been found at one location in the Cascade Range in N OR. *Torreyochloa* species closely resemble *Glyceria* in spikelets, habit, and habitat, and were formerly members of that genus. *Torreyochloa* differs from *Glyceria* in having leaf sheaths that are open, not fused in the front.

Torreyochloa pallida (Torr.) G.L. Church var. *pauciflora* (J. Presl) J.I. Davis
Weak Mannagrass

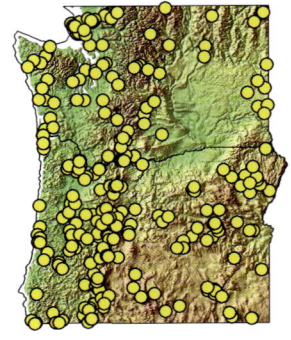

Rhizomatous perennial; **culms** 20–145 cm, 1.3–4.8 mm thick at the base. **Leaf sheaths** open; largest culm **leaf blades** 3.6–18 mm wide; **ligules** of largest leaves 3–9 mm, membranous, lacerate. **Panicles** narrowly to widely conic, ovoid, or obovoid, 1.2–7.5× as long as wide, (1)2–14 cm wide, branches spreading. **Spikelets** 3.6–6.9 mm with (3) 4–8 florets; **lower glumes** 0.7–1.6 mm; **upper glumes** 0.9–1.8 mm; **lemmas** 2.2–3.3 mm, with (5)7–9 prominent, ± parallel veins, awnless. **Anthers** 0.5–0.7 mm. **Habitat**: Marshes, bogs, and shores of lakes and streams, from sea level to high montane; not in alkaline habitats; widespread. **Comments**: Weak mannagrass resembles an unusually wide-leaved *Glyceria*, but *Glyceria* species have closed leaf sheaths. It is vegetatively similar in some ways to *Catabrosa aquatica* but has a much more erect habit.

Tripidium ravennae (L.) H. Sholz
Ravennagrass, Elephantgrass [E]

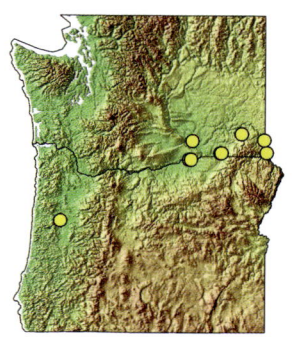

Enormous, cespitose perennial; **culms** 2–4+ m. **Leaf sheaths** open; **blades** 5–14 mm wide, with a patch of dense white hairs at base on upper surface; **ligules** 0.6–1.1 mm, membranous, ciliate, concealed by copious blade hairs. **Panicles** plumose, with many branches. **Spikelets** in sessile/pedicellate pairs, both spikelets fertile; **calluses** of spikelets with white hairs 4–6 mm; pedicel hairs a bit shorter than callus hairs. **Disarticulation** beneath spikelets; sessile spikelets falling with adjacent internode and pedicel. **Spikelets** 4–6 mm, ± dorsiventrally compressed, with 2 florets; **upper floret** fertile, with awned lemma, awns 2–5 mm; **lower floret** reduced, sterile, awnless. **Habitat**: Marshes, stream banks, roadsides, disturbed areas. Native to S Europe and W Asia. **Comments**: With fluffy panicles and towering culms *T. ravennae* resembles other giant grasses like *Cortaderia* and *Phragmites*, but the patch of dense hairs at the leaf blade base and the sessile/pedicellate pairs of spikelets are distinctive. Ravennagrass escapes from ornamental plantings and is extremely difficult to control. It is a noxious weed in several states.

Triplasis purpurea (Walter) Chapm. var. *purpurea*
Purple Sandgrass [E]

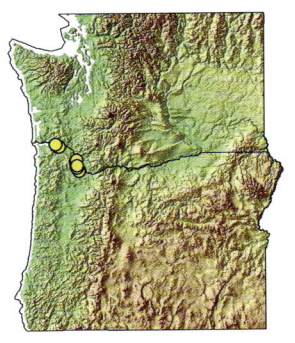

Tufted annual or perennial, occasionally rhizomatous; **culms** 14–100 cm, disarticulating at maturity. **Leaf sheaths** inflated, sometimes with tufts of hairs at the top; **blades** 1–5 mm wide, hairy; **ligules** of hairs to 1 mm. **Terminal panicles** often partially hidden in sheaths, with few spreading branches, 3–7 cm. **Axillary panicles** self-pollinating, completely hidden in the upper leaf sheaths. Terminal panicle **spikelets** 6.5–9 mm with 3–4 florets. **Glumes** about 2 mm. **Lemmas** 3–4 mm with short (< 1 mm), rounded lobes at the tip; awns < 2 mm, arising between the lobes. **Paleas** with a brush of fairly long hairs on the keels. **Anthers** about 2 mm. **Habitat**: Disturbed, sandy soils. **Comments**: Native from central and E US to Central America. Introduced along the lower Columbia River. At maturity, the culms break apart at the nodes, dispersing seed from inflorescences hidden in the upper sheaths.

Trisetum spicatum (L.) K. Richt.
Spike Trisetum

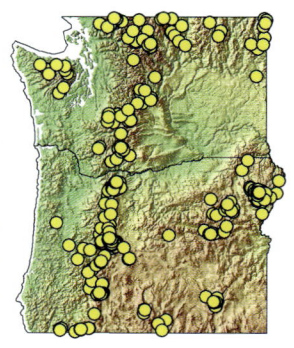

Cespitose perennial; **culms** 10–120 cm. **Leaf sheaths** hairy or glabrous; **blades** 1–5 mm wide, usually flat; **ligules** 0.5–4 mm, membranous, often truncate, erose. **Panicles** 20–30(50) cm, usually dense and spike-like, often shiny and silvery. **Spikelets** 5–7.5 mm, with 2–3 florets. **Lower glumes** 3–4(5.5) mm; **upper glumes** 4–7 mm, ≥ lowest floret, < 2× as wide as the lower glumes. **Lemmas** 3–6(7) mm, with 2 teeth tapering to tiny awn-like tips, usually < 1 mm, awns 3–8 mm, arising from the upper third of lemma back, bent and basally twisted. **Rachilla** hairy. **Callus hairs** to 1 mm. **Anthers** 0.7–1.4 mm. **Habitat**: Moist montane meadows, forests, and rocky slopes. **Comments**: The shiny inflorescence of *Koeleria macrantha* is narrower and usually awnless. *Trisetum canescens, T. cernuum,* and *T. wolfii* were moved to *Graphephorum*; it appears that *T. spicatum* is destined to become *Koeleria spicata,* leaving no native *Trisetums* in N America. Introduced *T. flavescens* has been reported from BC and ID and should be watched for in our area.

Triticum aestivum L.
Wheat [E]

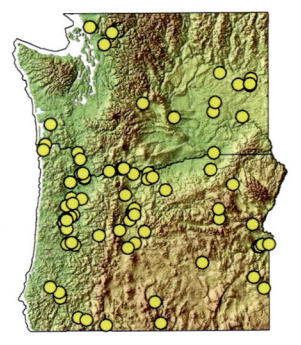

Tufted annuals; **culms** 14–150 cm. **Leaves** glabrous or hairy; **sheaths** open; **blades** 6–15(20) mm wide, flat; **auricles** prominent; **ligules** up to 1 mm, membranous. **Spikes** (3.5)6–18 × 0.8–2.1 cm, 2-sided, erect, with 1 spikelet per node. **Spikelets** 10–15 mm, with 3–9 florets, 2–5 florets seed-forming; **glumes** shorter than adjacent lemmas, hairy, truncate, short-awned; **lemmas** 10–15 mm, awns to 12 cm, straight or curved, sometimes awnless. **Anthers** 2.5 mm. **Habitat**: Disturbed sites, roadsides, cultivated fields. Originated in W and central Asia. **Comments**: Wheat is the most widely cultivated grain, domesticated at least 9,000 years ago. Both awned and awnless ("club wheat") forms are grown in the PNW. Club wheat has a thick, club-shaped spike. Roadside occurrences of wheat are waifs. Wheat is sometimes used for erosion control because it does not persist more than a few seasons. See *Aegilops cylindrica*, a serious weed in winter wheat.

Vahlodea atropurpurea (Wahlenb.) Fr. ex Hartm.
Mountain Hairgrass

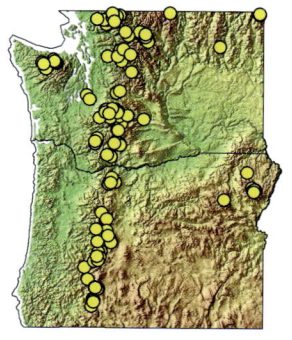

Glaucous, loosely cespitose perennial; **culms** 15–80 cm. **Leaf sheaths** open; **blades** 1–8.5 mm wide; primarily cauline; **ligules** 0.8–3.5 mm, truncate to obtuse, often torn. **Panicle** drooping with spikelets near ends of branches. **Spikelets** 4–7 mm; **glumes** 4–5.5(7) mm, enclosing 2 fertile florets (sometimes also with 1–2 vestigial florets). **Rachilla** ± glabrous, prolonged ≤ 0.5 mm beyond base of upper floret. **Lemmas** 1.8–3 mm, with long callus hairs, awned from midlength, awns 2–4 mm, twisted and geniculate. **Anthers** 0.5–1.2 mm. **Habitat**: Moist forest openings, seeps, streamsides, meadows, midmontane to alpine. **Comments**: Mountain hairgrass has the look of a *Poa*, but it has awned lemmas with long callus hairs. Formerly in the genus *Deschampsia*, which has longer, acute to acuminate ligules, basally disposed leaves, and hairy rachillas prolonged > 0.5 mm.

Ventenata dubia (Leers) Coss.
Ventenata, North Africa Grass [E]

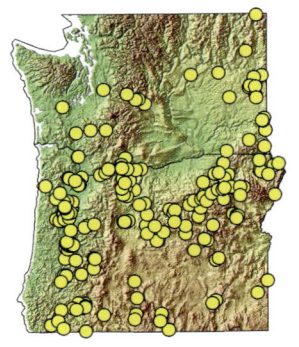

Slender annual; **culms** 15–75 cm; nodes dark. **Sheaths** open; blades 0.8–2.5 mm wide, flat or rolled; **ligules** 1–8 mm, membranous, lacerate. **Panicles** open, delicate; spikelets borne at branch ends; **spikelets** with 2–3 florets. **Glumes** (3)5–9-veined; lower glumes 4.5–6 mm; upper glumes 6–8 mm. **Lemmas** 5–7.5 mm; lowest lemma with straight awn to 4 mm arising from tip, remaining in spikelet after upper lemmas fall; upper lemmas awned from middle, awns 10–16 mm, bent, twisted. **Anthers** 1–2 mm. **Habitat**: Disturbed sites, roadsides, seasonally moist areas, often in heavy clay or shallow rocky soils. Native to N Africa, Europe. **Comments**: The delicate, mature panicles of a dense stand of *V. dubia* cast a distinctive reddish-silver glow recognizable from a moving car. Often confused with *Deschampsia danthonioides*, which has fewer glume veins, shorter lemmas, and both lemmas in a spikelet with a bent awn arising from below midlength. Invasive in crops and seasonal wetlands.

Zea mays L. ssp. *mays*
Corn, Maize [E]

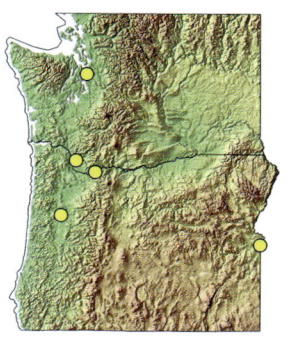

Robust annual; **culms** (1)2–4(6) m, (1)2–5 cm thick. **Leaf sheaths** open; **blades** 3–12 cm wide, margins entire. **Inflorescences** of 2 types: (1) **pistillate spikes** (ears) to 10 cm thick, surrounded by bracts (husks), arising in axils of leaves; **glumes** embedded in rachis (cob); (2) **staminate, terminal panicles** (tassels) with stiff, spreading branches; **spikelets** 9–14 mm; **lower glumes** loosely enclose the spikelet. **Habitat**: Ditches, roadsides, agricultural fields. Domesticated from Mexico and Central America. **Comments**: Domesticated about 7,000 years ago, corn is the world's third most important food crop. Waifs may grow in any sufficiently moist spot where kernels may fall. Corn silks are pollen tubes by which the female flowers are fertilized, which makes for good party conversation. The English word "corn" originally meant "grain" or wheat. European settlers in North America referred to the grain cultivated by the Native Americans as Indian corn, which became just "corn," except in Latin America, where it was called *maíz,* a Spanish word derived from "mahis," the name for corn used by the Indigenous Taíno people of the Caribbean region. Vegetative corn can be confused with *Sorghum bicolor*.

Zizania palustris L. var. *palustris*
Northern Wild Rice [E] (yellow dots)

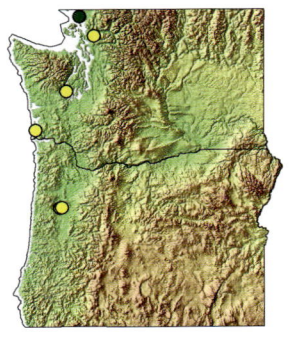

Aquatic annual; **culms** to 2 m, emergent. **Leaf sheaths** open; **blades** 3–21 mm wide; **ligules** 3–16 mm, membranous. **Upper panicle branches** appressed or ascending, with female flowers, **lower branches** spreading, with male flowers; **spikelets** with 1 floret. **Glumes** absent; **pistillate lemmas** 8–33 mm, hard or leathery, shiny, glabrous or hairy on veins, tips hairy, awns to 10 cm, aborted pistillate spikelets 0.6–2.6 mm wide. **Staminate lemmas** 6–17 mm, awns to 2 mm. **Habitat**: Lakes, marshes, streams. Native to E North America. **Comments**: Northern wild rice is a large aquatic grass grown as a food crop and planted for cover and food for wildlife. Most noncultivated populations in the PNW do not persist. Wild rice is culturally important to Indigenous people who harvest it using "beaters" to knock the grains into canoes.

Zizania aquatica L. var. *aquatica*
Southern Wild Rice [E] (green dots; not illustrated)

Differs from *Z. palustris* in having flexible lemmas with short, scattered hairs over the back, not concentrated at tips; upper panicle branches widely spreading at maturity.

BAMBOO

Bamboos are large, woody grasses. None are native in Oregon and Washington, and the ones brought in as ornamentals have not spread by seed. Thus, by one definition, bamboos should not be considered naturalized. However, tell that to someone who is trying to rid their property of a bamboo clone that arrived vegetatively. Established bamboo clones can be extremely difficult to eliminate. They persist in long-abandoned home sites, incrementally increasing in size. Not all bamboo clones are intentionally planted: some are carried downstream when floods tear rhizomes from riverbanks; some grow where deposited by earth-moving equipment in road construction; others establish from dumped yard debris. Bamboos can sneak under fences, sometimes even emerging though asphalt on nearby roads. However they start, they spread outward by rhizomes, outcompeting all adversaries. Even if one considers them only quasi-wild, as members of our wild habitats bamboos can no longer be ignored.

This treatment is not comprehensive. It includes only bamboo species recorded since the first edition of this field guide was published. We treat only species found "in the wild," or "feral," as we like to say. Generally, a bamboo must be in a location where it was not planted or appears to have been present a long time without horticultural care. All of our feral bamboos are "running bamboos." Their rhizomes have long internodes that allow the plants to spread widely. "Clumping bamboos" have very short rhizome internodes and appear to stay in well-behaved clumps. Given the wide variety of bamboos offered in the nursery trade, you may find other species escaping or persisting in the wild.

As members of the grass family, bamboos share many morphological features with our familiar grasses. But the features we use for identification focus on vegetative characters. Why? Because bamboos flower infrequently. Curiously, a particular species usually flowers within the same two- or three-year period all over the world. After flowering, nearly all the shoots die, although the rhizomes often survive to put up more shoots. If you find a flowering stand you may be able to identify it by asking at bamboo nurseries about which species are flowering at that time. Since the time between flowering can be as long as 65 or even 120 years, flowers are of little diagnostic importance for us. Fortunately, the vegetative characters show distinctive differences. We use culm and leaf characters to help us identify our feral bamboos.

Culms

Bamboo culms (canes in the nursery trade) have hollow internodes and solid nodes. Leaves and branches attach at the nodes. The overall growth form of culms may be straight, curved, arching, or zigzag.

Internodes vary in length and may be round or longitudinally flattened or grooved. The groove is the scar left behind from branch formation during the growth process, so it appears only on internodes above nodes with branches. (In a mature bamboo stand, shading can prevent branching on the lower culms and the lower internodes are not groovy.) Internode surfaces may be variously colored (green, yellow, gray, black, striped, or blotchy), shiny or dull, waxy or glaucous, smooth or rough. Caution: a layer of tiny lichens or fungi may make assessing culm roughness tricky. Culms may be distinctly rough in the lower six feet or so but smooth above; these count as rough. Roughness wears off with age. Internodes may be swollen below the nodes with a bulge or ridge, and they may have a waxy white ring that fades with age. Internodes of emerging shoots are covered with culm leaf sheaths, which either fall off, exposing the culm, or persist, covering all or part of the culm internode.

Nodes have useful traits. The lowest is the nodal ring or leaf sheath scar, what's left after the sheath falls. The bud that becomes the node starts within the sheath, so it is a bit narrower than the scar but usually becomes wider as it grows, flaring out to the nodal ridge, where branches originate. Depending on the species, the nodal ridge may be a very gentle swell of the culm or a clearly defined ridge.

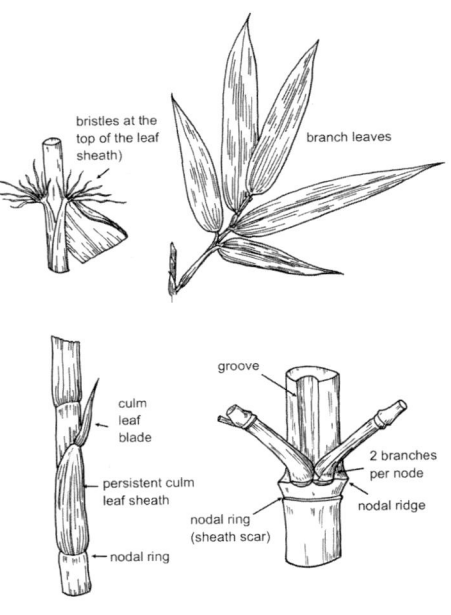

Leaves

Bamboos have two kinds of leaves: culm leaves and branch leaves. Culm leaves attach at the culm nodes. They have a sheath that wraps around the culm and a small blade at the top of the sheath. Culm leaves are either deciduous or persistent. The persistent ones are obvious from the old sheaths covering the culm internodes. The internodes become fully visible when the old sheaths fall off.

Branch leaves attach to the branches that arise from the nodes on the culm. Important characters are size and shape, and the presence or absence of long, hairlike bristles at the junction of the sheath and blade. The bristles can spread or be parallel to the branch and they may vary in color. They are fragile and wear off with age, making their absence an unreliable character. Always measure only healthy well-developed branch leaves.

Key to Bamboos

1a Branches 1 per node; internodes terete; culm sheaths persistent
  2a Branch leaves broadly oblong, 4–5(10)× as long as wide; culms blackish or mottled with black or dark brown
    3a Branch leaves to 25 cm, dark green, margins with a broad, white edge, veins on each side of midrib 5–9 .... **Sasa veitchii**
    3b Branch leaves to 40 cm, green with a yellow midrib, the margins sometimes becoming narrowly scarious and pale with age, veins on each side of midrib 8–14 .... **Sasa palmata** f. **nebulosa**
  2b Branch leaves linear-oblong to lanceolate, 6–16× as long as wide; culms variable in color but not mottled or black
    4a Plants to 5.5(6) m; lower surface of leaves glabrous ....***Pseudosasa japonica***
    4b Plants to about 1.5 m; lower surface of leaves with sparse hairs throughout...................... **Sasaella ramosa** (not treated)
1b Branches more than 1 per node; internodes terete or with a groove or flat side; culm sheaths deciduous
  5a Branches normally 2 per node; culm internodes with a groove or flat side extending from one node to the next
    6a Internodes widening abruptly about 1 cm below the sheath scar (feel for the swelling if it isn't obvious); the lower meter or so of some culms in each clump with 1 to several anomalously short internodes .... **Phyllostachys aurea**
    6b Internodes not abruptly widening below the sheath scar; lower internodes similar in length or gradually lengthening from base of culm
      7a Culms of older stems blackish or strongly marked with brown to blackish blotches; culms smooth, shiny ....***Phyllostachys nigra* ssp. *nigra***
      7b Culms of older stems green or yellow, sometimes striped green and yellow; culms smooth or rough, shiny or not
        8a Branch leaves shiny, 12–23 cm; culms shiny, green, not glaucous; branches perpendicular to the culm, the foliage appearing clearly layered.. **Phyllostachys vivax**
        8b Branch leaves dull, mostly 7–17 cm; culms dull, often glaucous; branches not perpendicular to the culm, the foliage not appearing layered.. ***Phyllostachys nigra* ssp. *henonis***
  5b Branches 3 or more per node; culm internodes terete, sometimes with a short groove not extending from one node to the next
    9a Culms smooth, green, becoming reddish in fall; leaves shiny; lower branches leafing out before the upper branches ....***Semiarundinaria fastuosa***
    9b Culms slightly rough, persistently glaucous, grayish; leaves dull; upper branches leafing out before the lower branches............... **Bashania fargesii**

Bashania fargesii (E.G. Camus) Keng f. & T.P. Yi
Windbreak Bamboo [E]

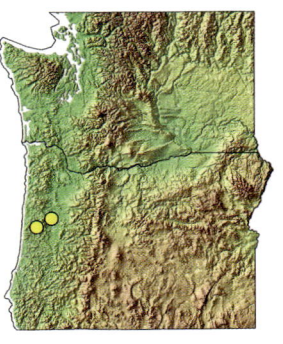

Culms to 7(9) m, to 5 cm in diameter, nearly straight; internodes to about 50 cm; terete or with short groove above origin of branches (groove not extending completely from one node to the next), very slightly rough, persistently glaucous, gray green. White waxy ring below nodes usually indistinct, blending into glaucous surface of internodes above and below, often wearing off. **Sheaths of culm leaves** semi-persistent, loosening on the sides but remaining attached in the middle, covering the base of the branches and falling tardily. **Branches** 3–8 per node, upper branches leafing out before the lower branches. **Branch leaves** 2–3 per distal branch; bristles at tops of sheaths usually lacking, or few, parallel to the sheath; longer leaf blades (14)19–26 cm, 1.4–3.1 cm wide, 6.5–10.5× as long as wide, slightly glaucous, dull. **Comments**: Similar to *Semiarundinaria fastuosa,* which has shiny leaves and green culms that turn reddish in autumn. The internodes of *Bashania* are the longest among our bamboos.

Phyllostachys aurea Carrière ex Rivière & C. Rivière
Golden Bamboo, Fishpole Bamboo [E]

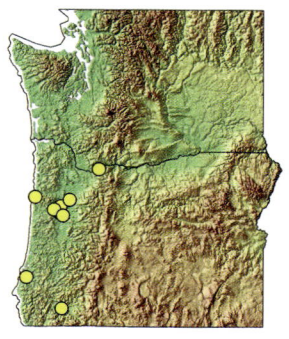

Culms to 10 m, 1–4 cm in diameter, straight or nearly so, erect; internodes grooved, smooth, green turning yellow (occasionally striped, rarely blotchy), usually expanded ~1 cm below the sheath scar (feel for subtle ridge or bulge); the lower meter or so of some culms in each clump with 1 to several anomalously short internodes. White waxy ring below node moderately prominent, yellowing in direct sunlight. **Sheaths of culm leaves** deciduous. **Branches** 2 per node. **Branch leaves** (1)2–3(7) per distal branch; bristles at tops of sheaths lacking or spreading, brown, rough; longer leaf blades 6–15 cm, 5–23 mm wide, 5–12.5× as long as wide, green or yellowish, not glaucous. **Comments**: *Phyllostachys aurea* is our most commonly planted bamboo. It is notoriously aggressive and considered invasive by the US Dept. of Agriculture.

Phyllostachys nigra (Lodd. ex Lindl.) Munro var. *henonis* (Mitford) Rendle
Black bamboo, Henon [E]

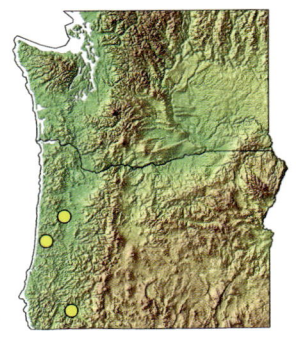

Culms to 14(20) m, to 9 cm diameter, nearly straight, often arching above; internodes grooved, hairy, and rough when young, then somewhat smooth but not slick when older, glaucous, olive green or grayish, not expanded below the sheath scar; lower internodes gradually lengthening from base of culm. White waxy ring below nodes absent or not prominent. **Sheaths of culm leaves** deciduous. **Branches** 2 per node. **Branch leaves** (1)2(4) per distal branch; bristles at tops of sheaths lacking, or spreading, brown, rough; longer leaf blades (5.5)7–17 cm, 0.8–2 cm wide, 6–14× as long as wide, dull, green, slightly glaucous. Foliage not appearing layered. **Comments**: Compare to *P. vivax* which has foliage that appears layered. *Phyllostachys nigra* var. *nigra* has shorter, slender culms that are so smooth they look polished. Its culms become black with age or develop brown blotches. There are no known populations of *P. n.* var. *nigra* in the wild in OR and WA.

Phyllostachys vivax McClure
Chinese Timber Bamboo [E]

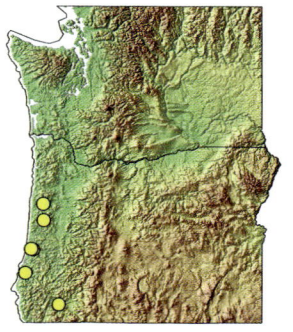

Culms to 12–15(21) m, to 9–13 cm diameter, not quite straight; internodes grooved, smooth, glossy, not glaucous on distal part of internodes, green to yellow-green; not expanded below the sheath scar; lower internodes similar in length or gradually lengthening from base of culm. White waxy ring below node absent or not prominent. **Sheaths of culm** leaves deciduous. **Branches** 2 per node. **Branch leaves** 1–5(6) per distal branch; bristles at tops of sheaths lacking, or spreading, brown, rough; longer leaf blades (9)12–23 cm, (1)1.5–2.5 cm wide, 7.3–13× as long as wide, shiny, green, not glaucous. Foliage appearing layered. **Comments**: *P. vivax* has the longest leaves of our quasi-wild *Phyllostachys*. *Phyllostachys vivax* and *P. nigra* var. *henonis* both can become massive after decades with adequate moisture W of the Cascades. They differ in general appearance, with Henon having dull, somewhat glaucous stems and slightly smaller, dull leaves.

Pseudosasa japonica (Siebold & Zucc. ex Steud.) Makino ex Nakai
Arrow Bamboo, Yadake [E]

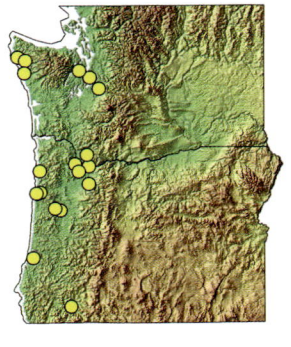

Culms to 5.5(6) m, 1–2.5 cm in diameter, straight; internodes to about 30 cm, terete, smooth, not glaucous, green to brownish or straw-colored. Nodes often oblique; white, waxy ring below node narrow, not prominent, often wearing off. **Sheaths of culm** leaves long-persistent, usually more than 75% as long as the internode. **Branches** 1 per node. **Branch leaves** (2)3–7 per distal branch; bristles at top of sheath usually lacking, or white, smooth, and parallel to the sheath; leaf blades linear-oblong to narrowly lanceolate, tapering to a point, longer blades 15–40 cm, 1.5–5.2 cm wide, 6.5–16× as long as wide, green, not glaucous above, lower surface glabrous. **Comments**: Perhaps the most commonly encountered bamboo found outside cultivation in the PNW; some spreading stands are the only evidence of a former homesite.

Sasa palmata (Burb.) E.G. Camus f. *nebulosa* J. Houz.
Broad-leaved Bamboo, Nebulosa [E] (yellow dots)

Culms 1.5–2.5(3.5) m, about 1 cm diameter; internodes terete, smooth, somewhat glaucous, blackish or straw-colored with purple or dark brown blotches. White, waxy ring below nodes usually present. **Sheaths of culm leaves** persistent, generally 75% as long as internode or shorter. **Branches** 1 per node. Sheaths of branch leaves minutely hairy toward the tip and base. **Branch leaves** 3–4 per distal branch; bristles at tops of sheaths spreading and rough, rarely absent; longer leaf blades broadly oblong, (17)24–36(40) cm, (2.5)4.5–8 cm wide, 4.5–5(10)× as long as wide, with 8–14 veins on each side of midrib, green with a yellow midrib, the margins sometimes becoming narrowly scarious with age. **Comments**: The broad leaves are used as food wrappers in Asian cultures.

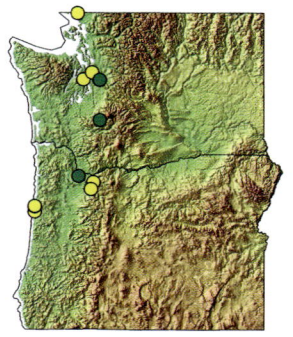

Sasa veitchii (Carrière) Rehder
Kuma Bamboo [E] (green dots)

Similar to *S. palmata*, but shorter (0.5–1.5 m), with shorter, dark green leaves with broad white margins, with 5–9 veins on each side of the midrib. Native to Japan.

Semiarundinaria fastuosa (Lat.-Marl. ex Mitford) Makino
Temple Bamboo, Narihira Bamboo [E]

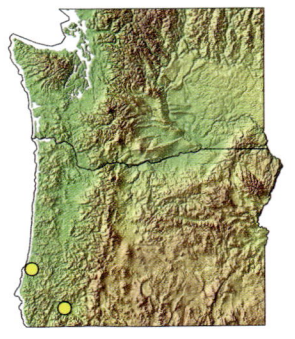

Culms to 6–7.5 m, to 4 cm diameter, nearly straight; internodes terete or with short groove above origin of branches (groove not extending completely from one node to the next), smooth and appearing polished, not glaucous, green turning reddish in autumn. White waxy ring below nodes not prominent, usually wearing off. **Sheaths of culm leaves** deciduous, but often hanging down from their bases for some time before falling off. **Branches** 3–8 per node, densely tufted; lower branches leafing out before the upper branches. **Branch leaves** 3–5 per distal branch, bristles at tops of sheaths usually parallel to sheath, or lacking; longer leaf blades (9)15–22 cm, 1.5–2(2.3) cm wide, 8–13× as long as wide, shiny, not glaucous. **Comments**: The densely tufted branches give a distinctive, columnar-bushy look to this bamboo.

GLOSSARY OF SELECTED GRASS TERMINOLOGY

Aborted – failing to develop into a functional part, often referring to anthers.

Abscission layer – a row of thin-walled cells that develop across the stem (or petiole) at the base of a leaf, flower, or fruit, forming a weak zone for detachment from the plant.

Acuminate – tapering gradually to a point, with the sides concave.

Acute – tapering to a point, with more or less straight sides, forming an angle of less than 90 degrees.

Angled – with sharply defined edges, in contrast to round or terete.

Annual – living and dying within 12 months. Annual plants lack previous year's leaves near the base. Their roots are usually (not always) easy to pull from the ground. Annuals may be stoloniferous or decumbent, but not rhizomatous.

Anther – the part of the stamen where pollen is produced.

Anthesis – flowering: when the stigmas emerge and the anthers release pollen.

Antrorse – directed forward or upward.

Apex – tip (plural: apices).

Apical – at or near the tip.

Apomixis – asexual reproduction involving the flower (see page 13). Adjective: apomictic.

Appressed – pressed closely against another structure.

Aristate – tipped with a long slender bristle.

Ascending – sloping upward, but at a wider angle than appressed.

Auricle – a small flap or lobe where the leaf blade meets the top of the leaf sheath.

Awl-shaped – tapering evenly from base to tip (like a leather punch).

Awn – a slender bristle attached at the lemma or glume tip or on the back of the lemma.

Axil – the angle between the upper side of a leaf or stem and the supporting stem or branch. Axillary panicles are found inside the leaf sheath, nestled against the culm. These panicles arise from the node where the sheath attaches and often remain hidden (cleistogamous) and the flowers self-pollinate. See p. 13.

Bidentate – with two teeth.

Bifid – with two lobes or teeth.

Bisexual floret – a floret with both male and female reproductive parts (functional stamens and ovary with associated style and stigma).

Blade – the flat, folded, or involute part of the leaf that does not wrap around the culm.

Bract – a modified, usually reduced, leaf. Glumes and lemmas are examples of bracts.

Bristle – a stiff hair.

Bulblet – a tiny bulb, usually produced in the inflorescence instead of sexual parts.

Callus – the hard, thickened base of the lemma, where the lemma attaches to the rachilla.

Callus hairs – a tuft of hairs at the base of the lemma.

Caryopsis – a one-seeded fruit, consisting of an embryo fused to the endosperm. The grass "seed," "fruit," or "grain" (plural: caryopses).
Cauline – attached to the stem (culm) rather than at the base of the plant.
Cespitose – with the shoots densely clustered, forming a bunchgrass.
Ciliate – with hairs along the edge (like eyelashes).
Ciliolate – minutely ciliate.
Cleistogamous – adjective to describe self-pollinating flowers hidden inside the sheath along the culm or at the base of the plant. See p. 13.
Cobwebby hairs – hairs that are curled and twisted but can be stretched out much longer than they first appear.
Collar – the back of the junction of the leaf blade and sheath; that junction on the outside of the leaf (compare to ligule).
Condensed – used to describe an inflorescence in which the flowers are crowded together, usually because the panicle branches are short. Also called dense or congested.
Congested – dense, crowded, usually referring to panicles.
Contracted – less crowded or dense than congested and condensed, but narrower than open, often because the panicle branches are not spreading, which can vary based on phenology related to flowering (anthesis).
Corm – a swelling of the underground part of the culm that serves as a storage organ. Corms are wrapped in scale leaves and lack the concentric rings seen in cross-sections of bulbs.
Crenulate – with fine rounded teeth; crenate refers to rounded teeth.
Crisp or **crisped** (hairs or leaf margins) – irregularly curled, crooked, wavy, or wrinkled.
Crown – in *Nasella*, a short, membranous, cylindrical extension of the lemma tip with hairs on its upper edge. Also, a fringe of dense, short hairs at the top of a callus.
Culm – the stem-like determinate shoot that produces the grass inflorescence.
Decumbent – lying on the ground, with the growing tip curving upward.
Decurrent – attached along the side. A decurrent leaf blade extends down the stem; a decurrent ligule extends down the blade.
Dehiscent – opening normally (in an anther, opening indicates functionality).
Depauperate – stunted or poorly developed, not reaching normal size.
Digitate – having veins, lobes, branches, etc., all arising from at or near a single point, like fingers from a palm.
Dioecious – with male and female flowers occurring on separate plants. See also monoecious.
Disarticulation – coming apart at maturity along a natural abscission layer (see p. 11).
Distal – toward the tip of the plant or structure (compare to proximal).
Divergent – spreading apart.
Dorsiventrally flattened (spikelet) – flattened from the front and the back so that when the spikelet lies naturally on the table, only one glume can be seen (see p. 9). The backs of the glumes and lemmas are obvious (compare to terete and laterally flattened).
Endemic – native and restricted to a specific area.
Entire – a smooth edge or margin that is not toothed or otherwise cut.
Erect – term used to describe panicle branches that extend upward narrowly. Compare to branches that extend upward and outward, ascending or spreading.
Erose – with a ragged edge.
Exceeding – sticking out beyond.

Exserted – sticking out beyond surrounding parts.
Extravaginal (shoot) – basal branching in which a young shoot grows more or less perpendicular to an older shoot, breaking through the sheath of the next older leaf. Extravaginal shoots are usually seen on rhizomatous plants (compare to intravaginal).
Fascicle – a cluster or bundle.
Fertile – having female or male flower parts, or both (opposite of sterile).
Flag leaf – the uppermost leaf of the culm, below the terminal inflorescence.
Flexuous – sinuous, curving alternately in different directions.
Floret – the individual grass flower and its associated bracts, the lemma and the palea.
Free – remaining separate, not joined or fused to another part.
Fruiting, in fruit – when the seeds mature.
Fusiform – shaped like a spindle, with the widest part in the middle and tapering to a point at both ends. Florets of some needlegrasses have this shape.
Gabbro – a geology term, see ultramafic, serpentine.
Gelatinous – becoming sticky-slimy when wet.
Geniculate – sharply bent, like a knee.
Glabrescent – becoming nearly or quite glabrous.
Glabrous – lacking hairs.
Glaucous – covered with a waxy powder that gives a whitish or bluish cast to the surface. The waxy powder can be rubbed off and may evaporate from plants dried over heat.
Glume – either of the two bracts at the base of the spikelet. The lower glume is adjacent to the lowest floret; the upper glume is adjacent to the second floret. Most grass spikelets have two glumes; *Lolium* has one; *Leersia* and *Zizania* have none.
Grain – caryopsis.
Gynodioecious – having pistillate (female) and bisexual flowers on separate plants in a population or species.
Gynomonoecious – having both pistillate (female) and bisexual flowers on the same plant.
Hirsute – with stiff, coarse hairs.
Hyaline – thin and translucent.
Indehiscent – not opening (e.g., an anther) and thus, by implication, sterile.
Inflorescence – the uppermost part of a flowering culm, consisting of all the spikelets and the part of the culm to which they are attached. Some grasses may have axillary inflorescences in addition to terminal ones.
Innovation – A vegetative shoot that arises from the base of a grass plant. Tiller is a synonym used in agricultural settings.
Internode – the part of the culm, rachis, or other structure between the nodes.
Intravaginal (shoot) – basal branching in which the young shoot grows up inside the leaf sheath, emerging at the sheath mouth, paralleling the older shoot, and not breaking through the sheath. Intravaginal shoots are often present on cespitose plants.
Inrolled (leaves) – with both margins rolled inward toward the midrib on the upper side. Involute.
Involute (leaves) – with both margins rolled inward toward the midrib on the upper side. Inrolled.
Keel – a sharp ridge like the keel of a boat. May be applied to glumes, lemmas, leaf sheaths, and blades. As an adjective, keeled means shaped like the keel of a boat, with the midrib projecting.

Lacerate – with a jagged margin that appears torn.
Lanceolate – lance-shaped: widest below the middle and tapering to both ends.
Lanceoloid – the 3-dimensional equivalent of lanceolate: widest below the middle and tapering at both ends, lance-shaped.
Laterally flattened (of a spikelet) – flattened from the sides (see p. 9).
Lateral veins – veins between the midrib and the marginal veins.
Leaf – grass leaves are made up of a sheath and blade; at the junction of those two parts is the collar (back side), ligule and auricles (inside and sides). See p. 7.
Lemma – the lower of the two bracts of the floret. This bract more or less encloses the palea and the flower. The lemma may be awned.
Ligule – a membranous flap or line of hairs at the junction of the leaf sheath and blade, on the inner side next to the culm.
Lodicule – highly modified remnant of the petals and sepals (perianth) at the base of the ovary. The lodicule's function is to open the floret, pushing the lemma and palea apart so the anthers and stigmas can emerge to offer and accept pollen, respectively.
Marginal veins – veins nearest the side edges (of a lemma, often).
Membranous – thin and pliable, like a membrane or skin, often somewhat transparent.
Midrib or **midvein** – the vein running along the middle of a leaf, glume, or lemma. It is usually more prominent than the other veins and may extend beyond the tip of the glume or lemma as an awn.
Monoecious – with separate male and female flowers on the same plant. See also dioecious.
Mucronate – with a mucro, a sharp, short, abrupt point at the tip.
Nerve – vascular bundle with its associated fibers; same as vein.
Node – the point of attachment of grass leaf sheaths on the culm, or spikelets on the rachis in spike inflorescences.
Nodulose – with pale, blunt bumps that are thickenings of fibers in the culms.
Oblanceolate – widest above the midpoint between attachment and the apex, and tapering to both ends; differs from lanceolate in the position of the widest point.
Obtuse – with a relatively blunt tip forming an angle of more than 90 degrees.
Ovate – egg-shaped, widest below the middle and broadly tapering to each end; wider than lanceolate.
Ovoid – the 3-dimensional equivalent of ovate; widest below the middle.
Palea – the upper of the two bracts of the floret, usually enclosed by the lemma.
Panicle – a branched inflorescence in which the branches are also branched.
Papilla – a short, rounded, blunt projection, like the little bumps on the tongue (plural: papillae).
Papillate – descriptive of a surface covered with small, raised projections called papillae.
Papillose – covered with or resembling papillae.
Pedicel – the stalk that supports one spikelet.
Pedicellate – a structure born on a stalk, having a pedicel (opposite of sessile).
Perennial – living for more than one year. Perennial grasses have remnants of the last year's leaves or shoots near the base and are usually hard to pull from the ground.
Petiole – the stem that attaches a leaf blade to the plant.
Pilose – with sparse, thin, spreading hairs.
Pistillate – having only female flowers.
Plumose – feathery, or bearing feather-like hairs or bristles.

Prow-shaped – shaped like the prow of a canoe (describes the leaf tip in *Poa*).

Proximal – toward the base of the plant or structure (compare to distal).

Pseudogamous – requiring pollen to stimulate embryo development but without fertilization; occurring in apomictic flowers of some grasses. See the description of apomixis in the section on Sex and Other Reproductive Options on p. 12–13.

Pseudopetiole – the narrowed portion of the leaf located between the ligule and the blade in many bamboos, and a few other Poaceae (*Pharus*).

Pseudovivipary – an asexual reproductive strategy in which bulblets or plantlets develop in place of seeds in the spikelets.

Puberulent – with minute hairs.

Pubescent – hairy (technically, with short hairs, but used more broadly).

Pulvinus – an enlargement at the base of a pedicel or inflorescence branch that swells to push the branch away from the main axis (plural: pulvini).

Raceme – an inflorescence in which each spikelet attaches to a pedicel that attaches directly to the rachis.

Rachilla – the axis of a grass spikelet; think of it as the miniature rachis within the spikelet, thus the diminutive ending "illa" on rachis. The parts of the rachilla between the florets are the rachilla internodes. A rachilla may also extend beyond the base of the terminal floret.

Rachis – the main axis of the inflorescence.

Reduced – smaller than a normal-sized structure of the same type. Reduced structures are often nonfunctional (vestigial anthers or rudiments).

Reflexed – angled downward.

Remote – distant.

Retrorse – directed backward or downward.

Rhizomatous – having rhizomes, see below. Rhizomatous plants are perennial.

Rhizome – a modified underground stem that usually grows horizontally (but under some conditions can grow vertically, as in deep sand). A rhizome has nodes and at each node there is a scale, which is a vestigial leaf.

Rib – a raised vein on a leaf, glume, or lemma. The central rib is called the midrib.

Riparian – describes the moist zone along a stream, river, or lake.

Rosette – a circle of tightly packed leaves from a central point, usually at ground level.

Rudiment – a sterile structure in the spikelet consisting of several small empty lemmas enfolding each other and lacking reproductive parts (found in *Melica*).

Rudimentary – describing an undeveloped or sterile floret (see rudiment).

Rugose – wrinkled or corrugated.

Scaber – a small hard projection on a plant surface (sheath, lemma, panicle branch, etc.), an assemblage of which makes the surface feel rough.

Scabridulous – rough, but a finer grade of sandpaper than scabrous.

Scabrous – rough, like the feel of sandpaper to the touch.

Scabrule – a tiny scaber.

Scarious – thin, dry, and membranous.

Sclerenchyma – supporting tissue composed of thickened, dry, hardened cells found in many plant parts, such as the white fiber bundles in leaf cross sections.

Seed – propagule; in grasses, this term is (mis)used for the caryopsis, a fruit.

Serpentine – see ultramafic.

Serrate – toothed along the margin.

Sessile – attached directly to the branch or inflorescence axis, without a pedicel (stalk). Grass taxonomists often use "sessile" when the pedicel is present but very short.

Sheath – the lower part of the leaf, which wraps around the culm or younger leaves.

Sinuous – curving back and forth.

Spike – an elongate, unbranched inflorescence in which the spikelets attach directly to the rachis, without pedicels (the spikelets are sessile on the rachis).

Spikelet – basic grass inflorescence unit: glumes and one or more florets.

Spreading – diverging widely from, such as a panicle branch from the inflorescence axis.

Spring (or summer) **annual** – a plant that germinates in the spring and reproduces that summer, then dies in the fall.

Stamen – the male organ of a flower, consisting of the anther, which contains the pollen, and a slender filament.

Staminate – consisting only of male flowers.

Sterile – not producing seeds or pollen.

Stipe – a short stalk formed from a narrowing of the structure that it supports, e.g., the little peg found at the base of the glumes in *Polypogon*.

Stolon – an aboveground horizontal stem that can produce roots, stems, and leaves.

Stoloniferous – having stolons, see above.

Subglobose – shaped like a sphere that's been squished a bit.

Subsessile – nearly sessile, with a very short stalk.

Subtending – located directly below and close to, often enclosing.

Subterete – not quite round in x-section, like a cylinder that's been squished a bit.

Terete – round in x-section (more or less cylindrical).

Terminal – at the upper or distal end or tip.

Tiller – innovation. A vegetative shoot.

Truncate – blunt, cut straight across, sometimes at an oblique angle.

Tufted – describes annual grasses that have multiple shoots from the base.

Ultramafic – substrates derived from igneous rock that are high in magnesium and iron; common examples are gabbro, serpentinite, and peridotite. Soils derived from this parent material are low in calcium and high in magnesium, with relatively high concentrations of nickel, chromium, and other heavy metals, and low levels of nitrogen and calcium. They are poor substrates for most plants. Botanists often refer to these soils as serpentine.

Umbel – an inflorescence with the branches spreading from a common point, somewhat like umbrella ribs.

Variegated – in grasses, usually leaves with green and white stripes.

Vein – vascular bundle with its associated fibers; same as nerve.

Vestigial – poorly developed and no longer functional, e.g., as in a vestigial anther that is a fraction of the normal size, appears shriveled, and produces no pollen.

Viviparous – with offspring developing while still attached to the parent plant; see the description on page 13.

Waif – a non-native plant species that only survives in the wild for a few generations before disappearing.

Wing – flat margin, especially of a glume or lemma.

Winter annual – a plant that germinates in the fall and reproduces the next spring, then dies in the summer.

REFERENCES

Electronic Resources

Grasses and More website: https://www.grassesandmore.net/. (Check this site for study guides, updates for corrections and additions to this field guide.)

Burke Herbarium, University of Washington: http://biology.burke.washington.edu/herbarium/imagecollection.php. (Photo gallery of Washington plants.)

CalFlora (http://www.calflora.org/) (Photo gallery of California plants.)

Consortium of Pacific Northwest Herbaria: http://www.pnwherbaria.org/. (Range maps, scanned images of specimens.)

Grass Manual on the Web: http://herbarium.usu.edu/webmanual. (Keys, descriptions, range maps for grasses treated in the Flora of North America.)

Jepson Interchange: http://ucjeps.berkeley.edu/interchange/. (Keys, descriptions, range maps for California plants; links to the CalFlora photo gallery.)

OregonFlora: http://www.oregonflora.org. (Atlas and photo gallery for Oregon plants.)

PLANTS Database: https://plants.usda.gov/. (Accepted plant names, synonyms; standard reference for federal agencies.)

Plants of the World Online: https://powo.science.kew.org/. (Accepted plant names, synonyms.)

Soreng, R. J., P. M. Peterson, K. Romanschenko, G. Davidse, J. K. Teisher, L. G. Clark, P. Barberá, L. J. Gillespie, F. O. Zuloaga. 2017. A Worldwide Phylogenetic Classification of Poaceae (Gramineae) II: An update and comparison of two 2015 classifications. Journal of Systematics and Evolution. https://onlinelibrary.wiley.com/doi/10.1111/jse.12262

World Flora Online: https://www.worldfloraonline.org/. (Accepted plant names, synonyms.)

Printed Resources

Anderton, L. K., and M. E. Barkworth, eds. 2009. *Grasses of the Intermountain Region.* Logan: Utah State University Press.

Baldwin, B. G., D. H. Goldman, D. J. Keil, R. Patterson, T. J. Rosatti, and D. H. Wilken, eds. 2012. *The Jepson Manual: Vascular Plants of California*, 2nd ed. Berkeley: University of California Press. (Keys, descriptions, illustrations; grass taxonomy is non-standard.)

Barkworth, M. E., L. K. Anderton, K. M. Capels, S. Long, and M. B. Piep. 2007. *Manual of Grasses for North America.* Logan: Utah State University Press. (A condensation of the *Flora of North America* grass treatment; keys, pictures, maps, minimal text.)

Barkworth, M. E., K. M. Capels, S. Long, L. K. Anderton, and M.B. Piep, eds. 2007. *Magnoliophyta: Commelinidae (in part): Poaceae, Part 1. Flora of North America North of Mexico*, vol. 24. New York: Oxford University Press. (Keys, drawings, descriptions, range maps. All North American grasses are included in this and the next listing, vol. 25.)

Barkworth, M.E., K. M. Capels, S. Long, and M. B. Piep. 2003. *Magnoliophyta: Commelinidae (in part): Poaceae, Part 2. Flora of North America North of Mexico*, vol. 25. New York: Oxford University Press.

Carex Working Group. 2017. *Identification Key to Grasses of Northern California and Northwestern Nevada.* Corvallis, OR: Carex Working Group.

Clark, L. G., and R. W. Pohl. 1954–1996. *Agnes Chase's First Book of Grasses: The Structure of Grasses Explained for Beginners.* 4th ed. Washington, DC: Smithsonian Institution Press.

Clarke, I. 2015. *Name Those Grasses: Identifying Grasses, Sedges and Rushes.* Victoria, Australia: Royal Botanic Gardens Victoria. (Exquisite line drawings and photographs, especially useful for our warm season and non-native species.)

Cronquist, A., A. H. Holmgren, N. H. Holmgren, J. L. Reveal, and P. K. Holmgren. 1977. *Intermountain Flora: Vascular Plants of the Intermountain West, U.S.A.* Vol. 6, The Monocotyledons. New York: Columbia University Press. (Keys, descriptions, excellent illustrations, outdated nomenclature.)

Hitchcock, C. L. and A. Cronquist. 2018. *Flora of the Pacific Northwest: an Illustrated Manual.* 2nd ed. D. E. Giblin, B.S. Legler, P. F. Zika, and R. G. Olmstead, eds. Seattle: University of Washington Press.

Hitchcock, C. L., A. Cronquist, M. Ownbey, and J. W. Thompson. 1969. *Vascular Plants of the Pacific Northwest. Part 1: Vascular Cryptogams, Gymnosperms, and Monocotyledons.* Seattle: University of Washington Press. (The classic reference for the PNW; lovely illustrations, a key to vegetative grasses; outdated nomenclature.)

Meyers, S. C., T. Jaster, K. E. Mitchell, and L. K. Hardison, eds. 2015. *Flora of Oregon. Vol. 1, Pteridophytes, Gymnosperms, and Monocots.* Botanical Research Institute of Texas, Fort Worth, Texas.

Stace, C. 1997. *New Flora of the British Isles.* 2nd ed. Cambridge: Cambridge University Press. (Excellent for exotic grasses, especially west of the Cascades.)

INDEX

*Species accounts are in **bold** type.*

A

Achnatherum
  hendersonii. See *Eriocoma hendersonii*
  hymenoides. See *Eriocoma hymenoides*
  lemmonii. See *Eriocoma lemmonii*
  lettermanii. See *Eriocoma lettermanii*
  nelsonii ssp. *dorei.* See *Eriocoma nelsonii* ssp. *dorei*
  nelsonii ssp. *nelsonii.* See *Eriocoma nelsonii*
  nevadense. See *Eriocoma nevadensis*
  occidentale ssp. *californicum.* See *Eriocoma occidentalis* ssp. *californica*
  occidentale ssp. *pubescens.* See *Eriocoma occicentalis* ssp. *pubescens*
  pinetorum. See *Eriocoma pinetorum*
  richardsonii. See *Eriocoma richardsonii*
  speciosum. See *Pappostipa speciosa*
  thurberianum. See *Eriocoma thurberiana*
  wallowaense. See *Eriocoma wallowaensis*
  webberi. See *Eriocoma webberi*
Aegilops, 10, 19, 24, 40, 96, 97
  cylindrica, **96,** 448
  triuncialis, **97**
Agropyron, 24, 25, 40, 93, 98
  caninum. See *Elymus trachycaulus* ssp. *trachycaulus*
  caninum. See *Elymus violaceus*
  cristatum, 40, **98**
  dasystachyum. See *Elymus lanceolatus* ssp. *lanceolatus*
  dasystachyum. See *Elymus lanceolatus* ssp. *riparius*
  desertorum. See *Agropyron cristatum*
  elongatum. See *Thinopyrum obtusiflorum*
  fragile, 40, **98**
  intermedium. See *Thinopyrum intermedium*
  latiglume, misapplied to *Elymus violaceus*
  repens. See *Elymus repens*
  scribneri. See *Elymus scribneri*
  smithii. See *Pascopyrum smithii*
  spicatum. See *Pseudoroegneria spicata*
  trachycaulum. See *Elymus violaceus*
  triticeum. See *Eremopyrum triticeum*
  violaceum. See *Elymus violaceus*
Agrostis, 4,13, 21, 22, 30, 31, 41, 99–117
  alba. See *Agrostis stolonifera*
  blasdalei, 42, **99**
  canina, 41, 43, **100**
  capillaris, 41, 43, **101**
  castellana, 41–43, **102**
  densiflora, 42, 99, **103**, 104, 113
  diegoensis. See *Agrostis pallens*
  exarata, 13, 41, 42, 103, **104**, 111, 113, 172, 323, 434
  gigantea, 41, 43, 100, 102, **105**, 115
  hallii, 42, **106**, 113
  hendersonii, 41, **107**
  howellii, 42, 100, **108**, 116, 171
  humilis. See *Podagrostis humilis*
  idahoensis, 44, **109**, 110, 112, 114, 117
  interrupta. See *Apera interrupta*
  mertensii, 42, 43, **110**, 108, 114
  microphylla, 41, 104, 107, **111**
  oregonensis, 43, 44, 109, **112**, 114
  pallens, 42, 44, 103, 106, **113**, 116
  palustris. See *Agrostis stolonifera*
  scabra, 44, 109, 110, 112, **114**
  semiverticillata. See *Polypogon viridis*
  spica-venti. See *Apera spica-venti*
  stolonifera, 41–43, 100, 102, 105, **115,** 402
  swalalahos, 42, 108, **116**, 171
  tenuis. See *Agrostis capillaris*
  variabilis, 42, **117**, 322, 397
  verticillata. See *Polypogon viridis*
Aira, 24, 36, 37, 44, 118–120
  caryophyllea var. *capillaris.* See *Aira elegans*
  caryophyllea, 44, **118,** 141
  elegans, 44, **118**
  elegantissima. See *Aira elegans*
  praecox, 24, 44, **120**
Alopecurus, 9, 21, 44, 53, 77, 121–127
  aequalis var. *aequalis*, 45, **121***,* 123, 124, 127
  arundinaceus, 44, **122***,* 126
  carolinianus, 45, **123***,* 124, 127
  geniculatus, 45, **124***,* 127
  myosuroides, 21, 44, **125**

pratensis, 44, **126,** 122, 348
saccatus, 45, 123, 124, **127**
Ammophila
  arenaria ssp. *arenaria.* See *Calamagrostis arenaria* ssp. *arenaria*
  breviligulata ssp. *breviligulata.* See *Calamagrostis breviligulata* ssp. *breviligulata*
Anemanthele lessoniana, 32
Anthoxanthum, 13, 22, 23, 45, 70, 128, 129
  aristatum ssp. *aristatum,* 45, 120, **128**
  hirtum. See *Hierochloë odorata*
  nitens. See *Hierochloë odorata*
  occidentale. See *Hierochloe occidentalis*
  odoratum, 4, 22, 45, 128, **129**
Apera, 31, 41, 45, 130
  interrupta, 41, 45, **130,** 274
  spica-venti, 41, 45, **130**
Aristida, 10–12, 21, 31, 45,131, 132
  adscensionis, 46, 131
  longiseta var. *robusta.* See *Aristida purpurea* var. *longiseta*
  oligantha, 46, **131,** 132
  purpurea var. *longiseta,* 46, **132**
Arrhenatherum, 6, 37, 40, 46, 133
  elatius ssp. *bulbosum,* 37, 40, 46, **133**
  elatius ssp. *elatius,* 8, 40, 46, **133**
Arundo donax, 34, **134**
Avena, 36, 46, 135–137
  barbata, 46, **135,** *136*
  fatua, 46, 135, **136,** *137*
  sativa, 46, 136, **137**
Avenella flexuosa, 37

B
bamboo, 453
  arrow, 460
  black, 458
  broad-leaved, 461
  Chinese timber, 459
  fishpole, 457
  golden, 457
  kuma, 461
  narihira, 462
  temple, 462
  windbreak, 456
barbgrass, 286
barley, 70
  cultivated, 70, 295
  foxtail, 294, 296
  hare, 296, 297
  low, 293
  meadow, 292
  Mediterranean, 295
  mouse, 296
  smooth, 296, 297
  wall, 296, 297

Bashania fargesii, 455, **456**
beachgrass, 50
  American, 167
  European, 167, 304, 373
beardgrass, 88
  annual, 402
  Asian, 399
  ditch, 400
  Mediterranean, 401
  water, 403
Beckmannia syzigachne, 27, 29, **138**
bentgrass, 41, 88
  alpine, 397
  Arctic, 396
  Blasdale's, 99
  California, 103
  colonial, 101
  creeping, 115
  dryland, 102
  dune, 113
  Hall's, 106
  Henderson's, 107
  highland, 102
  Howell's, 108
  mountain, 117
  northern, 110
  Saddle Mountain, 116
  seashore, 113
  small-leaf, 111
  spike, 104
  thin, 113
  Thurber's, 398
  velvet, 100
bermudagrass, 8, 192
blackgrass, 125
Blepharidachne, 11
  kingii, 18, **139**
bluegrass, 78
  alkali, 386, 388
  alpine, 353
  annual, 354
  arctic, 355
  big, 386, 388
  Bolander's, 356
  bulbous, 357
  Canada, 359
  Canby, 386, 388
  Chamber's, 358
  coastline, 360
  Cusick's, 362
  dune, 373
  few-flowered, 380
  fowl, 379
  glaucous, 366
  Howell's, 367
  Kentucky, 382

Leiberg's, 370
Letterman's, 372
little mountan, 361
looseflower, 369
Multnomah, 376
narrow-flower, 389
Nevada, 386, 388
nodding, 384
northern, 389
Pacific, 386
pale Cusick's, 362
pine, 386
Piper's, 381
Pringle's, 383
purple Cusick's, 362
rhizome, 385
rough, 391
San Francisco, 392
Sandberg's, 386, 388
seacliff, 392
seashore, 373
skyline, 362
Steens Mountain, 374
Suksdorf's, 390
timberline, 366
veiny, 378
Wallowa, 394
weak, 368
weeping, 375
Wenatchee, 361
western bog, 371
Wheeler's, 395
withered, 375
woodland, 377
bluestem, little, 416
Bouteloua
   dactyloides, 17, 19, 27
   gracilis, 19, 27
Brachypodium sylvaticum, 23, **140,** 166
bristlegrass, 91
   hooked, 423
   Japanese, 420
   yellow, 421
Briza, 38, 46
   maxima, 46, **141,** 142, 144
   media, 47, **141**
   minor, 47, **141**, 142
brome, 47
   Aleutian, 157
   Australian, 143
   barren, 163
   California, 158, 161, 263
   Chinook, 152
   Columbia, 166
   downy, 165
   foxtail, 155

   fringed, 146
   Japanese, 151
   maritime, 160
   meadow, 147
   mountain, 159
   Orcutt's, 153
   Pacific, 154
   poverty, 163
   rattlesnake, 144
   red, 155
   ripgut, 148
   rye, 156
   Sitka, 158, 161, 253
   smooth, 150
   soft, 149
   squarrose, 162
   sterile, 163
   Suksdorf's, 164
Bromus, 2, 11, 18, 32, 47, 143–166
   aleutensis. See *Bromus sitchensis* var. *aleutensis*
   arenarius, 50, **143,** 151
   breviaristatus. See *Bromus sitchensis* var. *marginatus*
   briziformis, 20, 49, 141, **144**
   carinatus. See *Bromus sitchensis* var. *carinatus*
   carinatus var. *marginatus.* See *Bromus sitchensis* var. *marginatus*
   carinatus var. *maritimus.* See *Bromus sitchensis* var. *maritimus*
   catharticus var. *catharticus,* 47, 49, **145**
   ciliatus, 48, **146,** 152, 154, 166
   commutatus, 50, 143, **147,** 149, 151
   diandrus, 49, **148,** 163
   hordeaceus, 50, 147, **149**
   hordeaceus ssp. *molliformis.* See *Bromus hordeaceus*
   hordeaceus ssp. *pseudothominei.* See *Bromus hordeaceus*
   hordeaceus ssp. *thominei.* See *Bromus hordeaceus*
   inermis, 48, **150,** 164
   japonicus, 50, 143, **151,** 162
   laevipes, 49, 146, **152,** 153, 154, 164, 166
   madritensis ssp. *rubens.* See *Bromus rubens*
   madritensis, 49, 155
   marginatus. See *Bromus sitchensis* var. *marginatus*
   maritimus. See *Bromus sitchensis* var. *maritimus*
   mollis. See *Bromus hordeaceus*
   orcuttianus, 48, 49, **153,** 164, 166
   pacificus, 47, 49, **154,** 166

richardsonii var. *pallidus*. See *Bromus ciliatus*
rigidus. See *Bromus diandrus*
rubens, 49, **155**
secalinus, 50, 147, **156**
sitchensis, 47, 145, 154, 157–161, 164
sitchensis var. *aleutensis,* 48, **157**
sitchensis var. *carinatus,* 48, **158,** 159, 160
sitchensis var. *marginatus,* 48, 157, 158, **159,** 164
sitchensis var. *maritimus,* 47, **160**
sitchensis var. *sitchensis,* 47, 157, 160, **161**
squarrosus, 50, 143, 151, **162**
sterilis, 49, 148, **163**
suksdorfii, 48, 153, **164,** 166
tectorum, 40, 49, 143, 148, 163, **165,** 263
vulgaris, 48, 49 140, 154, **166,** 314
brookgrass, 182
broomcorn, 335
browntop, 101
Buchloe dactyloides. See *Bouteloua dactyloides*

C
Calamagrostis, 13, 22, 30, 31, 50–52
 arenaria, 304
 arenaria ssp. *arenaria,* **167**
 breviligulata ssp. *breviligulata,* **167**
 breweri, **168**
 californica. See *Calamagrostis stricta* ssp. *inexpansa*
 canadensis, **169,** 170, 175, 176
 canadensis var. *canadensis,* **169**
 canadensis var. *langsdorffii,* **169**
 crassiglumis. See *Calamagrostis stricta* ssp. *inexpansa*
 densa. See *Calamagrostis koelerioides*
 howellii, **171**
 inexpansa. See *Calamagrostis stricta* ssp. *inexpansa*
 koelerioides, 172, **172,** 173, 175, 180
 neglecta. See *Calamagrostis stricta* ssp. *stricta*
 nutkaensis, 172, **173**
 purpurascens, 174, 175, 178
 rubescens, **175**
 stricta, **176**
 stricta ssp. *inexpansa,* **176,** 177
 stricta ssp. *stricta,* **176,** 177
 tacomensis, **178,** 180
 tweedyi, **179**
 utsutsuensis, **180**
Calamovilfa, 93
 longifolia var. *longifolia,* 30, **181**

canarygrass, 76
 annual, 345
 bulbous, 343
 California, 345
 hooded, 13, 346
 narrow, 343
 reed, 13, 76, 169, 344
Carex
 douglasii, 326
 filifolia, 257
 geyeri, 175
 praegracilis, 326
Catabrosa aquatica, 30, 34, **182,** 444
Cenchrus, 10, 13, 20, 28, 52, 183
 longisetus, 18, 52
 longispinus, 17, 28, 52, 183, **183**
 orientalis, 52
 setaceus, 52
 spinifex, 183
 villosus. See *Cenchrus longisetus*
cheatgrass, 2, 40, 165, 440
chess
 hairy, 147
 soft, 149
Chloris, 13
 verticillata, 27, **184**
Cinna latifolia, 30, 32, **185**
Coleanthus subtilis, 29, **186**
Coleataenia
 longifolia ssp. *rigidula*. See *Panicum rigidulum* ssp. *rigiduum*
 rigidula ssp. *rigidula*. See *Panicum rigidulum* ssp. *rigidulum*
cordgrass, 92
 alkali, 430
 dense-flowered, 429
 English, 428
 prairie, 432
 saltmeadow, 431
 smooth, 427
corn, 2, 11, 12, 19, 425, 451
Cortaderia, 13, 34, 53, 187, 188, 445
 jubata, 53, **187**
 selloana, 53, 187, **188**
Corynephorus canescens, 37, **189**
crabgrass, 2, 8, 12, 56
 hairy, 208
 smooth, 207
Crypsis, 11, 21, 22, 53, 93, 190 191, 411
 alopecuroides, 53, **190,** 191
 schoenoides, 53, **191**
 vaginiflora, 53, 139, **191**
cutgrass, rice, 299

Cynodon, 8
   dactylon, 27, 56, 184, **192**, 208
Cynosurus, 13, 19, 35, 54, 193, 194
   cristatus, 54, **193**
   echinatus, 54, **194**

D

Dactylis glomerata, 33, 195
dallisgrass, 12, 76, 341
Danthonia, 13, 22, 35, 37, 54, 90, 196–200
   americana. See *Danthonia californica*
   californica, 54, 129, **196**, 200
   californica var. *americana.* See *Danthonia californica*
   decumbens, 35, 54, **197**
   intermedia, 54, **198**, 199
   spicata, 54, 198, **199**
   unispicata, 18, 54, 196, 197, **200**
darnel, 2, 307
deergrass, 327
Deschampsia, 37, 55, 201–204, 449
   atropurpurea. See *Vahlodea atropurpurea*
   bolanderi, 20, 55, **201**, 286, 339
   caespitosa ssp. *holciformis.* See *Deschampsia cespitosa*
   cespitosa, 13, 55, **202**, 292
   cespitosa ssp. *beringensis.* See *Deschampsia cespitosa*
   cespitosa ssp. *cespitosa.* See *Deschampsia cespitosa*
   danthonioides, 36, 37, 55, **203**, 204, 450
   elongata, 55, **204**
Dichanthelium, 9, 29, 55, 75, 205, 206
   acuminatum ssp. *fasciculatum,* 55, **205**
   lanuginosum. See *Dichanthelium acuminatum* ssp. *fasciculatum*
   oligosanthes ssp. *scribnerianum,* 55, 205, **206**
   scribnerianum. See *Dichanthelium oligosanthes* ssp. *scribnerianum*
Digitaria, 8, 28, 56, 184, 192, 207, 208, 341
   ischaemum, 56, **207**
   sanguinalis, 56, 207, **208**
Diplachne, 28, 37, 38
   fusca, 56
   fusca ssp. *fascicularis,* 28, 40, 56, **209**
   fusca ssp. *uninervia,* 40, 56, **209**
Distichlis, 12
   spicata, 35, 197, 198, **210**
   stricta. See *Distichlis spicata*
dogtail, 13, 54
   bristly, 194
   crested, 193
   hedgehog, 194

dropseed, 93
   puffsheath, 438
   rough, 436
   sand, 437

E

Echinochloa, 10, 28, 56, 138, 211, 212
   crus-galli, 56, **211**, 212
   muricata var. *microstachya,* 56, 211, **212**
Ehrharta erecta, 26, **213**
elephantgrass, 445
Eleusine, 27
   indica, 27, 192
   tristachya, 27
Elymus, 2, 4, 24–26, 40, 56–59, 71, 214–227, 306, 442
   alaskanus. See *Elymus violaceus*
   arenicola. See *Leymus flavescens*
   canadensis var. *canadensis,* 59, **214**
   caput-medusae. See *Taeniatherum caput-medusae*
   cinereus. See *Leymus cinereus*
   condensatus var. *condensatus.* See *Leymus condensatus*
   curvatus, 58, **215**
   elymoides, 11, 58, 216, 222, 224, 294
   elymoides ssp. *brevifolius,* 58, **216**, 217
   elymoides ssp. *californicus.* See *Elymus elymoides* ssp. *elymoides*
   elymoides ssp. *elymoides,* 58, **216**, 217
   elymoides ssp. *hordeoides,* 24, 58, **216**, 217, 222
   flavescens. See *Leymus flavescens*
   glaucus ssp. *glaucus,* 58, 59, **218**, 219
   glaucus ssp. *jepsonii.* See *Elymus glaucus* ssp. *glaucus*
   glaucus ssp. *virescens,* 58, 59, **218**
   hirsutus, 58, **219**
   lanceolatus, 57, **220**, 223, 225, 303, 306, 340
   lanceolatus ssp. *lanceolatus,* 57, 221, **220**
   lanceolatus ssp. *psammophilus,* 57, **220**, 221, 303
   lanceolatus ssp. *riparius,* 57, **220**, 221
   mollis. See *Leymus mollis* ssp. *mollis*
   multisetus, 58, 216, **222**
   ponticus. See *Thinopyrum obtusiflorum*
   repens, 57, **223**
   scribneri, 57, **224**
   smithii. See *Pascopyrum smithii*
   spicatus. See *Pseudoroegneria spicata*
   trachycaulus ssp. *trachycaulus,* 58, 220, **225**, 226
   triticoides. See *Leymus triticoides*
   violaceus, 57, **226**

virescens. See *Elymus glaucus* ssp. *virescens*
virginicus var. *submuticus.* See *Elymus curvatus*
wawawaiensis, 26, 57, **227**, 404
Elytrigia
    elongata. See *Thinopyrum obtusiflorum*
    intermedia. See *Thinopyrum intermedium*
    pontica. See *Thinopyrum obtusiflorum*
    repens. See *Elymus repens*
    spicata. See *Pseudoroegneria spicata*
Eragrostis, 11, 35, 59, 228–235
    cilianensis, 599, **228**, 243
    curvula, 59, **229**
    hypnoides, 59, **230**
    lutescens, 59, **231**
    mexicana ssp. *virescens,* 60, **232**
    minor, 59, **233**
    pectinacea, 232, 233, 235
    pectinacea var. *miserrima,* 60, **234**
    pectinacea var. *pectinacea,* 60, **234**
    pilosa var. *pilosa,* 59, **235**
Eremopyrum triticeum, 25, **236**
Eriocoma, 32, 33, 60, 237–249, 338
    hendersonii, 41, 61, **237**
    hymenoides, 61, **238**, 412
    lemmonii ssp. *lemmonii,* 61, **239**, 240
    lettermanii, 60
    nelsonii, 61, **240**, 243
    nelsonii ssp. *dorei,* 61, **240**, 241
    nelsonii ssp. *nelsonii,* 61, **240**, 241, 243
    nevadensis, 62, 240, **242**, 243
    occidentalis, 240, **243**
    occidentalis ssp. *californica,* 61, 240–242, **243**, 244
    occidentalis ssp. *pubescens,* 61, 62, **243**, 244
    pinetorum, 61, **245**, 249
    richardsonii, 61, **246**
    speciosa. See *Pappostipa speciosa*
    thurberiana, 60, **247**, 257, 338
    wallowaensis, 61, 237, **248**
    webberi, 61, **249**

F
false oat, 69
    Wolf's, 285
falsebrome, 2, 140, 166
feathergrass, 75
    Mexican, 331
feathertop, 52
fescue, 5, 62–64
    alpine, 251
    bearded, 268
    brome, 252
    California, 253
    Chewing's, 261
    Colorado, 251
    confused, 258
    covar sheep, 271
    crinkle-awn, 269
    desert, 259
    Elmer's, 255
    foxtail, 260
    green, 272
    hair, 256
    hard, 270
    Idaho, 247, 257, 328
    inland Roemer's, 264
    meadow, 415
    mountain rough, 254
    Oregon, 264
    rattail, 252, 260
    red, 266
    Rocky Mountain, 267
    Roemer's, 2, 264
    sand, 250
    sheep, 270
    sixweeks, 263
    small, 259
    spike, 300
    tall, 90, 414
    Washington, 273
    western, 262
Festuca, 4, 12, 13, 18, 24, 28, 33, 38, 39, 62, 90, 250–273
    altaica. See *Festuca campestris*
    ammobia, 63, 66, **250**, 266
    arida. See *Festuca microstachys*
    arundinacea. See *Schedonorus arundinaceus*
    brachyphylla ssp. *brachyphylla,* 63, 67, **251**, 267
    brachyphylla ssp. *coloradensis,* 63, 66, **251**, 267
    brevipila. See *Festuca trachyphylla*
    bromoides, 64, **252**, 259, 260
    californica, 65, 66, **253**, 254
    campestris, 66, **254**
    capillata. See *Festuca filiformis*
    confinis. See *Leucopoa kingii*
    confusa. See *Festuca microstachys*
    eastwoodiae. See *Festuca microstachys*
    elatior var. *arundinacea.* See *Schedonorus arundinaceus*
    elatior var. *elatior.* See *Schedonorus pratensis*

elmeri, 65, **255**, 268
filiformis, 63, 66, **256**
glauca, 258
grayi. See *Festuca microstachys*
idahoensis var. *idahoensis.* See *Festuca idahoensis*
idahoensis var. *roemeri.* See *Festuca roemeri*
idahoensis, 63, 67, 254, **257**, 262, 264, 267, 270–272
kingii. See *Leucopoa kingii*
lemanii, 63, 64, 68, **258**, 270, 271
megalura. See *Festuca myuros*
microstachys, 64, 252, **259**, 263
myuros, 64, 252, **260**
nigrescens, 64, 67, 68, **261**, 264, 266
occidentalis, 63, 67, **262**
octoflora, 64, **263**
ovina, misapplied to *Festuca lemanii, F. trachyphylla, F. valesiaca*
ovina ssp. *saximontana.* See *Festuca saximontana*
ovina var. *brachyphylla.* See *Festuca brachyphylla*
ovina var. *capillata.* See *Festuca filiformis*
pacifica. See *Festuca microstachys*
perennis. See *Lolium perenne*
pratense. See *Schedonorus pratensis*
reflexa. See *Festuca microstachys*
roemeri, 64, 65, 67, 257, 261, 262, **264**, 265, 270
roemeri var. *klamathensis,* 63, 65, **264**
roemeri var. *roemeri,* 63, 65, **264**
rubra ssp. *commutata.* See *Festuca nigrescens*
rubra, 6–68, 250, 261, 262, **266**, 272, 273
saximontana, 63, 67, 251, **267**
saximontana var. *purpusiana,* 67, **267**
saximontana var. *saximontana,* 67, **267**
scabrella. See *Festuca campestris*
subulata, 65, 255, **268**, 269, 284
subuliflora, 255, 268, **269**, 284
temulenta. See *Lolium temulentum*
trachyphylla, 63, 64, 68, 257, 258, **270**, 271
valesiaca, 63, 64, 68, 257, 258, 270, **271**
viridula, 33, 63, 65, 66, **272**
washingtonica, 63, 65, 67, **273**
fountaingrass, 52
foxtail, 9, 91
  black, 122
  creeping meadow, 122
  giant, 420
  giant green, 424
  green, 424
  meadow, 9, 44, 126
  Pacific meadow, 127
  shortawn, 121
  slender meadow, 125
  tufted, 123
  water, 124
  yellow, 421

G
Gastridium
  phleoides, 21, **274**, 346, 402
  ventricosum, misapplied to *Gastridium phleoides*
Glyceria, 13, 34, 68, 89, 94, 182, 275–282, 443, 444
  borealis, 67, **275**
  canadensis, 279
  canadensis var. *canadensis,* 69, **276**
  declinata, 69, **277**, 280
  elata, 69, **278**, 282
  erecta. See *Torreyochloa*
  fluitans, 69, **280**, 281
  grandis, 276
  grandis var. *grandis,* 69, **279**
  leptostachya, 69, 275, **280**, 281
  maxima, 68, 276, 279
  × *occidentalis,* 69, 277, **280**, 281
  striata, 69, 278, **282**
glyceria, low, 277
goatgrass, 2, 10, 40
  barbed, 97
  jointed, 96, 97
goosegrass. Common name for *Eleusine*
grama, blue. Common name for *Bouteloua gracilis*
Graphephorum, 37, 69, 283–285, 447
  canescens, 69, 268, **283**, 284, 285
  cernuum, 70, 268, **284**, 283, 285
  wolfii, 24, 37, 39, 69, **285**
grass
  American barnyard, 212
  alkali, 89
  barnyard, 12, 56, 211
  beardless rabbitsfoot, 403
  big quaking, 141
  blowout, 412
  buffalo. Common name for *Bouteloua dactyloides*
  bunny tail. Common name for *Lagurus ovatus*
  cattail, 91, 421
  common saltmarsh, 407
  European alkali, 405
  eyelash, 139

fairgrounds, 417
Garrison, 122
hairy wallaby, 413
Harding, 76, 343
hare's tail. Common name for *Lagurus ovatus*
Johnson, 91, 426
King's Eyelash, 139
Lemmon's alkali, 406
little quaking, 141
moss, 186
mud, 186
nit, 274
North Africa, 450
Nuttall's alkali, 409
orchard, 195
Pacific alkali, 408
Pampas, 53, 188
perennial quaking, 141
pheasant tail. Common name for *Anemanthele lessoniana*
poverty, 413, 439
purple Pampas, 187
quaking, 46
rabbitsfoot, 88, 402
rattlesnake, 141
rescue, 145
rosette, 55
Scribner's, 201
Scribner's rosette, 206
smooth alkali, 410
tapered rosette, 205
tumble windmill, 184
vanilla, 289
wallaby, 90, 413
western alkali, 411
whitetop, 418
winter, 332

H
Hainardia cylindrica, 20, **286**, 339
hairgrass, 44, 55
　annual, 203
　elegant, 118
　gray, 189
　mountain, 449
　silver, 118
　slender, 204
　spike, 120
　tufted, 202
hardgrass, 417
heathgrass, common, 197
Henon, 458, 459
Hesperochloa kingii. See *Leucopoa kingii*

Hesperostipa, 60
　comata ssp. *comata,* 33, **287**
Hierochloë, 13, 36, 38, 70, 128, 288, 289
　occidentalis, 70, **288**
　odorata, 70, **289**
Holcus, 36, 70, 290, 291
　lanatus, 70, **290**, 291, 344
　mollis ssp. *mollis,* 70, 290, **291**
Hordeum, 24, 70, 216, 292–297, 306, 440
　brachyantherum, 71, **292**
　depressum, 71, **293**, 295
　glaucum. See *Hordeum murinum* ssp. *glaucum*
　gussoneanum. See *Hordeum marinum* ssp. *gussoneanum*
　jubatum ssp. *jubatum,* 71, **294**
　leporinum. See *Hordeum murinum* ssp. *leporinum*
　marinum ssp. *gussoneanum,* 71, **295**
　murinum, 71, 295, 296, 297
　murinum ssp. *glaucum,* **296–297**
　murinum ssp. *leporinum,* **296–297**
　murinum ssp. *murinum,* **296–297**
　vulgare ssp. *vulgare,* 70, **295**, 296

J
Jarava speciosa. See *Pappostipa speciosa*
jubatagrass, 187
junegrass, 298

K
knotgrass, 76, 342
Koeleria
　cristata. See *Koeleria macrantha*
　macrantha, 23, 24, 37, 39, 172, **298,** 433, 434, 447
　nitida. See *Koeleria macrantha*
　pyramidata. See *Koeleria macrantha*
　spicata. See *Trisetum spicatum*

L
Lagurus ovatus, 18
Leersia, 9, 13, 94, 466
　oryzoides, 29, **299**
Leptochloa. See *Diplachne*
　fascicularis. See *Diplachne fusca* ssp. *fascicularis*
　fusca ssp. *fascicularis.* See *Diplachne fusca* ssp. *fascicularis*
　fusca ssp. *uninervia.* See *Diplachne fusca* ssp. *uninervia*
　uninervia. See *Diplachne fusca* ssp. *uninervia*

Leucopoa, 62
 kingii, 39, 64, 65, **300**
Leymus, 24, 25, 37, 40, 57, 71, 301–306
 cinereus, 71, **301,** 302, 442
 condensatus, 23, 71, 301, **302**
 flavescens, 23, **303**
 multicaulis, 306
 mollis ssp. *mollis,* 25, 71, 167, **304**
 racemosus, 71, **305**
 triticoides, 26, 71, 302, 303, **306,** 340
Lolium, 9, 20, 22, 62, 72, 90, 296, 307, 308, 466
 arundinaceum. See *Schedonorus arundinaceus*
 multiflorum. See *Lolium perenne* ssp. *multiflorum*
 perenne ssp. *perenne,* 72, **307–308**
 perenne ssp. *multiflorum,* 72, **307–308**
 temulentum ssp. *temulentum,* 72, **307–308**
lovegrass, 59
 India, 235
 little, 233
 Mexican, 232
 sixweeks, 231
 teal, 230
 tufted, 234
 weeping, 229

M
Macrobriza maxima. See *Briza maxima*
maize, 451
mannagrass, 68, 94
 American, 279
 boreal, 275
 Canadian, 276
 false, 94
 fowl, 282
 narrow, 280
 northwestern, 280
 rattlesnake, 276
 small floating, 275
 spiked false, 443
 tall, 278
 water, 280
 waxy, 277
 weak, 444
matgrass, 328
meadow foxtail, 9, 44, 126
 creeping, 122
 Pacific, 127
 slender, 125
medusahead, 2, 97, 440
melic, 72
 awned, 309
 hairy, 316

 Harford's, 313
 rock, 316
 silky–spike, 316
 Smith's, 314
Melica, 6, 13, 20, 33, 72, 309–317, 469
 aristata, 72, **309**
 bulbosa, 73, **310,** 315
 ciliata, 72, **316**
 fugax, 73, **311**
 geyeri var. *geyeri,* 73, **312**
 harfordii, 72, 73, **313**
 smithii, 73, 268, 309, **314**
 spectabilis, 73, 310, **315**
 stricta var. *stricta,* 73, **316**
 subulata, 73, 268, 310, 312, **317**
Milium vernale, 29, 30
millet, 2, 12
 common, 335
 proso, 335
 wild, 335
milo, 425
Miscanthus, 9
 sinensis, 26, 27, **318**
Molinia
 arundinacea. See *Molinia caerulea*
 caerulea, 35, 139, **319**
moorgrass, purple, 319
Muhlenbergia, 21, 22, 30, 31, 32, 38, 73, 320–327
 andina, 74, **320,** 323, 324
 asperifolia, 74, **321,** 324, 435
 filiformis, 74, **322,** 326
 foliosa ssp. *ambigua.* See *Muhlenbergia mexicana* var. *filiformis*
 glomerata, 74, 320, **323,** 324
 mexicana, 74, 320, 323
 mexicana var. *filiformis,* 75, **324**
 mexicana var. *mexicana,* 75, **324**
 minutissima, 74, **325**
 multiflora. See *Redfieldia flexuosa*
 richardsonis, 74, 322, **326**
 rigens, 74, **327**
 squarrosa. See *Muhlenbergia richardsonis*
 uniflora, 74, **325**
muhly, 73
 alkali, 321
 bog, 325
 foxtail, 320
 least, 325
 mat, 326
 pull-up, 322
 spike, 323
 western. See *Muhlenbergia filiformis*
 wirestem, 324

Munroa squarrosa, 18
muttongrass, Vasey's, 365

N
Nardus, 286
  stricta, 19, **328**
Nassella, 32, 60, 75, 329–331
  lepida, 75, **329**
  pulchra, 75, **330**
  tenuissima, 75, **331**
nebulosa, 461
needle-and-thread, 287
needlegrass, 32, 60, 75, 465
  California, 243
  Columbia, 240
  desert, 338
  Dore's, 240
  foothills, 329
  Lemmon's, 239
  Mexican, 331
  Nevada, 242
  pine, 245
  purple, 330
  Richardson's, 246
  Thurber's, 247
  Webber's, 249
  western, 243

O
oatgrass, 13, 54
  California, 196
  hairy, 413
  nodding, 284
  poverty, 199
  tall, 6, 46, 133
  timber, 198
oats, 2, 46, 137
  cultivated, 137
  false, 69
  slender wild, 135
  wild, 136
oniongrass, 6, 13, 72, 310
  Alaska, 317
  Geyer's, 312
  little, 311
  purple, 315
Oryzopsis 60
  asperifolia, 19, 32, **332**
  exigua. See *Piptatherum exiguum*
  hendersonii. See *Eriocoma hendersonii*
  hymenoides. See *Eriocoma hymenoides*
  webberi. See *Eriocoma webberi*

P
panicgrass, 9, 55, 75
  fall, 334
  hairy, 205
  redtop, 4, 336
  Scribner's, 206
  western, 205
Panicum, 9, 11–13, 29, 55, 75–76, 206, 333–337, 435
  capillare ssp. *capillare,* 76, 321, **333**
  dichotomiflorum ssp. *dichotomiflorum,* 76, **334**
  miliaceum ssp. *miliaceum,* 76, **335**
  miliaceum ssp. *ruderale,* 76, **335**
  oligosanthes var. *scribnerianum.* See *Dichanthelium oligosanthes* ssp. *scribnerianum*
  rigidulum ssp. *rigidulum,* 4, 28, 75, **336**
  scribnerianum. See *Dichanthelium oligosanthes* ssp. *scribnerianum*
  virgatum, 76, **337**
Pappostipa, 60
  speciosa, 32, **338**
Parapholis, 286
  incurva, 20, **339**
Pascopyrum, 9, 24, 40
  smithii, 26, 220, 306, **340**
Paspalum, 13, 28, 76
  dilatatum, 11, 12, 76, **341**
  distichum, 76, **342**
Pennisetum, 52
  villosum. See *Cenchrus longisetus*
  orientale. See *Cenchrus orientalis*
  setaceum. See *Cenchrus setaceus*
Phalaris, 22, 29, 30, 31, 36, 76, 343–346
  angusta, 77, **343**
  aquatica, 76, 77, **343**, 346
  arundinacea, 13, 76, 77, 299, **344**
  californica, 77, **345**
  canariensis, 77, **345**
  paradoxa, 13, 77, **346**
  tuberosa var. *stenoptera.* See *Phalaris aquatica*
Phleum, 9, 21, 77
  alpinum, 77, **347**
  pratense, 77, 126, 347, **348**
Phragmites, 349, 445
  americanus. See *Phragmites australis* ssp. *americanus*
  australis, 34, 40, 77
  australis ssp. *americanus,* 78, **349**
  australis ssp. *australis,* 78, **349**
  communis. See *Phragmites australis* ssp. *australis*

phragmites. See *Phragmites australis* ssp. *australis*
Phyllostachys
  aurea, 455, **457**
  nigra var. *henonis,* 455, **458**, 459
  nigra var. *nigra,* 455, 458
  vivax, 455, 458, **459**
pinegrass, 175
Piptatheropsis, 60
  exigua, 32, **350**
Piptatherum exiguum. See *Piptatheropsis exigua*
Pleioblastus viridistriatus. See *Sasaella ramosa*
Pleuropogon, 10, 18, 20, 78, 351–352
  oregonus, 78, **351**
  refractus, 78, 351, **352**
Poa, 2, 4, 7, 8, 12, 13, 17, 33, 39, 78, 89, 298, 353–395, 449, 468
  alpina ssp. *alpina,* 82, **353**
  ampla. See *Poa secunda* ssp. *juncifolia*
  annua, 79, **354,** 368, 417
  arctica ssp. *arctica,* 79, 86, **355**
  bolanderi, 80, **356**, 367
  bulbosa, 13, 79, **357**
  chambersii, 81, 85, **358**, 374, 378, 385, 395
  compressa, 80, **359**
  confinis, 81, 86, **360**, 382
  curtifolia, 82, 83, **361**
  cusickii, 13, 84, 362–365, 374, 383
  cusickii ssp. *cusickii,* 84, **362–363**
  cusickii ssp. *epilis,* 84, **362–363**
  cusickii ssp. *pallida,* 84, **362, 364**
  cusickii ssp. *purpurascens,* 81, 84, 86, 87, 358, **362, 364**
  epilis. See *Poa cusickii* ssp. *epilis*
  fendleriana ssp. *longiligula,* 81, 82, 362, **365**
  glauca, 353, 379
  glauca ssp. *glauca,* 87, **366**
  glauca sp. *rupicola,* 84, **366**
  gracillima var. *gracillima.* See *Poa secunda* ssp. *secunda*
  gracillima var. *multnomae.* See *Poa* × *multnomae*
  howellii, 80, 356, **367**
  iconia, 79, **357**
  incurva. See *Poa secunda* ssp. *secunda*
  infirma, 79, 354, **368**
  interior, 78–79
  juncifolia. See *Poa secunda* ssp. *juncifolia*
  laxiflora, 85, **369**, 375
  leibergii, 82, 362, **370**
  leptocoma, 88, 369, **371**, 380, 384, 394
  lettermanii, 84, **372**, 390
  longiligula. See *Poa fenderleriana* ssp. *longiligula*
  macrantha, 80, 360, **373**
  mansfieldii, 12, 81, 358, 366, **374**
  marcida, 85, 369, **375**
  × *multnomae,* 83, **376**, 378
  nemoralis, 85, 87, 366, **377**, 379
  nervosa var. *nervosa.* See *Poa nervosa*
  nervosa var. *wheeleri.* See *Poa wheeleri*
  nervosa, 81, 83, 85, 369, 376, **378**, 395
  nevadensis. See *Poa secunda* ssp. *juncifolia*
  palustris, 85, 87, 377, 378, **379**, 382
  paucispicula, 88, 371, **380**, 384
  piperi, 86, 87, **381**, 385
  pratensis, 85–87, 360, 374, 378, 379, **382**, 385, 391, 395
  pratensis ssp. *alpigena,* 86, 355
  pringlei, 82, 362, **383**
  reflexa, 88, 371, 380, **384**, 394
  rhizomata, 86, 358, 381, 382, **385**
  rupicola. See *Poa glauca* ssp. *rupicola*
  sandbergii. See *Poa secunda* ssp. *secunda*
  scabrella. See *Poa secunda* ssp. *secunda*
  secunda, 78, 80, 83, 359, 361, 370, 376, 386–88, 392
  secunda ssp. *juncifolia,* 382, **386–88**
  secunda ssp. *juncifolia* var. *ampla,* 388
  secunda ssp. *juncifolia* var. *juncifolia,* 388
  secunda ssp. *juncifolia* var. *nevadensis,* 388
  secunda ssp. *secunda,* 83, **386–88**, 389
  secunda ssp. *secunda* var. *scabrella,* 388
  secunda ssp. *secunda* var. *secunda,* 388
  stenantha var. *stenantha,* 83, 388, **389**
  suksdorfii, 84, 372, **390**, 394
  supina, 368
  trivialis ssp. *trivialis,* 86, 87, 382, **391**
  unilateralis, 19, 23, 392, 393
  unilateralis ssp. *pachypholis,* 83, **392–393**
  unilateralis ssp. *unilateralis,* 82, **392–393**
  wallowensis, 78, 87, 353, 366, **394**
  wheeleri, 81, 358, 362, 374, 378, 382, 385, **395**
Podagrostis, 31, 32, 41, 45, 88, 168, 396–398
  aequivalvis, 43, 88, **396**
  humilis, 42, 88, 117, 322, **397**, 398
  humilis, also misapplied to *Podagrostis thurberiana*
  thurberiana, 43, 88, 396, **398**
Polypogon, 11, 21, 31, 88–89, 399, 403
  fugax, 89, **399**, 400
  interruptus, 89, 399, **400**
  maritimus, 89, **401**
  monspeliensis, 89, 194, 400, 401, **402**
  viridis, 31, 89, **403**

pricklegrass, 53, 190, 191
    foxtail, 190
    modest, 191
    swamp, 191
proso millet
Psathyrostachys juncea, 25
Pseudoroegneria spicata, 26, 40, 57, 227, **404**
Pseudoroegneria spicata. Also misapplied to *Elymus wawawaiensis*
Pseudosasa japonica, 455, **460**
Puccinellia, 38, 68, 89, 405–411
    distans, 89, **405**
    lemmonii, 90, **406**
    maritima, 89, **407**
    nutkaensis, 90, **408**, 409
    nuttalliana, 90, 408, **409**
    pauciflora, See *Torreyochloa pallida* var. *pauciflora*
    pumila, 89, **410**
    simplex, 89, **411**

Q
quackgrass, 223

R
ravennagrass, 445
Redfieldia flexuosa, 38, **412**
redtop, 105
    Idaho, 109
    Oregon, 112
reed
    American common, 349
    common, 77
    giant, 134
    Old World common, 349
reedgrass, 502
    bluejoint, 169
    Brewer's, 168
    Cascade, 179
    fire, 172
    Howell's, 171
    northern, 176
    Pacific, 173
    pine, 172
    purple, 174
    Rainier, 178
    slimstem, 176
    Steens Mountain, 180
    Tweedy's, 179
rice, 2, 9
ricegrass, 32, 60, 237
    Henderson's, 237, 248
    Indian, 60, 238
    little, 350

roughleaf, 332
    Wallowa, 248
rivergrass, common, 418
rye, 419
    cereal, 419
ryegrass, 72
    annual, 307
    bearded, 307
    perennial, 307
Rytidosperma, 35, 54, 90
    biannulare, 90, **413**
    penicillatum, 90, **413**

S
sacaton, alkali, 321, 435
saltgrass, 12, 210
sandbur, 52
    longspine, 183
    mat, 183
sandgrass, 13
    purple, 445
sandreed, prairie, 181
Sarcobatus vermiculatus, 210
Sasa
    palmata f. *nebulosa,* 455, **461**
    veitchii, 455, **461**
Sasaella ramosa, 455
Schedonorus, 38, 39, 62, 64, 90, 414, 415
    arundinaceus, 90, 126, **414**, 415
    pratensis, 90, **415**
Schizachyrium, 9
Schizachyrium scoparium var. *scoparium,* 27, **416**
Sclerochloa dura, 19, 34, 35, **417**
Scolochloa festucacea, 38, **418**
scratchgrass, 321
Secale cereale, 25, **419**
Secar, 227
semaphoregrass, 78
    nodding, 352
    Oregon, 351
Semiarundinaria fastuosa, 455, 456, **462**
Setaria, 9, 10, 12, 13, 20, 28, 91, 420–424
    faberi, 91, **420**
    italica, 28, 91, **424**
    pumila ssp. *pallide-fusca,* 91, **421**
    pumila ssp. *pumila,* 91, **421–422**
    verticillata, 91, **423**
    viridis var. *major,* 91, **424**
    viridis var. *viridis,* 91, 421, **424**
shattercane, 91, 425
sicklegrass, curved, 339
silvergrass, 9
    Chinese, 318

Sitanion
- *elymoides*. See *Elymus elymoides* ssp. *elymoides*
- *hordeoides*. See *Elymus elymoides* ssp. *hordeoides*
- *hystrix* var. *brevifolium*. See *Elymus elymoides* ssp. *brevifolius*
- *hystrix* var. *californicum*. See *Elymus elymoides* ssp. *elymoides*
- *hystrix* var. *hordeoides*. See *Elymus elymoides* ssp. *hordeoides*
- *hystrix* var. *hystrix*. See *Elymus elymoides* ssp. *elymoides*
- *hystrix*. See *Elymus elymoides*
- *jubatum*. See *Elymus multisetus*

sloughgrass, American, 138

Sorghum, 29, 34, 91, 425, 426
- *bicolor*, 17, 91, **425**, 451
- *halepense*, 91. **426**

sorghum, 12, 91, 425

Spartina, 19, 27, 92, 93, 427–432
- *alterniflora*, 92, **427**, 431
- *anglica*, 92, **428**
- *densiflora*, 92, **429**, 431
- *gracilis*, 92, **430**, 432
- *michauxiana*. See *Spartina pectinata*
- *patens*, 92, 429, **431**
- *pectinata*, 92, 430, **432**
- × *townsendii*, 92, 428

Sphenopholis, 23, 39, 92, 433, 434
- *intermedia*, 93, **433**
- *obtusata*, 93, 298, 433, **434**

Sporobolus, 11–13, 28, 30, 92, 93, 181, 190, 435–439
- *airoides*, 93, 321, **435**
- *alopecuroides*. See *Crypsis alopecuroides*
- *alterniflorus*. See *Spartina alterniflora*
- *anglicus*. See *Spartina anglica*
- *compositus* var. *compositus*, 93, **436**
- *cryptandrus*, 93, **437**
- *densiflorus*. See *Spartina densiflora*
- *hookerianus*. See *Spartina gracilis*
- *michauxianus*. See *Spartina pectinata*
- *neglectus*, 93, **438**, 439
- *niliacus*. See *Crypsis vaginiflora*
- *pumilus*. See *Spartina patens*
- *rigidus* var. *rigidus*. See *Calamovilfa longifolia* var. *longifolia*
- *schoenoides*. See *Crypsis schoenoides*
- × *townsendii*. See *Spartina* × *townsendii*
- *vaginiflorus* var. *vaginiflorus*, 93, 438, **439**

sprangletop, 56
- bearded, 209
- Mexican, 209

squirreltail
- barley, 216
- big, 222
- bottlebrush, 216
- common, 216
- longleaf, 216

stinkgrass, 228

Stipa
- *arundinacea*. See *Anemanthele lessoniana*
- *californica*. See *Eriocoma occidentalis* ssp. *californica*
- *columbiana* var. *columbiana*. See *Eriocoma nelsonii* ssp. *dorei*
- *columbiana* var. *nelsonii*. See *Eriocoma nelsonii* ssp. *nelsonii*
- *comata*. See *Hesperostipa comata*
- *lemmonii* var. *lemmonii*. See *Eriocoma lemmonii* ssp. *lemmonii*
- *lepida*. See *Nassella lepida*
- *nevadensis*. See *Eriocoma nevadensis*
- *occidentalis* var. *californica*. See *Eriocoma ocidentalis* ssp. *californica*
- *occidentalis* var. *nelsonii*. See *Eriocoma nelsonii* ssp. *nelsonii*
- *pinetorum*. See *Eriocoma pinetorum*
- *pulchra*. See *Nassella pulchra*
- *speciosa*. See *Pappostipa speciosa*
- *tenuissima*. See *Nassella tenuissima*
- *thurberiana*. See *Eriocoma thurberiana*
- *webberi*. See *Eriocoma webberi*
- *williamsii*. See *Eriocoma nelsonii* ssp. *nelsonii*

sweetgrass, 70, 128
- California, 288
- hairy, 289

switchgrass, 337

T

Taeniatherum caput-medusae, 24, 263, **440**

Thinopyrum, 9, 24, 25, 93
- *elongatum*. See *Thinopyrum obtusiflorum*
- *intermedium* ssp. *barbulatum*, **441**
- *intermedium* ssp. *intermedium*, 94, **441**
- *obtusiflorum*, **442**
- *ponticum*. See *Thinopyrum obtusiflorum*

thintail, 286

threeawn, 10, 45–46, 131, 132
- oldfield, 131
- prairie, 131
- red, 132
- six-weeks, 131

ticklegrass, 114

timothy, 9, 77, 348
- alpine, 347

Torreyochloa, 38, 68, 89, 94, 182, 443, 444
   erecta, 94, **443**
   pallida var. *pauciflora,* 94, **444**
Tripidium ravennae, 9, 29, 34, **445**
Triplasis, 11, 13
   purpurea var. *purpurea,* 35, **446**
Trisetum, 51
   canescens. See *Graphephorum canescens*
   cernuum. See *Graphephorum cernuum*
   flavescens, 447
   spicatum, 24, **447**
   wolfii. See *Graphephorum wolfii*
trisetum, 69
   nodding, 284
   spike, 447
   tall, 283
Triticum aestivum, 21, 25, **448**

V

Vahlodea atropurpurea, 37, **449**
veldtgrass, panic, 213
velvetgrass, 70, 169
   common, 290
   creeping, 291
ventenata, 2, 450
Ventenata dubia, 32, 36, 203, **450**
vernalgrass, 45
   annual, 128
   sweet, 13, 129
Vulpia
   bromoides. See *Festuca bromoides*
   microstachya. See *Festuca microstachys*
   myuros. See *Festuca myuros*
   octoflora. See *Festuca octoflora*

W

wedgegrass, 92
   prairie, 434
   slender, 433
wheat, 2, 3, 12, 29, 30, 96, 125, 307, 419, 448, 451
wheatgrass, 24, 40, 56, 93
   annual, 236
   arctic, 226
   bluebunch, 404. Also misapplied to *Elymus wawawaiensis*
   crested, 40, 98
   intermediate, 441
   pubescent, 441
   sand-dune, 220
   Scribner's, 224
   Siberian, 40, 98
   slender, 225
   Snake River, 227
   streambank, 220
   tall, 442
   thickspike, 220
   western, 340
whorlgrass, water, 182
wild rice, 12, 94
   northern, 452
   southern, 452
wildrye, 56, 71
   awnless, 215
   beardless, 306
   blue, 218
   Canada, 214
   coastal blue, 218
   giant, 302
   Great Basin, 301
   mammoth, 305
   northwestern, 219
   yellow, 303
windgrass, 45
   common, 130
   interrupted, 130
witchgrass, 75, 333
woodreed, slender, 185

Y

yadake, 460
Yorkshire fog, 290

Z

Zea mays ssp. *mays,* 11, 19, 17, **451**
Zizania, 9, 12, 29, 94, 466
   aquatica var. *aquatica,* 94, **452**
   palustris var. *palustris,* 94, **452**

ABOUT THE AUTHORS

Cindy Talbott Roché earned a PhD from the University of Idaho in plant science, and a BS and MS at Washington State University in Forest Management and Range Ecology. She has lived and worked in both Washington and Oregon. Her field experience includes agency employment and consulting work. She has been working with native and weedy grasses for forty-five years; in the past ten years she has taught numerous grass identification workshops. From her background in writing and editing natural resource publications as well as illustrating grasses for the Flora of North America, she brings a patient attention to detail to the field guide text and illustrations.

Richard E. Brainerd holds an MS from Oregon State University. He has been a botanical consultant in the Pacific Northwest for over thirty years, conducting rare plant and weed surveys, wetland delineations, natural resource inventories, and plant taxonomic research. His strengths combine taxonomic expertise with field experience. He enjoys working with difficult grasses such as *Bromus* and *Poa*, as well as with nongrasses like sedges, rushes, and willows.

Barbara L. Wilson holds a PhD from Oregon State University, where she studied the taxonomy of *Festuca*. She founded the *Carex* Working Group and has taught sedge and grass identification workshops for many years. She brings a strong background in grass taxonomy to this project. She, Otting, and Brainerd have published several new species of grasses. The three members of the former *Carex* Working Group (Wilson, Brainerd, Otting) collaborated on the keys in this book.

Nick Otting has an MS from Oregon State University. He spends the field seasons studying the flora of the shrub steppe and the mountains east of the Cascades. He dedicated many off-seasons to aggressively editing and testing grass identification keys with a view toward making grasses ever more accessible for students, and of course, toward opening the doors of this— to him—high art and practice of grass identification.

Robert C. Korfhage received his MS from Washington State University in range and wildlife ecology. He enjoyed a long career with the Bureau of Land Management as wildlife biologist (Wyoming), planning/environmental coordinator (Idaho and California), and natural resource manager (Oregon). Using his passion for photography, he captured many of the images and edited all of the photos for the field guide.